International Nuclear Science, Technology and Engineering Conference (iNuSTEC 2020)

Held online due to COVID-19

IOP Conference Series: Materials Science and Engineering
Volume 1106

Selangor, Malaysia
17 - 19 November 2020

Editors:

Faridah Mohamad Idris
Nasri A. Hamid
Mohd Syukri Yahya
Hassan Mohamed
Rogemah Ramli
Madihah Mujaini
Julie Andrianny Murshidi
Mohd Idzat Idris
Abu Hassan Husin
Khaidzir Hamzah
Abdul Aziz Mohamed

ISBN: 978-1-7138-3399-4
ISSN: 1757-8981

Printed from e-media with permission by:

Curran Associates, Inc.
57 Morehouse Lane
Red Hook, NY 12571

Some format issues inherent in the e-media version may also appear in this print version.

Institute of Physics
Dirac House, Temple Back
Bristol BS1 6BE UK

Phone: 44 1 17 929 7481
Fax: 44 1 17 920 0979

techtracking@iop.org

Additional copies of this publication are available from:

Curran Associates, Inc.
57 Morehouse Lane
Red Hook, NY 12571 USA
Phone: 845-758-0400
Fax: 845-758-2633
Email: curran@proceedings.com
Web: www.proceedings.com

TABLE OF CONTENTS

ENERGY AND SUSTAINABILITY

MATERIAL AND PROCESS

ENVIRONMENT AND INFRASTRUCTURE

IOP Conf. Series: Materials Science and Engineering 1106 (2021) 011001 doi:10.1088/1757-899X/1106/1/011001

IOP Conference Series: Materials Science and Engineering

International Nuclear Science, Technology and Engineering Conference 2020 (iNuSTEC2020)

"Sustainable Nuclear Science and Technology for Well-Being of Mankind"

Universiti Tenaga Nasional
Selangor, Malaysia
17-19 November 2020

Editors

Faridah Mohamad Idris
Malaysian Nuclear Agency (Nuklear Malaysia)
Nasri A. Hamid
Universiti Tenaga Nasional (UNITEN)
Mohd Syukri Yahya
Universiti Tenaga Nasional (UNITEN)
Hassan Mohamed
Universiti Tenaga Nasional (UNITEN)
Rogemah Ramli
Universiti Tenaga Nasional (UNITEN
Madihah Mujaini
Universiti Tenaga Nasional (UNITEN
Julie Andrianny Murshidi
Malaysian Nuclear Agency (Nuklear Malaysia)
Mohd Idzat Idris
Universiti Kebangsaan Malaysia (UKM)
Abu Hassan Husin
Universiti Teknologi MARA (UiTM)
Khaidzir Hamzah
Universiti Kebangsaan Malaysia (UTM)
Abdul Aziz Mohamed
Malaysian Nuclear Society (MNS)

All papers have been peer-reviewed

Supporting/Sponsoring Organisations

Ministry of Science, Technology and Innovation Malaysia; Ministry of Education Malaysia; Ministry of Energy and Natural Resources; Malaysian Nuclear Agency; ASME Malaysia Section; Malaysian Welding and Joining Society; Reactor Interest Group; Asia-Oceania Neutron Scattering Association, Japan Atomic Energy Society.

IOP Conf. Series: Materials Science and Engineering 1106 (2021) 011001 doi:10.1088/1757-899X/1106/1/011001

PREFACE

It is my pleasure to welcome all the distinguished guests, key note speakers and participants today for the three-day virtual conference of the "International Nuclear Science, Technology and Engineering Conference 2020 (iNuSTEC2020)".

I wish a special welcome to all the eminent speakers from Malaysia and nuclear experts from overseas such as the Republic of Korea, the United States of America and Germany to iNuSTEC2020, and thus making this conference a truly international conference in spirit. I'm very glad and thankful that you are willing to spend your precious time to be with us and sharing your knowledge and vast experiences with us, especially for the young budding nuclear scientists and engineers in Malaysia.

I was informed that this is the 3rd time that the Universiti Tenaga Nasional (UNITEN) is organizing iNuSTEC with MNS. The first was in 2010, the second in 2017. NuSTEC2020 is especially more significant and challenging this year as the conference is held online, in line with the Conditional Movement Control Order (CMCO) SOP's due to COVID-19 pandemic. UNITEN students from the Skuad Media UNITEN (SMU) played a big role in bringing this conference online to you. Congratulations to Skuad Media UNITEN!

Indeed, nuclear science and technology development is closely intertwined with the world wide web (www), networking and computing development. The demand from the former drives the latter to the fore front. It was the scientists who work at CERN, Geneva, who first invented the world wide web (www) to share the vast nuclear data across the world. The high-performance computers are used extensively on grid to run complex nuclear simulations.

This conference is jointly organized by the Malaysian Nuclear Society (MNS) in collaboration with several local universities and vendors. I wish to thank all the committee members for their tireless effort in ensuring the success of this virtual online conference. Due to the changing time caused by the COVID-19, all participants are urged to stay safe and maintain social distancing.

For this iNuSTEC2020 conference, there will be more than 70 papers to be presented by the keynote speakers, invited speakers and participants. Like what we have every year, the conference hosts several other events and for this year we are very proud to host the Radiation Safety Refresher Course for Radiation Protection Officer (RPO) in collaboration with Arey Solutions Sdn Bhd., and the Nuclear Debate 2020. Both events are held online, and for the Nuclear Debate 2020, it involved students from the local universities.

The theme of this year conference is "Sustainable Nuclear Science and Technology for Well-Being of Mankind". Indeed, I'm very convinced that this conference is an ideal platform for scientists, researchers and stakeholders in nuclear science, technology and engineering to share experiences and exchange knowledge in related areas. The Nuclear Science, Technology and Engineering Conference (NuSTEC) has come a long way since 2010 and has surpassed the various challenges that the nuclear science and technology has faced over the years and has adapted to the changing public perception. From having this conference with just 20 - 30 participants back in 2010, we have come a long way in making the conference a huge success with almost 100 participants and exhibitors participating in NuSTEC annually.

I would like to take this opportunity to share the Universiti Tenaga Nasional's involvement in nuclear education. In providing students with sound knowledge in nuclear, the College of Engineering (COE), Universiti Tenaga Nasional (UNITEN) initiated the introduction of several technical electives courses in nuclear engineering and technology within the Mechanical Engineering Department. To ensure the effective implementation of the nuclear technical elective courses, cooperation and collaboration with several other institutions such as the Malaysian Nuclear Agency and the Universiti Kebangsaan Malaysia were established. Since there is no nuclear facilities at UNITEN, cooperation with the Malaysian Nuclear Agency (MNA) is vital to ensure that students have the opportunity to use MNA facilities. Exposing students to the TRIGA Mark II research reactor and NDT laboratories at MNA help students to understand the physics and

IOP Conf. Series: Materials Science and Engineering 1106 (2021) 011001 doi:10.1088/1757-899X/1106/1/011001

applications of nuclear technologies better. Besides, research activities are also conducted at MNA facilities for the benefits of staff and students. In strengthening the cooperation, a Memorandum of Understanding (MOU) between UNITEN and MNA was formalized in December 2010.

Currently, the College of Engineering is offering two nuclear courses, the "Introduction of Nuclear Engineering" and "Nuclear Policy, Security and Safeguards" as technical elective courses for all engineering students at UNITEN. The number of students enrolled for the two courses maintained between 50 – 80 students every semester. In addition, in every academic year there are always UNITEN students that undergo industrial training and conduct final year projects in nuclear. Furthermore, at post graduate level, there are couple of students who specialize in nuclear science and technology while doing masters and PhD. It is hopeful that these students will be the ambassadors in supporting the establishment of nuclear power in Malaysia in the near future.

I would like to express my gratitude and congratulations to all participants who have been actively involved during the three days of the digital conference, achieving this conference's objectives.

Thank you.

Prof. Dato' Ts. Dr. Mohd Zamri Bin Yusoff
Deputy Vice-Chancellor (Student Affairs, Alumni and Management)
Universiti Tenaga Nasional (UNITEN)
Advisor of iNuSTEC2020

IOP Conf. Series: Materials Science and Engineering 1106 (2021) 011001 doi:10.1088/1757-899X/1106/1/011001

MESSAGE FROM:
PRESIDENT, MALAYSIAN NUCLEAR SOCIETY (MNS)

On behalf of the Malaysian Nuclear Society (MNS), I would like to express our thanks to all participants of the International Nuclear Science, Technology and Engineering Conference 2020 (iNuSTEC2020). MNS is honoured to receive supports from distinguished international and national researchers and experts in the field of nuclear science, technology and engineering as International Advisory and National Advisory Panels for iNuSTEC2020. We are indeed pleased to have Organizing Committees of iNuSTEC2020 and the Secretariats working alongside to ensure the success of iNuSTEC2020 which was conducted in the teleconference mode. At the forefront of nuclear science and engineering, MNS collaborates with universities, government agencies and related NGOs to bring the success of iNuSTEC2020. I believe MNS members shall strive to make it better and stronger in future.

In conjunction with this conference, MNS is also honoured to organise Radiation Safety Refresher for RPO on 18-19 Nov 2020 with the MNS corporate member Arey Solutions Sdn Bhd for radiation protection officers registered with Atomic Energy Licencing Board (AELB) of Malaysia. MNS is also honoured to organise Inter Varsity Nuclear Debate 2020 on 19 Nov 2020 in collaboration with Skuad Media UNITEN. These events will help bring university students especially the young Z nuclear generation to discuss issues related to nuclear landscape and the mainstream of development. There are 60 papers received and about 34 papers are accepted to be published in the proceeding.

This conference brings together experienced and young researchers to discuss and debate latest development in all areas of nuclear science, engineering and technology applications for friendly purposes at international level. This would spur and catalyse further research and development leading to the sustainable and improvement of the quality of life without compromising the quality of the environment in Malaysia and in the world.

We would like to express our gratitude to UNITEN for hosting iNuSTEC2020 and Inter Varsity Nuclear Debate 2020 and, to all the personnel involved for their unwavering efforts to make this event, which is held online, a reality. I wish to also record my appreciations and thanks to all sponsors for their generous support. May all these efforts bring benefits to all of us and I congratulate everyone involved in making this conference successful event.

Thank you.

Assoc. Prof. Dr. Abdul Aziz Mohamed
President, MNS

IOP Conf. Series: Materials Science and Engineering 1106 (2021) 011001 doi:10.1088/1757-899X/1106/1/011001

CONTENTS

PAPERS

PROCEEDING EDITORIAL COMMITTEE

IOP Conf. Series: Materials Science and Engineering 1106 (2021) 011001 doi:10.1088/1757-899X/1106/1/011001

iNuSTEC2020 ORGANISING COMMITTEE

Patron:
Prof. Dato' Ir Dr. Kamal Nasharuddin Bin Mustapha,
Vice-Chancellor, UNITEN

Advisor:
Prof. Dato' Ts. Dr. Mohd. Zamri Bin Yusoff,
Deputy Vice-Chancellor (Student Affairs, Alumni and Management)
UNITEN

Chairperson:
Assoc. Prof. Dr. Nasri A Hamid
Head of Nuclear Energy and Engineering Group (GrEEN),
UNITEN

Co-chairperson:
Dr. Abdul Aziz Mohamed
President,
Malaysian Nuclear Society (MNS)

Deputy Chairpersons:
Dr. Mohd Syukri Bin Yahya, *UNITEN*
Ts. Dr. Hassan Bin Mohamed, *UNITEN*

Secretary:
Dr. Faridah Mohamad Idris, *MNS/Nuklear Malaysia*
Rogemah Binti Ramli, *UNITEN*

Treasurer:
Dr. Julie Andrianny Bt. Murshidi, *MNS/Nuklear Malaysia*

Deputy Treasurer:
Madihah Binti Mujaini, *UNITEN*

INTERNATIONAL ADVISORY COMMITTEE

Prof. Emiretus Dr. Noramly Muslim, *MNS Honorary Fellow*
Prof. Dr. Sung-Min Choi, *KAIST, Republic of Korea*
Prof. Dr. Edy Giri Rachman Putra, *BATAN, Indonesia*
Dr. Craig Marianno, *Texas A&M, United States of America*
Prof. Dr. Nikolay Kardjilov, *Helmholtz-Zentrum Berlin, Germany*
Prof. Dato' Dr. Abd Aziz Bin Tajuddin, *Al Bukhary International University (AIU), Malaysia*
Assoc. Prof. Ir. Dr. Mokhtar Awang, *ASME Malaysia Section*

NATIONAL ADVISORY COMMITTEE

Ts. Dr. Siti Aia'shah Binti Hashim, *Nuklear Malaysia*
Prof. Dr. Rosalina Abdul Salam, *Universiti Sains Islam Malaysia*
Prof. Dr. Hjh. Farida Zuraina Mohd Yusof, *Universiti Teknologi MARA*
Prof. Dr. Ismail Yusoff, *Universiti Malaya*
Prof. Dr. Nahrul Khair Bin Alang Md Rashid, *Universiti Teknologi Malaysia*
Prof. Dr. Izham Bin Datuk Zainal Abidin, *Universiti Tenaga Nasional*
Assoc. Prof. Dr. Khaidzir Hamzah, *Universiti Teknologi Malaysia*
Assoc. Prof. Dr. Abu Hassan Husin, *Universiti Teknologi MARA*
Ts. Saharuddin Bin Kambari, *Malaysian Welding and Joining Society (MWJS)*
Assoc. Prof. Dr. Faizal Mohamed, *Universiti Kebangsaan Malaysia*

IOP Conf. Series: Materials Science and Engineering 1106 (2021) 011001 doi:10.1088/1757-899X/1106/1/011001

TECHNICAL COMMITTEE

Technical and Logistics Committee
Muhammad Amir Asraf Bin Ramli, *MNS/Arey Solutions*
Ikmal Azrai Bin Abdul Aziz, *MNS/Arey Solutions*

Skuad Media UNITEN (SMU)
Nayli Aliah binti Mohd Afandi, *UNITEN*
Zafran Akmal bin Zainal Hisham, *UNITEN*
Zidaniel Zikri bin Zulkifli, *UNITEN*
Ashwin Anbalagan, *UNITEN*
Ahmad Azizul Iqram bin Musa, *UNITEN*
Muhammad Zahirulhaq bin Sharani, *UNITEN*
Muhammad Darwisy bin Abd Aziz, *UNITEN*
Kamaluddin Abdullah, *UNITEN*
Kamarudin Yaacob, *UNITEN*
Muhammad Ferdaus Bin Abdullah, *UNITEN*
Hazirah Adlina binti Mohammad Hazani, *UNITEN*
Ariff bin Mohd Sa'ad, *UNITEN*

Scientific Committee
Dr. Mohd Syukri Yahya, *MNS/UNITEN*
Ts. Dr. Hassan Mohamed, *MNS/UNITEN*
Safwan Shalbi, *INE MNS/UMP*
Rogemah Ramli, *MNS/UNITEN*
Dr. Julia Abdul Karim, *MNS/Nuklear Malaysia*

Publicity Committee
Dr. Faridah Mohamad Idris, *MNS/Nuklear Malaysia*
Dr. Julie Andrianny Murshidi, *MNS/Nuklear Malaysia*
Khairul Izzat Bin Abdul Aziz, *MNS/Reltech Lab*
Dr. Hafizal Yazid, *MNS/Nuklear Malaysia*

Souvenirs Committee
Madihah Mujaini, *MNS/UNITEN*
Dr. Juniza Md Saad, *MNS/UPM*
Azraf Bin Azman, MNS/*Nuklear Malaysia*

Malaysian Nuclear Society (MNS)
Khair'ah Binti Yazid@Khalid, *Nuklear Malaysia*
Anas Bin Mohamad Fauzi, *UNITEN*
Dr. Julia Bt. Abdul Karim, *Nuklear Malaysia*
Dr. Muhammad Rawi B. Mohamed Zin, *Nuklear Malaysia*
Khairul Azhar Bin Abdul Halim, *Nuklear Malaysia*
Mohd Rizal B. Mamat @ Ibrahim, *Nuklear Malaysia*
Anwar B. Abdul Rahman, *Nuklear Malaysia*
Shalina Bt. Sheik Muhamad, *Nuklear Malaysia*
Nazaratul Ashifa Bt. Abdullah Salim, *Nuklear Malaysia*
Dr. Mazleha Maskin, *Nuklear Malaysia*

Arey Solutions
Muhammad Amir Asraf Bin Ramli
Ikmal Azrai Bin Abdul Aziz

Universiti Teknologi Malaysia (UTM)
Dr. Khaidzir Hamzah
Dr. Jasman Zainal
Dr. Zulhairun Bin Abdul Karim

Universiti Putra Malaysia (UPM)
Dr. Juniza Md Saad

Universiti Teknologi Mara (UiTM)
Assoc. Prof. Dr. Abu Hassan Husin

Universiti Sains Islam Malaysia (USM)
Dr. Mohd Hafez bin Mohd Isa

Universiti Malaya (UM)
Prof. Dr. Hasan Abu Kassim
Assoc. Prof. Dr. Norhasliza Yusof

Universiti Tun Hussien Onn Malaysia (UTHM)
Assoc. Prof. Dr. Zaidi Embong

Institute of Nuclear Engineers (INE)
Safwan Shalbi
Muhammad Aliff Ashraff Rosdi
Asyraf Arif Bin Abu Bakar
Abdul Hannan Damahuri

Malaysian Nuclear Professional Chapter MNS (MNPC-MNS)
Ts. Razali Bin Harun, *REM*
Khairul Izzat Bin Abdul Aziz, *Reltech Lab*
Mohd Taha Bin Hj. Suja', *REM*
Muhammad Amir Asraf Bin Ramli, *Arey Solutions*
Amir Syafiq Bin Azhar, *Reltech Lab*
Ts. Dr. Mohd Idzat Bin Idris, *UKM*
Mohamad Aminudin Bin Said, *MOH*
Muhammad Khairul Azfar, *ANM*
Ikmal Azrai Bin Abdul Aziz, *Arey Solutions*

RPO REFRESHER COURSE ORGANISING COMMITTEE

Advisor
Assoc. Prof. Dr. Abdul Aziz Bin Mohamed, *MNS*

Technical
Ts. Razali Bin Harun, *REM*
Muhammad Amir Asraf Bin Ramli, *Arey Solutions*
Khairul Izzat Bin Abdul Aziz, *Reltech Lab*
Ikmal Azrai Bin Abdul Aziz, *Arey Solutions*

DEBAT NUKLEAR ANTARA VARSITI 2020 ORGANISING COMMITTEE

Advisors
Assoc. Prof. Dr. Abdul Aziz Bin Mohamed, *MNS*
Assoc. Prof. Dr. Nasri A Hamid, *UNITEN*

Chairperson
Dr. Mohd Syukri Yahya, *MNS/UNITEN*

Deputy Chairperson
Ts. Dr. Hassan Mohamed, *MNS/UNITEN*

Logistic
Madihah Mujaini, *MNS/UNITEN*
Rogemah Ramli, *MNS/UNITEN*
Azraf Azman, *MNS/Nuklear Malaysia*
Dr. Hafizal Yazid, *MNS/Nuklear Malaysia*

Speakers
Dr. Mohd Syukri Yahya, *MNS/UNITEN*
Dr. Juniza Md Saad, *MNS/UPM*

Judges
Ang Wei Eng, *Texas A&M University*
Syafiqah Syakira binti Saiful Yazan
Dr. Juniza Md Saad, *MNS/UPM*
Dr. Shamsul Amri bin Sulaiman, *Tenaga Nasional Berhad*
Azrudi bin Mustapha, *MyPOWER*
Sheriffah Noor Khamseah Al-Idid binti Dato Syed Ahmad Idid

SKUAD MEDIA UNITEN (SMU) COMMITTEE MEMBERS

Director
Nayli Aliah binti Mohd Afandi

Head of Photographer
Ashwin Anbalagan

Head of Social Media Executives
Ahmad Azizul Iqram bin Musa

Head of Videographer
Muhammad Zahirulhaq bin Sharani

Videographers
Zafran Akmal bin Zainal Hisham
Muhammad Darwisy bin Abd Aziz

Graphic Designers
Ariff bin Mohd Sa'ad
Hazirah Adlina Binti Mohammad Hazani

Photographers
Zidaniel Zikri bin Zulkifli
Kamaluddin Abdullah
Kamarudin Yaacob
Muhammad Ferdaus Bin Abdullah

IOP Conf. Series: Materials Science and Engineering 1106 (2021) 011001 doi:10.1088/1757-899X/1106/1/011001

INUSTEC 2020 AWARDS AND RECOGNITION

Congratulations All Winners

BEST THESIS AWARD
Abdul Hanan B Damanhuri, *UNITEN*
"Effect of Thorium Fuel Addition in the Configuration of PUSPATI TRIGA Reactor"

DEBAT NUKLEAR ANTARA VARSITI 2020

Champion:
Universiti Tenaga Nasional
Mohamad Hakimi bin Mohamed Kaberi (*Best Speaker Award*)
Muhammad Syazran Haikal bin Azlan
Ahmad Thareeq bin Mohd Isa
Muhammad Anas Amjad bin Abdullah

1st Runner Up:
Universiti Teknologi MARA (UiTM)
Muhammad Harits Ikhwan bin Khairul Azman
Asyraff Danial bin Abdullah
Ameer Syahiran bin Azhan

Consolation Prize 1:
Universiti Kebangsaan Malaysia (UKM)
Gilbert Raphael Bong
Mohammad Danial bin Sakaria
Sharendravarman a/l Balakrisnan

Consolation Prize 2:
Universiti Teknologi Malaysia (UTM)
Muhammad Khairul Fahmi bin Ibrahim
Muhammad Haikal bin Jamil
Muhammad Amir Safwan bin Zamani Ahmad

NUCLEAR QUIZ CHALLENGE

Winners: Ang Wei Eng (*TAMU, USA*), Muhd Amirul Shazwan Shalihin, Siti Hajar Abdullah
1st Runner Up: Nuraslinda Anuar (*TAMU, USA*), Muhammad Ferdaus Abdullah, Nur Amira Kamarulzaman
2nd Runner Up: Nor Arymaswati Bt. Abdullah (*Nuklear Malaysia*), Nuraslinda Anuar (*TAMU, USA*), Norfarizan binti Mohd Said (*Nuklear Malaysia*)

IOP Conf. Series: Materials Science and Engineering 1106 (2021) 011001 doi:10.1088/1757-899X/1106/1/011001

ACKNOWLEDGEMENT

iNuSTEC2020 is jointly organized by Malaysian Nuclear Society (MNS)
and Universiti Tenaga Nasional (UNITEN)

In Collaboration with

Malaysian Nuclear Agency (Nuklear Malaysia)
Universiti Kebangsaan Malaysia (UKM)
Universiti Teknologi MARA (UiTM)
Universiti Islam Antarabangsa Malaysia (IIUM)
Universiti Teknologi Malaysia (UTM)
Universiti Malaya (UM)
Universiti Sains Islam Malaysia (USIM)
Universiti Tun Hussien Onn Malaysia (UTHM)
Arey Solutions Sdn Bhd

Co-organized by
National Centre for Particle Physics (NCPP)
Institute of Nuclear Engineers (INE MNS)
Kelab Sains Nuklear (KESAN)
Malaysian Nuclear Professional Chapter (MNPC MNS)

Supported by
Reltech Lab Sdn. Bhd.
ASME Malaysia Section
Malaysian Welding and Joining Society
Majlis Perbandaran Sepang

IOP Conf. Series: Materials Science and Engineering 1106 (2021) 011001 doi:10.1088/1757-899X/1106/1/011001

MNS COUNCIL MEMBERS 2020/2021

Advisor / Past President	Prof. Dr. Nahrul Khair Bin Alang Md Rashid (*UTM*)
President	Dr. Abdul Aziz Bin Mohamed (*UNITEN*)
Vice President	Assoc. Prof. Dr. Faizal Bin K.P. Kunchi Mohammed (*UKM*)
Secretary	Dr. Faridah Bt Mohamad Idris (*Nuklear Malaysia*)
Deputy Secretary	Anas Bin Mohamad Fauzi (*UNITEN*)
Treasurer	Dr. Julie Andrianny Bt. Murshidi (*Nuklear Malaysia*)
Deputy Treasurer	Khair'ah Binti Yazid@Khalid (*Nuklear Malaysia*)

Members
Ts. Dr. Hassan Bin Mohamed (*UNITEN*)
Dr. Julia Bt. Abdul Karim (*Nuklear Malaysia*)
Dr. Shukri Mohd (*Nuklear Malaysia*)
Dr. Muhammad Rawi B. Mohamed Zin (*Nuklear Malaysia*)
Dr. Mohd Syukri Yahya (*UNITEN*)

Auditors
Khairul Azhar Bin Abdul Halim (*Nuklear Malaysia*)
Ikmal Azrai Bin Abdul Aziz (*Arey Solutions*)

MNS Chapters
RadTech (*Radiation Tech*)
XApp (*X-ray App*)
NuEnergy (*Nuclear Energy*)
Neutron Photon (*Neutron- Photon*)
Student Chapter (*UKM, UiTM, UNITEN, UTM - NESS*)
Women in Nuclear, Malaysia (*WiN Malaysia*)
Institute of Nuclear Engineers (*INE MNS*)
Malaysian Nuclear Professional Chapter (*MNPC MNS*)

MNPC-MNS COMMITTEE MEMBERS 2020/2021

Advisor / President MNS
Dr. Abdul Aziz Bin Mohamed

President
Ts. Razali Bin Harun (*REM*)

V. President
Khairul Izzat Bin Abdul Aziz (*Reltech Lab*)

Secretary
Mohd Taha Bin Hj. Suja' (*REM*)

Visit us at http://www.nuklearmalaysia.org

IOP Conf. Series: Materials Science and Engineering 1106 (2021) 011001 doi:10.1088/1757-899X/1106/1/011001

With Best Compliments

RELTECH LAB SDN BHD, MALAYSIA

REM CONSULTANTS SDN BHD, MALAYSIA

AREY SOLUTIONS SDN BHD, MALAYSIA

MAJLIS PERBANDARAN SEPANG

ASME MALAYSIA SECTION

MALAYSIAN WELDING AND JOINING SOCIETY

Nuclear Energy and Engineering Group (GrEEN), UNITEN

IOP Conf. Series: Materials Science and Engineering 1106 (2021) 011001 doi:10.1088/1757-899X/1106/1/011001

Congratulation to MNS and UNITEN from

Peer review declaration

All papers published in this volume of *IOP Conference Series: Materials Science and Engineering* have been peer reviewed through processes administered by the Editors. Reviews were conducted by expert referees to the professional and scientific standards expected of a proceedings journal published by IOP Publishing.

All papers published in this proceeding have been peer-reviewed, where single-blind peer-reviewed was practiced for all papers published in this proceeding. An average number of 2 reviewers was assigned for each of the papers published. The reviewed papers and comments by the reviewers were transmitted to the authors by the editors. The authors were requested to resubmit their papers as per comments by the reviewers.

Out of 60 papers submitted, 39 were accepted for publication in the proceeding after being reviewed and resubmitted by authors. Table 1 provides statistics of the papers involved in this proceeding. Selected papers published were checked against plagiarism using *Turnitin* software, courtesy of Universiti Tenaga Nasional (UNITEN) as co-organizer of iNuSTEC2020.

Among the criteria used by the editorial committee in accepting the papers are:

- The work by the author(s) is original/novel and relevant to the nuclear community
- It has notably scientific merit and motivation, with the contents of the paper stated are correct and accurate together with references
- The terminology used in the works are acceptable/in accordance with the nuclear community as a whole
- The presentation of the papers are in accordance with the practiced standard of IOP

Table 1. Statistics of the paper review process in iNuSTEC2020

No.	Item	
	Conference submission management system	*Website and Email*
	Number of submissions received	*: 60 papers*
	Number of submissions accepted	*39 papers*
	Acceptance Rate ((Number of Submissions Accepted) / (Number of Submissions Received)) X 100)	*65% acceptance rate*
	The average number of reviews per paper:	*2 reviews*
	The total number of reviewers involved:	*48 reviewers*
	Any additional info on the review process (i.e., plagiarism check system):	*Turnitin for selected papers*

IOP Conf. Series: Materials Science and Engineering 1106 (2021) 011002 doi:10.1088/1757-899X/1106/1/011002

Contact person for queries:	*Dr. Faridah Mohamad Idris (email: faridah_idris@nuclearmalaysia.gov.my)*

A multipronged core power control strategy for Reaktor TRIGA PUSPATI

Mohd Sabri Minhat[1,2], Nurul Adilla Mohd Subha[1, a)], Fazilah Hassan[1], Abdul Rashid Husain[1], Anita Ahmad[1], Fatimah Sham Ismail[1], Norikhwan Hamzah[3]

[1]School of Electrical Engineering, Faculty of Engineering, Universiti Teknologi Malaysia, 81310 Skudai, Malaysia.
[2]Reactor Instrumentation and Control Engineering, Reactor Technology Centre, Malaysian Nuclear Agency, 43000 Kajang, Malaysia.
[3]School of Mechanical Engineering, Faculty of Engineering, Universiti Teknologi Malaysia, 81310 Skudai, Malaysia

[a)]nuruladilla@utm.my

Abstract. At present, the power tracking performance of nuclear Reaktor TRIGA PUSPATI (RTP) is considered unsatisfactory performance due to relatively long settling time during transient and a chattering noise during steady-state power output. Application of the conventional Feedback Control Algorithm (FCA) as a power control technique is proven to be inadequate to keep the core power output stable and within tight multiple parameter constraints for the safety demand of the RTP. Hence, the present study proposed a multipronged core power control strategy improvement through manipulation of the current Control Rod Selection Algorithm (CRSA), Control Rod Speed Design (CRSD), and Power Change Rate Constraint (PCRC) which are part of the core power control design. In this paper, the profiling and analysis of the multipronged core power control strategy are presented. The model for core power control consists of mathematical models of the reactor core, FCA controller, and a series of multipronged models. The mathematical models of the reactor core are based on the point kinetics model, thermal-hydraulic model, and reactivity model. The reactor model is integrated with the FCA controller and a combination of CRSA-CRSD-PCRC models. The power tracking performance of the proposed control strategy and conventional FCA is compared via computer simulation. Overall, the results show the multipronged FCA offers a wider options for optimum operation of the TRIGA reactor.

1. Introduction

The Reaktor TRIGA PUSPATI (RTP), made by General Atomics is the only nuclear research reactor in Malaysia. The reactor is mainly used for research and development (R&D), isotope production, education, and training purposes. One of the main safety concern in nuclear operation is the core power control system design which is vital for the safe operation of the reactor and minimize any possibilities of failures or malfunctions of the reactor control system which may lead to abnormal behavior of the reactor. Thus, the development of safe, rapid, and efficient core power control for nuclear plants is attracting ongoing research interest [1].

In general, there is no universal solution to resolve all the problems related to power tracking performance in nuclear reactor operation. Hence, a comprehensive countermeasures strategy needs to be identified towards improving efficiency and enhancing safety. One of the powerful methods is known as a multipronged strategy that has been implemented for optimisation problems in non-nuclear applications [2][3]. Surprisingly, no previous study has investigated a multipronged approach in a nuclear power tracking system.

The core power control is physically designed based on common safety parameter constraints such as step reactivity limiting [4], reactor period limiting [5], power change rate-limiting [6], and control rod speed limiting [7]. To date, the effects of the combination of all these safety parameter constraints in the single-core power control system have not been extensively studied. Generally, these distinct parameters can be categorized into three-pronged elements in core power control design; Control Rod Selection Algorithm (CRSA), Control Rod Speed Design (CRSD), and Power Change Rate Constraint (PCRC).

The RTP uses a conventional core power control known as Feedback Control Algorithm (FCA) [8] for the power maneuvering up to 1 MWth based on the conventional CRSA (cCRSA), conventional CRSD (cCRSD), and conventional PCRC (cPCRC). In our recent works, the new CRSA based on Single Control Absorbing Rod (SCAR) algorithm [9], new CRSD based on different saturation types and changes in the maximum rod speed limiter values [10], and new PCRC based on fuzzy approach [11] have been designed separately to provide better results than conventional methods. However, the results produced are not fully optimized and do not have a simple analytic link to the choice of the prongs. Hence, in this paper, the three-pronged and combination prongs of the strategies in [9-11] are studied, analyzed, and profiled to further improve the power tracking performance of the RTP.

The current core power control operates within tight multiple parameter constraints for the safety demand of the RTP. The power tracking performance measurement is based on the sudden change of power demand which currently leads to significant power output delayed known as settling time during transient and a chattering noise during steady-state. The effect of all prongs is evaluated through the power maneuvering performance of the core power control and actuation signal of the Control Rod Drive Mechanism (CRDM).

This paper is organized as follows. The modelling of RTP in state space form is presented in Section 2 and the RTP core power control system is briefly described in Section 3. The proposed multipronged core power control strategy is presented in Section 4. The results and discussion on the implementation of a multipronged core power control strategy are given in Section 5. Finally, conclusions are given at the end of the paper.

2. State space Reaktor TRIGA PUSPATI model

2.1. State Space Model

The RTP model is built according to the point kinetics with six groups of delayed neutrons, reactivity feedback due to control rod movement, fuel temperature, and moderator temperature changing are considered. The equations of the RTP reactor core model are shown as [9]:

$$\left.\begin{aligned}
\frac{d\psi}{dt} &= \frac{\rho-\beta}{\Lambda}\psi + \textstyle\sum_{i=1}^{6}\lambda_i\eta_i \\
\frac{d\eta_i}{dt} &= \frac{\beta_i}{\Lambda}\psi - \lambda_i\eta_i \quad i = 1,\ldots,6 \\
\frac{dT_m}{dt} &= \frac{\Omega}{\mu_m}T_f - \frac{(\Omega+2M)}{\mu_m}T_m + \frac{2M}{\mu_m}T_{in} \\
\frac{dT_f}{dt} &= \frac{N_0}{\mu_f}\psi - \frac{\Omega}{\mu_f}T_f + \frac{\Omega}{\mu_f}T_m \\
\mu_m &= M_m C_m \\
\mu_f &= M_f C_f \\
M &= C_m\Gamma \\
\dot{\rho}_r &= G_r Z_r \\
\rho &= \rho_r + \alpha_m T_m + \alpha_f T_f \\
N &= \psi N_0
\end{aligned}\right\} \quad (1)$$

where, ψ is relative neutron density, ρ is total reactivity, β is the total fraction of effective delayed neutron, Λ is mean neutron generation time, λ_i is decay constant of the i-th group of delayed neutron precursor, η_i is the i-th group of normalized precursor concentration, β_i is the i-th group of delayed neutron, T_m is average temperature of coolant, N_o is nominal core power, M_m is moderator total mass, C_m is moderator specific heat capacity, μ_m is heat capacity of moderator, T_f is average temperature of fuel, Ω is global heat transfer coefficient, Γ is coolant mass flow rate, T_{in} is average inlet temperature of coolant, M_f is fuel total mass, C_f is fuel specific heat capacity, μ_f is heat capacity of fuel, ρ_r is reactivity due to control rod movement, G_r is reactivity worth of the control rod, Z_r is velocity of the control rod, α_m is reactivity due to change in temperature moderator, α_f is reactivity due to change in temperature fuel and, N actual core power.

The linearized model is widely used in TRIGA reactor modelling [12] and can be represented in state-space form as:

$$\left.\begin{aligned}
\dot{x} &= Ax + Bu \\
y &= Cx + Du \\
A &= \begin{bmatrix}
-\frac{\beta}{\Lambda} & \lambda_1 & \lambda_2 & \lambda_3 & \lambda_4 & \lambda_5 & \lambda_6 & \frac{\alpha_f}{\Lambda}\psi_0 & \frac{\alpha_m}{\Lambda}\psi_0 & \frac{\psi_0}{\Lambda} \\
\frac{\beta_1}{\Lambda} & -\lambda_1 & 0 & 0 & 0 & 0 & 0 & 0 & 0 & 0 \\
\frac{\beta_2}{\Lambda} & 0 & -\lambda_2 & 0 & 0 & 0 & 0 & 0 & 0 & 0 \\
\frac{\beta_3}{\Lambda} & 0 & 0 & -\lambda_3 & 0 & 0 & 0 & 0 & 0 & 0 \\
\frac{\beta_4}{\Lambda} & 0 & 0 & 0 & -\lambda_4 & 0 & 0 & 0 & 0 & 0 \\
\frac{\beta_5}{\Lambda} & 0 & 0 & 0 & 0 & -\lambda_5 & 0 & 0 & 0 & 0 \\
\frac{\beta_6}{\Lambda} & 0 & 0 & 0 & 0 & 0 & -\lambda_6 & 0 & 0 & 0 \\
\frac{N_0}{\mu_f} & 0 & 0 & 0 & 0 & 0 & 0 & -\frac{\Omega}{\mu_f} & \frac{\Omega}{\mu_f} & 0 \\
0 & 0 & 0 & 0 & 0 & 0 & 0 & \frac{\Omega}{\mu_m} & \frac{(\Omega+2M)}{\mu_m} & 0 \\
0 & 0 & 0 & 0 & 0 & 0 & 0 & 0 & 0 & 0
\end{bmatrix} \\
B &= [0 \;\; 0 \;\; 0 \;\; 0 \;\; 0 \;\; 0 \;\; 0 \;\; 0 \;\; 0 \;\; G_r]^T \\
C &= [1 \;\; 0 \;\; 0 \;\; 0 \;\; 0 \;\; 0 \;\; 0 \;\; 0 \;\; 0 \;\; 0] \\
D &= [0]
\end{aligned}\right\} \quad (2)$$

where $\boldsymbol{A}$ is the state matrix, $\boldsymbol{B}$ is the input matrix, $\boldsymbol{C}$ is the output matrix, $\boldsymbol{D}$ is the disturbance matrix, core state $\boldsymbol{x} = [\boldsymbol{\delta N}\ \boldsymbol{\delta\eta_1}\ \boldsymbol{\delta\eta_2}\ \boldsymbol{\delta\eta_3}\ \boldsymbol{\delta\eta_4}\ \boldsymbol{\delta\eta_5}\ \boldsymbol{\delta\eta_6}\ \boldsymbol{\delta T_m}\ \boldsymbol{\delta T_f}\ \boldsymbol{\delta\rho}]^{\boldsymbol{T}}$, core input $\boldsymbol{u} = [\boldsymbol{z_r}]$, and core output $\boldsymbol{y} = [\boldsymbol{N}]$.

3. RTP core power control system

The RTP feedback core power control system for power maneuvering is illustrated in Figure 1. The term input refers to a Power Demand (PDM) and the output is the neutron power at the core. The core power is measured by an ex-core neutron detector and Neutron Measurement System (NMS) as signal processing. The NMS provides two signals which are core power (*N*) and the rate of power change (Log Rate). The error deviation in percentage between the PDM and the core power output is subtracted with conventional Power Change Rate Constraint (cPCRC) based on the constant gain rate (*G2*), which then are used as the inputs for the signal filter and Proportional-Integral (PI) controller. The controller output in form of the control rod velocity is fed to the Control Rod Speed Design (CRSD) to constraint the reactivity insertion rate in the core. The conventional Control Rod Selection Algorithm (cCRSA) is used to determine the sequence of control rod movement at Control Rod Drive Mechanism (CRDM) based on rod position balancing pattern to regulate the reactor power. Using the said configuration, the core power control has 1% full power chattering error with a large settling time in the case of a sudden change in power demand.

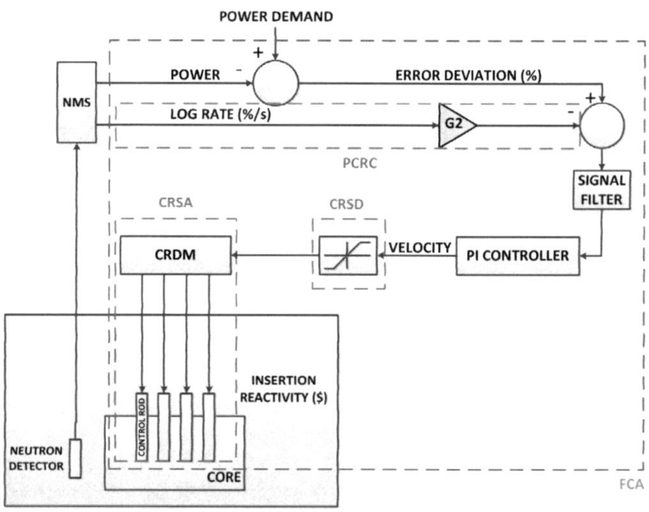

Figure 1. Block Diagram of Conventional Three-pronged in Core Power Control System of the RTP.

The FCA consist of PCRC, signal filter, PI controller, CRSD, and CRSA. The PI controller with a signal filter is designed as follows [9]:

$$\left.\begin{aligned} u_c = &\ G_3 E_{fi} + G_4 \int_{t_0}^{t} E_{fi} dt \\ E = &\left[G_1 log\left(\frac{PDM}{N}\right)\right]_{\pm 1} - \alpha G_2 \frac{1}{N}\frac{dN}{dt} \\ E_{fi}(k) = &\ 1.47197E(k) + 0.882\left[E_{fi}(k-1) - E(k-1)\right] \end{aligned}\right\} \quad (3)$$

where u_c is the output signal from the controller, G_1, αG_2, G_3, G_4 , are controller tuning gain for FCA, and E_{fi} is input filter calculation based on the error signal (*E*).

Profiling and analysis of three-pronged control strategy of CRSA, PCRC and CRSD, all possible combination of conventional and the new three-pronged control strategy need to be taken into account. All parameters in Eq. (3) are considered as constant throughout this study.

4. Multipronged core power control strategy

In this section, the three-pronged control strategies in core power control system are presented based on our works in [9-11] which consists of conventional and new CRSA, CRSD, and PCRC.

In [9], the CRSA model is represented as follows:

$$\left.\begin{aligned} \rho_r(s) &= G_r(iCRSA)Z_r(s) \\ cCRSA &= \left(\frac{1.25s+0.02752}{s^3+0.02202s^2+2.519e-16s}\right) \\ nCRSA &= \left(\frac{5s+0.3854}{s^3+0.0771s^2+6.3158e-18s}\right) \end{aligned}\right\} \tag{4}$$

where the velocity of the control rod (Z_r) is calculated from the controller u_c and the control rod position dynamics in both cCRSA and nCRSA models are derived based on the set of actual input and output data from the CRDM using System ID. In [9], the nCRSA is modelled to improve the pattern control rod movement by providing the most responsive reactivity during transient region.

In the CRSD, the maximum allowable control rod speed and the saturation models are determined to produce optimum tracking performance without exceeding the predetermined safety limit stated in the Final Safety Analysis Report (FSAR) [13]. The cCRSD and nCRSD are both represented by the hard saturation models following our recent work in [10], at 2 mm/s and 3 mm/s respectively. The model is defined as:

$$u_{CRSD} = f_{sat}(u_c) = \begin{cases} -u_{max}, & u_c < -u_{max} \\ u_c, & -u_{max} \le u_c \le u_{max} \\ u_{max}, & u_c > u_{max} \end{cases} \tag{5}$$

where u_{max} is the maximum velocity control rod permitted. Hence, the application of saturation prevents power output overshoot for safety concern.

In conventional PCRC (cCPRC), the constant gain is used which represented by *G2* in Figure 1. However, this approach is not satisfying to introduce a different level of penalizing value on the control rod speed signal to improve the drivability of the CRDM. Thus, to solve the problem, the nPCRC based on Fuzzy Logic with straight lines membership function is designed for the RTP as

$$\hat{u}_{PCRC} = \left[1 - \left(\frac{u_a w_1 + u_b w_2 + u_c w_3 + u_{ab} w_4}{u_a + u_b + u_c + u_{ab}}\right)\right] u_{CRSD} \tag{6}$$

where u_a, u_b, u_c and u_d are designed based on linear membership functions, and w_i (for $i = 1, 2, 3, 4$) is an adjustable weighting parameter that can be varied at different levels of the penalty based on the rate of power change as shown in [11].

IOP Conf. Series: Materials Science and Engineering 1106 (2021) 012001 doi:10.1088/1757-899X/1106/1/012001

The three-pronged control strategy is unified into a single model with PI controller in Eq. (3) as follows:

$$\rho_r = \underbrace{G_r(iCRSA)\left[1-(1-\alpha)\left(\frac{u_a w_1+u_b w_2+u_c w_3+u_{ab} w_4}{u_a+u_b+u_c+u_{ab}}\right)\right]\underbrace{\left[\underbrace{G_3 E_{fi} + G_4\int_{t_0}^{t} E_{fi}dt}_{u_c}\right]_{\pm u_{max}}}_{u_{CRSD}}}_{\hat{u}_{PCRC}} \quad (7)$$

Referring to Eq. (7), switching function is employed to select either cPCRC or nPCRC where α is set to 1 for cPCRC and α= 0 for nPCRC. The integration of the three new prongs is illustrated in Figure 2.

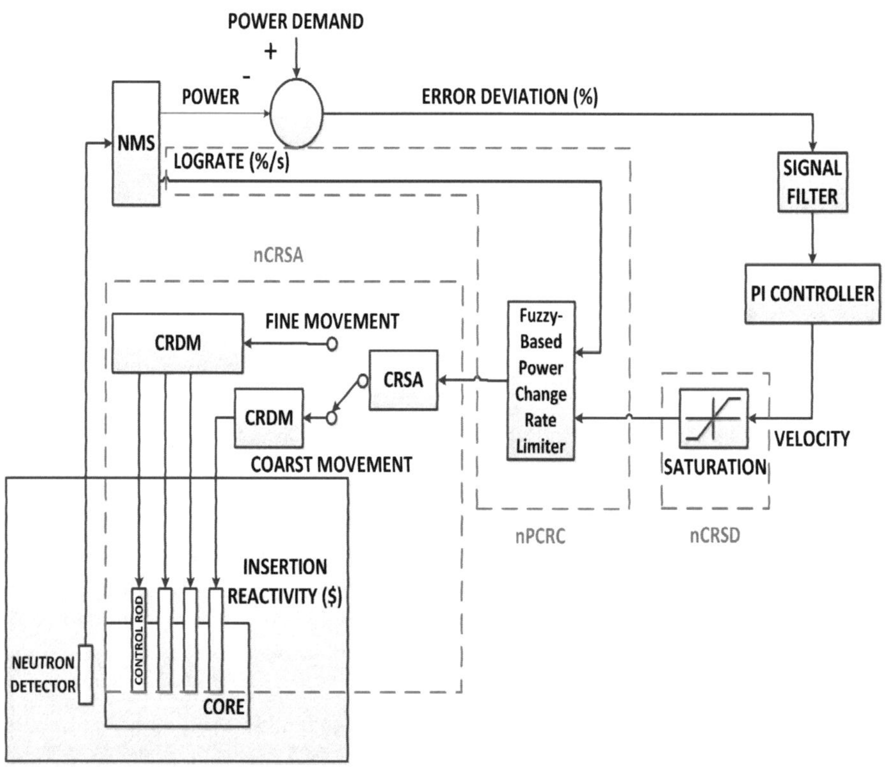

Figure 2. Block Diagram of the Multipronged Core Power Control.

The improved version of FCA using the single-prong control strategy has been published in [9-11] with the sequence prong as shown in Figure 3. The design of nPCRC prong based on [11] requires minimum changes of original structure FCA.

The combination of new two-prongs and the new three-prongs which defines the new multiprong control strategy. The input signal starts with an error deviation that leads to actuation signal at the CRDM. The new prong component in the block diagram is marked with green colour.

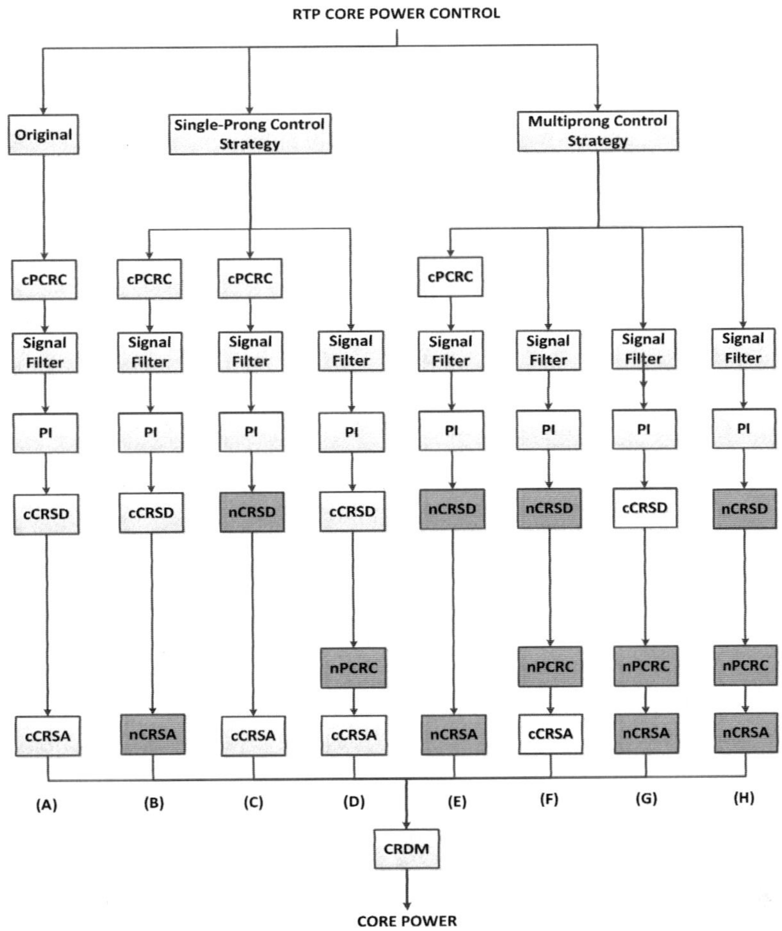

Figure 3. Block Diagram of RTP Core Power Control Design Framework.

In this study, for easier to reference the strategy name from Figure 3 will need to be defined with an alphabet in Table 1. To investigate the capability of the proposed control strategy in Table 1, the strategy is simulated in the next section.

Table 1. Defined strategy name for types of core power control system in RTP

Type		*Label*
Original	FCA-cCRSA-cCRSD-cPCRC	(A)
New Single-Prong Control Strategy	FCA-**nCRSA**-cCRSD-cPCRC	(B)
	FCA-cCRSA-**nCRSD**-cPCRC	(C)
	FCA-cCRSA-cCRSD-**nPCRC**	(D)
New Multiprong Control Strategy	FCA-**nCRSA-nCRSD**-cPCRC	(E)
	FCA-cCRSA-**nCRSD-nPCRC**	(F)
	FCA-**nCRSA**-cCRSD-**nPCRC**	(G)
	FCA-**nCRSA-nCRSD-nPCRC**	(H)

5. Results and discussion

The linearized model of the TRIGA reactor is modelled using Eq. 2 with the existing FCA in Eq. 3. The validation of the RTP simulation model using the experimental data has been reported in our previous work [9-10]. The FCA parameters are set to 12.3, 0.08, 10 and 1e-7 for *G1*, *G2*, *G3*, and *G4*, respectively.

The combination of the conventional, new single-prong and, the new multiprong control strategy in Figure 3 produce eight types of core power control system. The performance for each combination is profiled and analyzed to determine its suitability for the system. To observe the effect of nPCRC, the rate of power change is set to ±12.5%FP/s as maximum allowable limit set by the Final Safety Analysis Report.

In the simulation, an initial low power of 10% FP is set for $0 \leq t < 200$s and 75% nominal core power is set for $200\text{s} \leq t \leq 700$s to simulate condition of sudden power demand. The performance of the tested prongs are evaluated through the power maneuvering of the core power control and actuation signal of the CRDM as shown in Figure 4 and Figure 5 respectively.

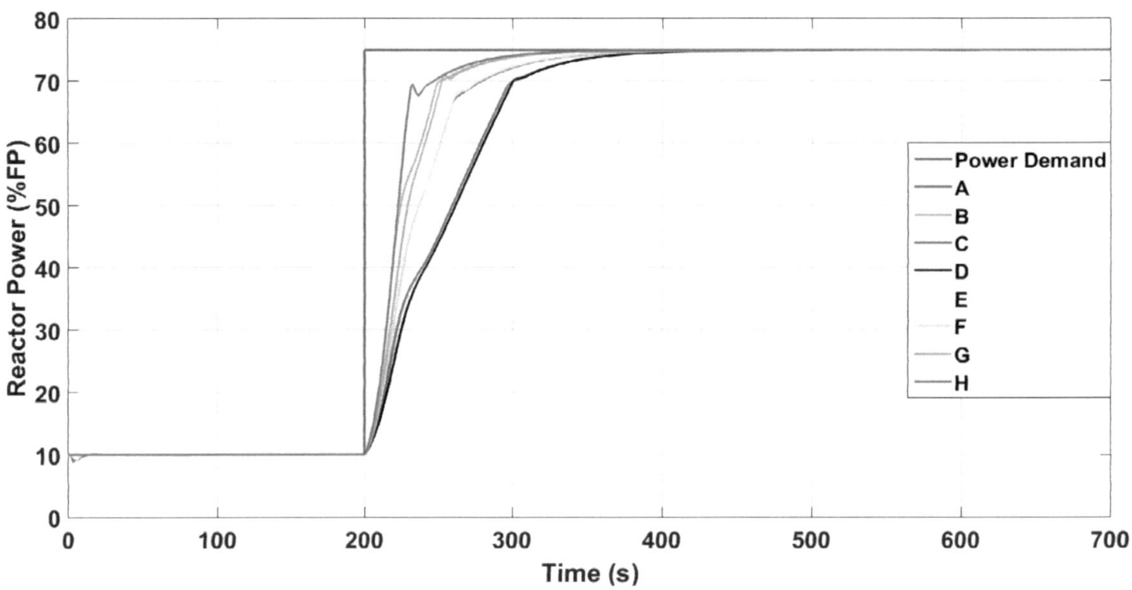

Figure 4. The power tracking performance with a single and multiprong control strategy approach.

The performance of the new single-prong and multiprong control strategy are benchmarked with the existing (A) control strategy represented by the red line in Figure 4 and Figure 5. The quantitative performance characteristic of the existing, new single-prong, and new multiprong control strategy are presented in Table 2. In addition, workload and energy released are calculated based on the accumulation signal and area under the graph.

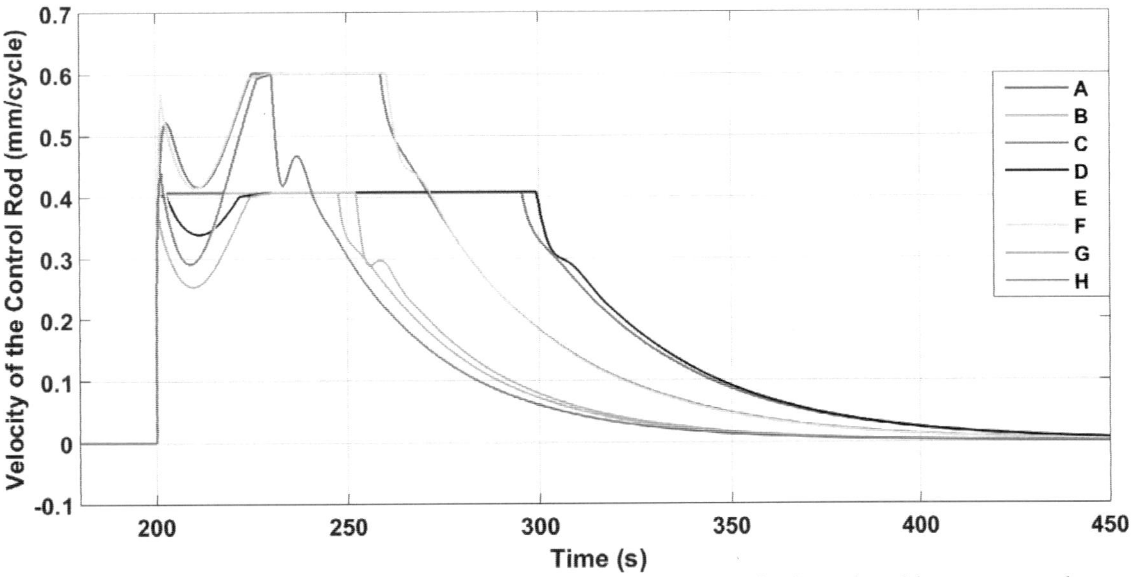

Figure 5. The comparison of velocity control signal between single and multiprong control strategy approach.

Generally, Figure 4 shows that the new multiprong control strategy which is indicated by yellow, turquoise, brown, grey lines able to significantly improve the power tracking performance than the original FCA. However, prong combination such as (F) which represented by the turquoise line generate limited effect on tracking control performance that is inferior to the single-prong control strategy such as (B) which indicated by green line. All multipronged control strategy able to improve the smoothness of the velocity signal which is beneficial in reducing the possibilities of mechanical drive damage at CRDM as shown in Figure 5. Also, minimise the power output overshoot for wide-range operation at RTP is guaranteed.

Table 2. Transient and steady-state response for multipronged core power control at 750kW.

Type		*Settling Time, T_s (s)*	*Rise Time, T_r (s)*	*Percent Overshoot, P_{os} (%)*	*Chattering Error, Δe_{ce} (%)*	*Work Load (mm/cycle) / Energy Released (kW-h)*
Original	(A)	112.5	84.0	0.001076	0.027155	105.56 / 5334.69
New Single-Prong Control Strategy	(B)	62.0	40.0	0.000639	0.002947	60.00 / 5774.26
	(C)	90.0	59.5	0.000868	0.015175	105.63 / 5590.44
	(D)	113.5	86.0	0.001308	0.026213	105.55 / 5299.06
New Multiprong Control Strategy	(E)	57.5	31.5	0.000562	0.002544	60.00 / 5830.51
	(F)	88.5	58.8	0.001067	0.013630	105.63 / 5594.64
	(G)	63.5	42.0	0.000863	0.002940	59.98 / 5723.76
	(H)	55.5	23.0	0.000744	0.002263	60.01 / 5837.95

By referring to Table 2, the (B) design provides the most responsive reactivity compared to other single-prong control strategy with reduced settling and rise times. In addition, it is also capable of significantly reducing the workload and optimizes energy release from the reactor core. By changing the control rod speed in (C), this prong can still produce a small settling time and chattering error better performance compare to (D). However, (C) still unable to reduce non-smooth control surface during transient. The higher control rod speed can overcome the wind-up problem and provide high reactivity insertion rate. However, it may not be the best velocity signal to reduce the workload without the help from nCRSA prong.

The core power tracking performance based on the application of fuzzy logic for PCRC such as (D) produced result almost similar to the original FCA (A) except for small chattering error due to less sensitivity effect introduced by nPCRC prong. Therefore, (D) is not sufficient to improve the current tracking performance without combination with other prong.

In the case of multiprong control strategy, the combination of nCRSA and increase of the control rod speed (nCRSD) label as (E) able to boost up the tracking performance to be at par with the single-prong strategy (B). The (E) is the best solution to eliminate the unsmooth control surface during the transient region. Based on the data from Table 2, the (H) is the best performing control strategy in term of reducing settling time, rise time, and minimum chattering error with high optimal energy release. However, there is still room for improvement in term of the smooth control surface and actuation signal performances during transient which mainly caused by the effect nPCRC.

Based on several predefined performance criteria [10], the performance of all multipronged types are summaries in Table 3. Overall, all multipronged types produced small settling time and chattering error which fulfil the two main criteria in designing the core power system. However, the (E) type is advantageous since it able to produce better result in controlling the reactor power in all aspects. With nPCRC, (H) control strategy is a better choice in the case for a long period of operation which produce less chattering error, better drivability of the CRDM, and better optimization for high energy release in the core.

Table 3. Profiling of CRSA, CRSD and PCRC for core power control in TRIGA

Type		*Small settling time*	*Fast response*	*Smooth control surface during transient*	*Smooth actuation signal during transient*	*Small chattering error during steady-state*	*Small actuation signal during steady-state*	*High Energy Released*
New Single-Prong Control Strategy	(B)	√				√	√	
	(C)	√				√		
	(D)					√		
New Multiprong Control Strategy	(E)	√	√	√	√	√	√	√
	(F)	√				√		
	(G)	√				√	√	
	(H)	√	√			√	√	√

6. Conclusion

A study of the multipronged core power control strategy for the TRIGA reactor is presented to provide greater level of operational safety with a realistic design towards solving practical problems. The new multiprong control strategy is analyzed and profiled to improve the current tracking performance of core power control at RTP. Instead of using new single-prong control strategy with nCRSA, nCRSD, and nPCRC, the new multiprong control strategy consider other alternative to offer wide solutions in optimizating the core power control performance reactor. The significant finding from this study is that the new multipronged core power system able to provide better understanding about the relationship between element in the core power control system and provide insight for future research on this area.

7. References

[1] Zarei M 2020 Progress in nuclear energy A physically based PID controller for the power maneuvering of nuclear reactors *Prog. Nucl. Energy* **127** 103431.

[2] Shen J, Duan H, Zhang B, Wang J, Ji J S, Wang J, Pan L, Wang X, Zhao K, Ying B, Tang S, Zhang J, Liang C, Sun H, Lv Y, Li Y, Li T, Li L, Liu H, Zhang L, Wang L and Shi X 2020 Prevention and control of COVID-19 in public transportation: Experience from China *Environ. Pollut.* **266.**

[3] Jalil S A, ElKabbash M, Zihao li, Zhang J, Singh S, Zhan Z and Guo C 2020 Multipronged heat-exchanger based on femtosecond laser-nano/microstructured Aluminum for thermoelectric heat scavengers *Nano Energy* **75** 104987.

[4] Benítez-Read J S, Ruan D, Najera-Hernandez M, Perez-Clavel B, Pacheco-Sotelo J O and Lopez-Callejas R 2005 Comparison between a continuous and discrete method for the aggregation and deffuzification stages of a TRIGA reactor power fuzzy controller *Prog. Nucl. Energy* **46** 309–20.

[5] Pérez-Cruz J H and Poznyak A 2008 Neural control for power ascent of a TRIGA reactor *Proc. Am. Control Conf.* **2190** (5).

[6] Eom M, Chwa D and Baang D 2015 Robust Disturbance Observer-Based Feedback Linearization Control for a Research Reactor Considering a Power Change Rate Constraint *IEEE Trans. Nucl. Sci.* **1**(12).

[7] Kim J H, Park S H and Na M G 2014 Design of a model predictive load-following controller by discrete optimization of control rod speed for PWRs *Ann. Nucl. Energy* **71(**343) 51.

[8] Minhat M S, Selamat H and Mohd Subha N A 2018 Adaptive control method for core power control in TRIGA Mark II reactor *IOP Conf. Ser. Mater. Sci. Eng.* **298.**

[9] Minhat M S, Mohd Subha N A, Hassan F and Mohamad Nordin N 2020 An improved control rod selection algorithm for core power control at TRIGA PUSPATI Reactor *J. Mech. Eng. Sci.* **14(**6362) 79

[10] Minhat M S, Adilla N, Subha M, Hassan F, Ahmad A and Rashid A 2020 Progress in Nuclear Energy Profiling and analysis of control rod speed design on core power control for TRIGA reactor *Prog. Nucl. Energy* **128** 103481

[11] Minhat M S, Adilla N, Subha M and Hassan F 2020 Application of fuzzy logic for power change rate constraint in core power control at Reaktor TRIGA PUSPATI *IOP Conf. Ser. Mater. Sci. Eng.* **785**

[12] Cammi A, Ponciroli R, Borio A, Magrotti G, Prata M, Chiesa D and Previtali E 2013 A zero dimensional model for simulation of TRIGA Mark II dynamic response *Prog. Nucl. Energy* **68(**43) 54.

[13] Malaysian Nuclear Agency 2020 *Safety Analysis Report for PUSPATI TRIGA Reactor.*

Acknowledgments
This work was supported by a Fundamental Research Grant Scheme (FRGS) grant (FRGS/1/2017/TK04/UTM/02/55) from the Ministry of Higher Education, Malaysia. The author would like to thank all individuals who have contributed to this paper especially to Dr. Nurul Adilla Binti Mohd Subha from Universiti Teknologi Malaysia and to the Malaysian Nuclear Agency under the Ministry of Science, Technology, and Innovation, Malaysia (MOSTI) for the support.

An analytical approach of the chromaticity correction for compact hadron river

Leo Kwee Wah[1,a)], Ken Takayama[2], Toshikazu Adachi[2], Tadamichi Kawakubo[2] and Tanuja Dixit[3]

[1]Nuclear Agency, Bangi 43000 Kajang. Selangor. Malaysia
[2]High Energy Accelerator Research Organization (KEK), Tsukuba, Ibaraki, Japan.
[3]Society for Applied Microwave Electronics Engineering & Research (SAMEER), Mumbai, India

[a]leo@nuclearmalaysia.gov.my

Abstract. A design of a compact hadron driver for future cancer therapies based on the induction synchrotron concept is given. In order to realize a slow extraction technique in a fast cycling synchrotron, which allows the energy sweep beam scanning, the zero momentum-dispersion *D(s)* region and high flat *D(s)* region are necessary. The present design lattice which meets both requirements. In this study, details of the lattice parameters is introduced and an analytical solution for chromaticity correction are described and discussed.

1. Introduction

A lattice design of a compact hadron driver for future cancer therapies as shown in Figure 1 has been identified [1]. Whereby the lattice has the two fold symmetry with a circumference of 52.8 m, 2 m long dispersion free straight section, and 3 m long large flat dispersion straight section [1][2]. The lattice parameter is design based on the linear term of motion equation. Hence, the individual common components will be treated as a 3 X 3 matrices as the following equations [3][4].

For drift space,

$$\text{Matrix Drift}, M_{Drift} = \begin{bmatrix} 1 & L & 0 \\ 0 & 1 & 0 \\ 0 & 0 & 1 \end{bmatrix} \quad (1)$$

For focusing magnet,

$$\text{Matrix Focusing}, M_F = \begin{bmatrix} \cos[\sqrt{k}L] & 1/\sqrt{k}\sin[\sqrt{k}L] & 0 \\ -\sqrt{k}\sin[\sqrt{k}L] & \cos[\sqrt{k}L] & 0 \\ 0 & 0 & 1 \end{bmatrix} \quad (2)$$

For defocusing magnet,

$$\text{Matrix Defocusing}, M_D = \begin{bmatrix} \cosh\left[\sqrt{|k|}L\right] & 1/\sqrt{|k|}\sinh\left[\sqrt{|k|}L\right] & 0 \\ \sqrt{|k|}\sinh\left[\sqrt{|k|}L\right] & \cosh\left[\sqrt{|k|}L\right] & 0 \\ 0 & 0 & 1 \end{bmatrix} \quad (3)$$

For bending magnet,

$$\text{Matrix Bending}, M_B = \begin{bmatrix} \cos[\sqrt{k}L] & \frac{1}{\sqrt{k}}\sin[\sqrt{k}L] & \frac{1}{\sqrt{k}}(1-\cos[\sqrt{k}L]) \\ -\sqrt{k}\sin[\sqrt{k}L] & \cos[\sqrt{k}L] & \sin[\sqrt{k}L] \\ 0 & 0 & 1 \end{bmatrix} \quad (4)$$

where:

$k = k - \text{value of the related magnets}$
$L = \text{Length of the related magnets}$

For perturbation term, which can be treated as nonlinear term and it is an essential term must be evaluated, so that the beam can beam accelerated inside the accelerator ring with the designed lattice. Therefore, the chromaticity as one of perturbation term caused by the momentum deviation of the particles must be considered and will be discussed at the following section.

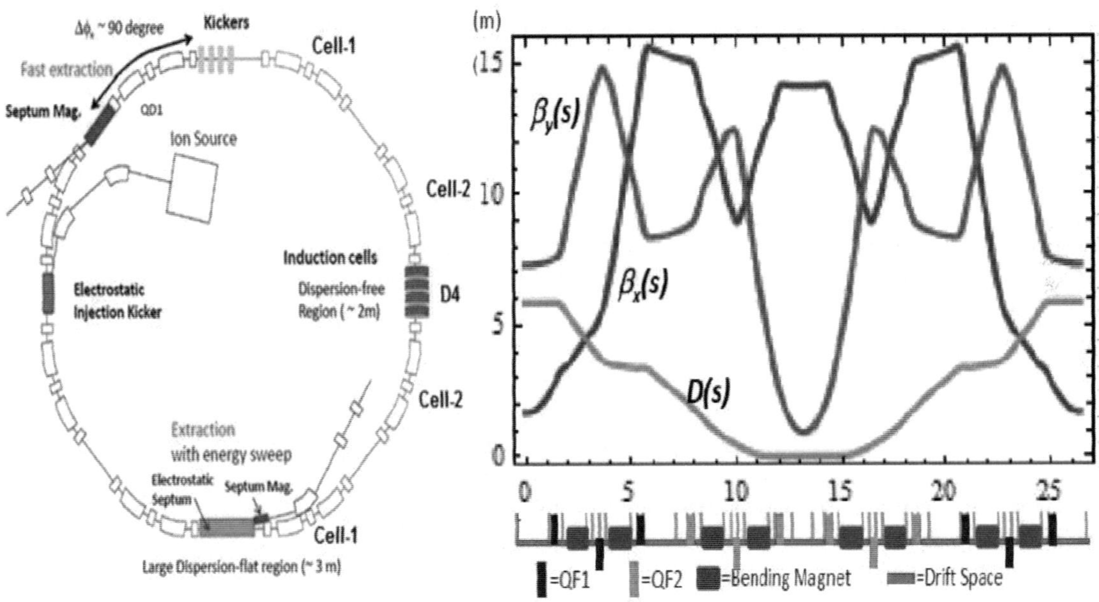

Figure 1. Outline of the ring and its lattice [1].

..

A lattice configuration of half ring with two fold symmetry is shown in Figure 2. The configuration is the combination of two cells with a drift space as Cell A-Cell B- Drift Space 4- Cell B-Cell A.

IOP Conf. Series: Materials Science and Engineering 1106 (2021) 012002 doi:10.1088/1757-899X/1106/1/012002

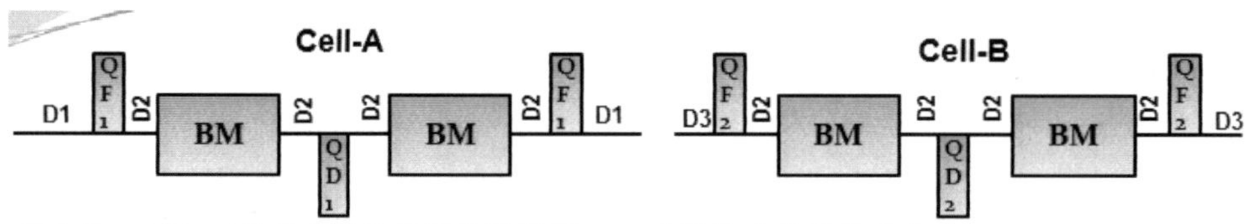

Configuration =Cell-A X Cell-B X Drift Space 4 X Cell-B X Cell-A

Drift Space 4 = Free Dispersion Region

A/Z=2, Energy = 200 MeV/U Total length of 26.4 m

Drift Space	S [m]
D1	1.5
D2	0.3
D3	0.5
D4	1.0

Bending Magnet	Bending Angle, Θ[°]	Edge Angle, Θ[°]	Curvature, ρ [m]	Magnet Length, l_B [m]
BM	22.5	11.25	2.8662	1.1256

Quad Magnet	Length, l_Q[m]	K-Value [m^{-2}]
QF1	0.30	0.75568
QD1	0.30	-0.53774
QF2	0.30	0.40173
QD2	0.30	-0.8957

Figure 2. Configuration of the lattice.

Therefore, the 3 X 3 transfer matrix of the lattice and the Courant-Snyder parameters as the following.

For horizontal axis,

$$\text{Transfer matrix of Cell } A, M_{Hcell\,A} = M_{Drift1}M_{F1}M_{Drift2}M_{B}M_{Drift2}M_{D1}M_{Drift2}M_{B}M_{Drift2}M_{F1}M_{Drift1} \quad (5)$$

$$\text{Transfer matrix of Cell } B, M_{Hcell\,B} = M_{Drift3}M_{F2}M_{Drift2}M_{B}M_{Drift2}M_{D2}M_{Drift2}M_{B}M_{Drift2}M_{F2}M_{Drift3} \quad (6)$$

$$\text{Transfer matrix of the lattice}, M_x = M_{Hcell\,A}M_{Hcell\,B}M_{Drift4}M_{Hcell\,B}M_{Hcell\,A} \quad (7)$$

$$= \begin{bmatrix} Mx_{11} & Mx_{12} & Mx_{13} \\ Mx_{21} & Mx_{22} & Mx_{23} \\ Mx_{31} & Mx_{32} & Mx_{33} \end{bmatrix} \quad (8)$$

For vertical axis,

$$\text{Transfer matrix of } Cell\ A, M_{Vcell\,A} = M_{Drift1}M_{D1}M_{Drift2}M_{B}M_{Drift2}M_{F1}M_{Drift2}M_{B}M_{Drift2}M_{D1}M_{Drift1} \quad (9)$$

$$\text{Transfer matrix of } Cell\ B, M_{Vcell\,B} = M_{Drift3}M_{D2}M_{Drift2}M_{B}M_{Drift2}M_{F2}M_{Drift2}M_{B}M_{Drift2}M_{D2}M_{Drift3} \quad (10)$$

$$\text{Transfer matrix of } the\ lattice, M_y = M_{Vcell\,A}M_{Vcell\,B}M_{Drift4}M_{Vcell\,B}M_{Vcell\,A} \quad (11)$$

$$= \begin{bmatrix} My_{11} & My_{12} & My_{13} \\ My_{21} & My_{22} & My_{23} \\ My_{31} & My_{32} & My_{33} \end{bmatrix} \quad (12)$$

From Equation (9) and (12), the Courant-Snyder parameters can be derived as:

For horizontal axis,

$$\beta_x(s) = (Mx_{11})^2\beta_x(0) - 2\,Mx_{11}Mx_{12}\alpha_x(0) + (Mx_{12})^2\ \gamma_x(0) \tag{13}$$

$$\alpha_x(s) = Mx_{11}Mx_{21}\beta_x(0) + (\,Mx_{12}Mx_{21} + Mx_{11}Mx_{22})\alpha_x(0) - Mx_{12}Mx_{22}\ \gamma_x(0) \tag{14}$$

$$\gamma_x(s) = (Mx_{21})^2\beta_x(0) - 2\,Mx_{21}Mx_{22}\alpha_x(0) + (Mx_{22})^2\ \gamma_x(0) \tag{15}$$

For vertical axis,

$$\beta_y(s) = (My_{11})^2\beta_y(0) - 2\,My_{11}My_{12}\alpha_y(0) + (My_{12})^2\ \gamma_y(0) \tag{16}$$

$$\alpha_y(s) = My_{11}My_{21}\beta_y(0) + (\,My_{12}My_{21} + My_{11}My_{22})\alpha_y(0) - My_{12}My_{22}\ \gamma_y(0) \tag{17}$$

\

$$\gamma_y(s) = (My_{21})^2\beta_y(0) - 2\,My_{21}My_{22}\alpha_y(0) + (My_{22})^2\ \gamma_y(0) \tag{18}$$

Whereby the initial condition of Courant-Snyder parameters can be identified as:

$$\beta_{x,y}(0) = \beta_{x,y}(C) \tag{19}$$

$$\alpha_{x,y}(0) = \alpha_{x,y}(C) \tag{20}$$

$$\gamma_{x,y}(0) = \gamma_{x,y}(C) \tag{21}$$

where:

$C =$ The end postioin of the ring

For Equation (19), (20) and (21), the beta function along the ring can be identified and it will be used to analysis the chromaticity at the following section.

2. Chromaticity and chromaticity correction

Perturbations of beam are occurred in the beam transportation even in the absence of magnet field and alignment errors. The momentum deviations of particle from the ideal design energy so-called as chromaticity which causes the perturbations in the solutions of the equations of motion [3].

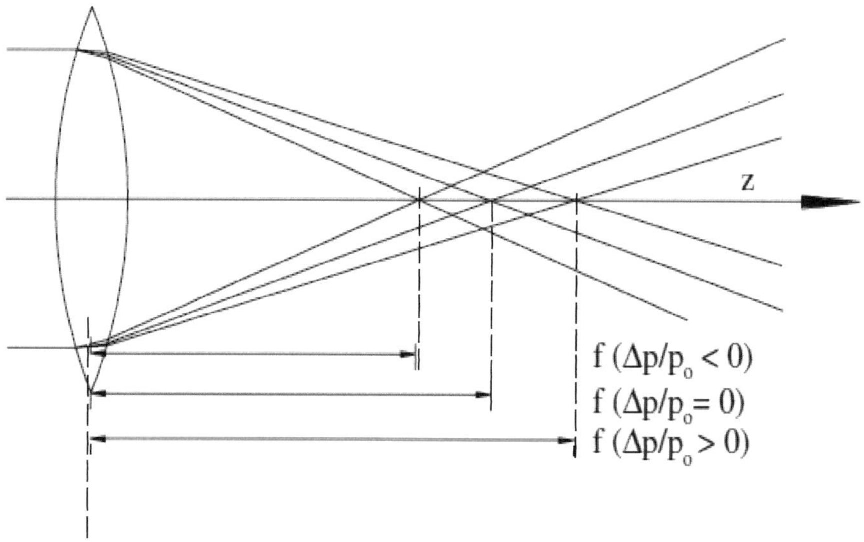

Figure 3. Chromaticity focusing errors.

As shown in Figure 3, the lowest order chromatic perturbation is caused by the variation of the focal length of the quadrupoles with energy which the different momentum deviation is occurred. This kind of error is well known from light optics, where a correction of this chromatic aberration by using different kinds of glasses for the lenses in a composite focusing system. For the chromatic perturbations correction in beam optics, the particles with different energies can be separated by the introduction of a dispersion function. Once the particles are separated by energy, different focusing corrections depending on the energy of the particles are introduced. As well-known that higher energy particles are focused less than ideal energy particles and lower energy particles are over focused. For such phenomena correction, a correction magnet: sextupoles with the property as focusing for higher energy particles and defocusing for lower energy particles can be utilized as shown in Figure 4.

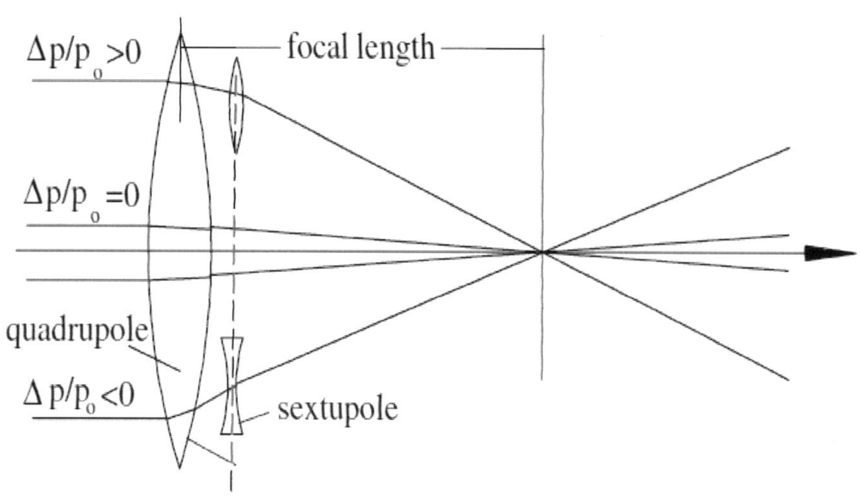

Figure 4. Chromaticity correction with sextupoles.

The variation of tunes with energy is called the chromaticity and is defined by [3]:

$$\xi = \frac{\Delta v}{\Delta p/p_0} \tag{22}$$

where:

$\Delta v =$ Tune shift

$\frac{\Delta p}{p_0} =$ momentum deviation

Equation (22) is applicable for both circular and open beam lines. Using the definition of the chromaticity and with the derivation, the natural chromaticity at the quadrupoles of the circular accelerators for horizontal and vertical axis can be defined as [3][5]:

$$\xi_x = -\frac{1}{4\pi}\int_0^C k_x(s)\beta_x(s)ds \tag{23}$$

$$\xi_y = -\frac{1}{4\pi}\int_0^C k_y(s)\beta_y(s)ds \tag{24}$$

where:

$k_{x,y}(s) = k -$ value of horizontal or vertical quadrupole magnets as function of position

$\beta_{x,y}(s) =$ Horizontal or vertical beta as function of position

To correct both the horizontal and the vertical chromaticity, two different groups of horizontal and vertical sextupoles are required. For chromaticity correction by sextupoles, the required field strength and number of sextupoles need to be optimized. Such of chromaticity correction for individual horizontal and vertical sextupoles can be defined as:

$$\xi_{sextx} = \frac{1}{4\pi}\int_0^C 2G(s)\eta(s)\beta_x(s)ds \tag{25}$$

$$\xi_{sexty} = \frac{1}{4\pi}\int_0^C G(s)\eta(s)\beta_y(s)ds \tag{26}$$

where:

$G(s) =$ Strength of sextupoles magnets as function of position

$\beta_{x,y}(s) =$ Horizontal or vertical beta as function of position

From Equation (23), (24), (25) and (26), we're try to find the strength and number of the sextupoles magnets by compensation of the natural chromaticity as defined follow:

$$\xi_x + \sum_m \xi_{sextx,m} = 0 \tag{27}$$

$$\xi_y + \sum_m \xi_{sexty,m} = 0 \tag{28}$$

where:

$m =$ Number of sextupole magnets

3. Result and discussion

In general, the approach based on Equation (27) and (28) is used to evaluate the number of sextupoles and its strength and to determine the best combination to correct the chromaticities without causing any sextupoles magnetic design problems. For such evaluation, momentum deviation is assumed as 0.01 and three cases have been evaluated as follow.

Case A

In this lattice, the two sextupoles with the different field strength as G(s) and F(S) are inserted as shown in Figure 5. It is located at the position with the different Courant-Snyder parameters.

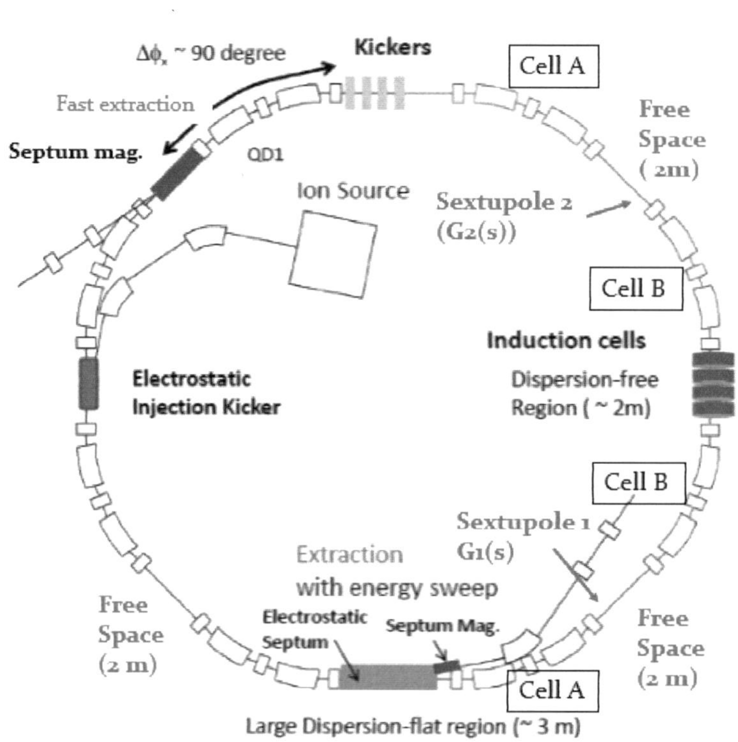

Figure 5. Case A: Two sextupoles are inserted into the ring.

By considering the sextupoles configuration as shown in "Case A" , the location and required parameters of the Sextupoles are fixed and the field strength can be identified from Equation (24) and (25) as shown in Table 1.

Table 1. Parameters of sextupoles for Case A.

Sextupole	Length of Sextupoles, l [m]	Initial Position, S[m]	Field Strength, $G_i[m^{-3}]$
Sextupole 1	0.2	6.1512	28.4157
Sextupole 2	0.2	18.8536	-38.3670

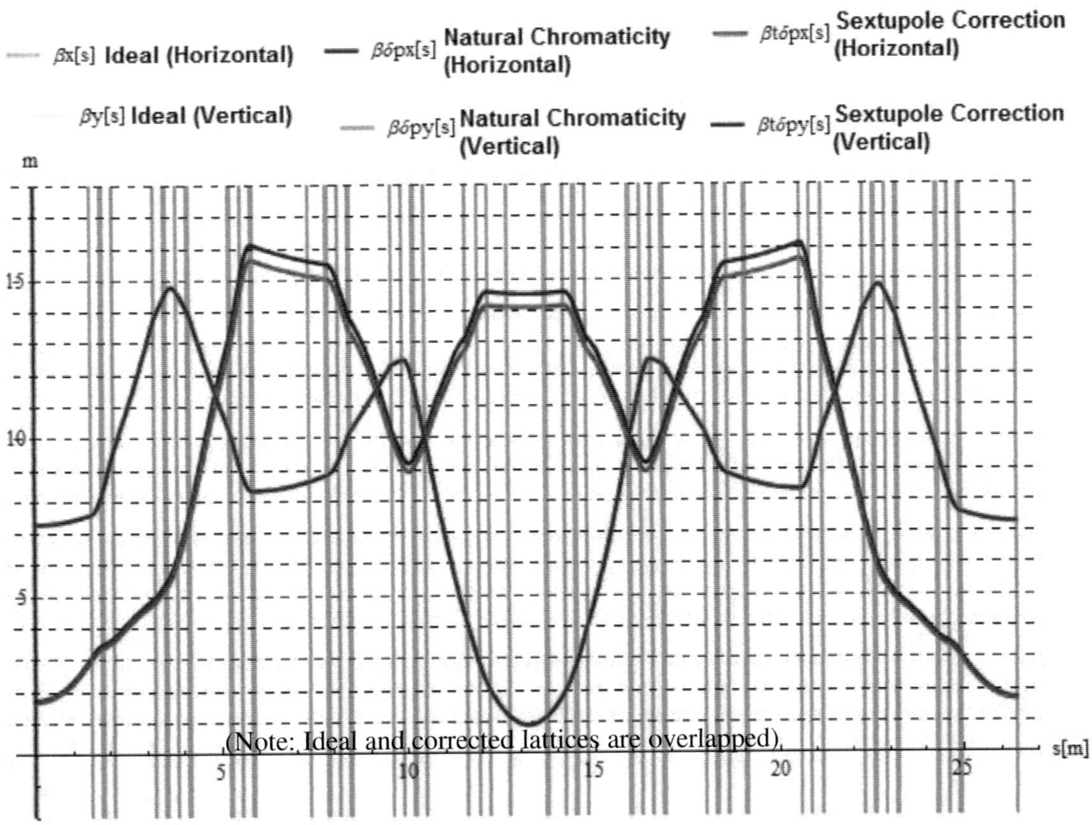

Figure 6. Horizontal and vertical lattice: Ideal, natural chromaticity and corrected lattice by two sextupoles with $\Delta p/p=0.01$.

From Figure 6, the field strengths of $G_{1,2}(s)$ 28.4157 and -38.3670 can compensate the chromaticity perturbation ideally.

Case B

In this lattice, four sextupoles are inserted as shown in Figure 7. Sext1 and Sext3 have same strength and located at the position with the same Courant-Snyder parameters. Also Sext2 and Sext4 have same strength and located at the position with the same Courant-Snyder parameters.

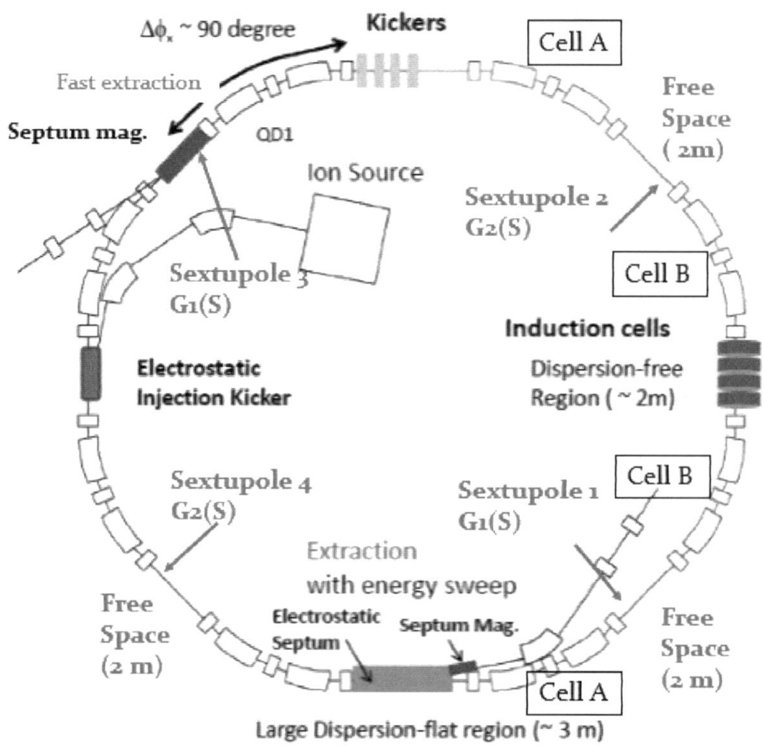

Figure 7. Case B: Four sextupoles are inserted into the ring.

By considering the sextupoles configuration as shown in "Case B" , the locations and parameters of the Sextupoles are fixed and the field strength can be identified from Equation (24) and (25) as shown in Table 2.

Table 2. Parameters of sextupoles for Case B.

Sextupole	Length of Sextupoles, l [m]	Initial Position, S[m]	Field Strength, $G_i[m^{-3}]$
Sextupole 1	0.2	6.1512	14.0728
Sextupole 2	0.2	18.8536	-19.1835
Sextupole 3	0.2	34.0560	14.0728
Sextupole 4	0.2	45.7584	-19.1835

.

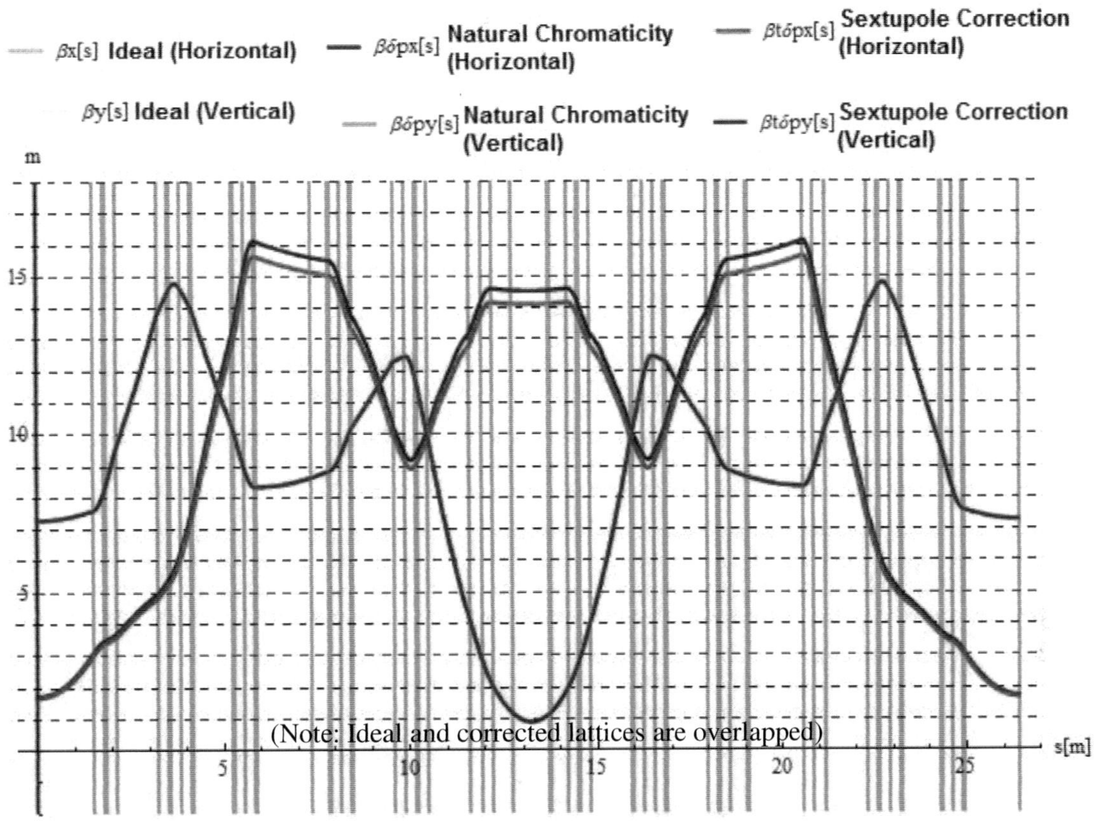

Figure 8. Horizontal and vertical lattice: Ideal, natural chromaticity and corrected lattice by four sextupoles with $\Delta p/p=0.01$

From Figure 8, the field strengths of four Sextupoles with $G_{1,3}(s)$ as 14.0728 and $G_{2,4}(s)$ as -19.1835 can compensate the chromaticity perturbation ideally.

Case C

In this lattice, the four sextupoles are inserted as shown in Figure 9. Sext1 and Sext3 have same strength and located at the position with the same Courant-Snyder parameters. Also Sext2 and Sext4 have same strength and located at the position with the same Courant-Snyder parameters

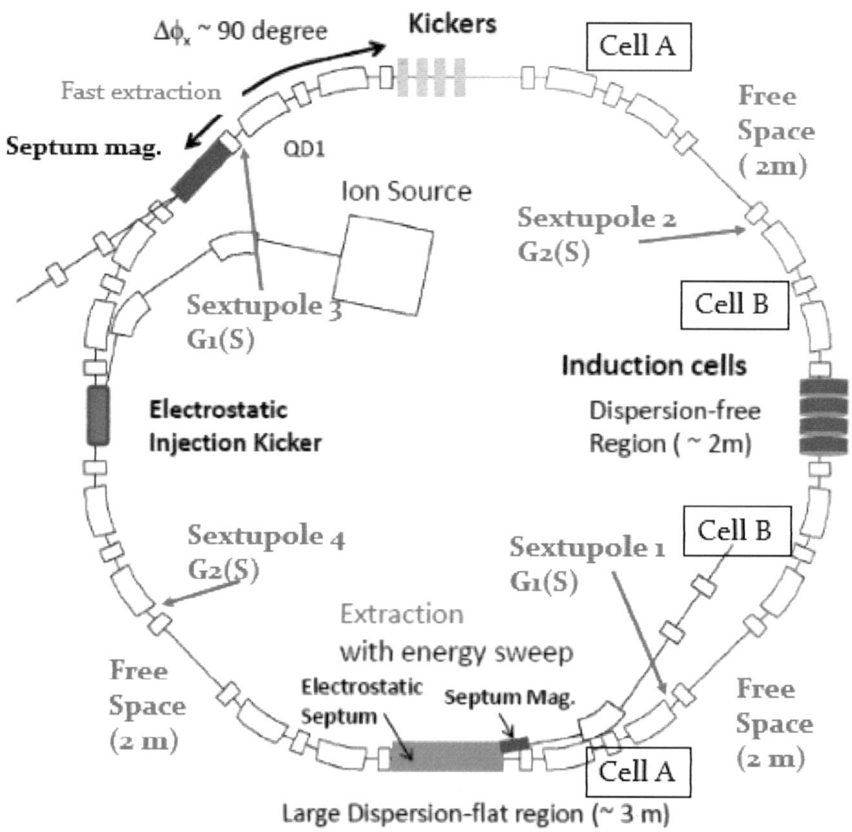

Figure 9. Case C: Four sextupoles are inserted into the ring

By considering the sextupoles configuration as shown in "Case C" , the locations and parameters of the sextupoles are fixed and the field strength can be identified from Equation (24) and (25) as shown in Table 3.

Table 3. Parameters of sextupoles for Case C

Sextupole	Length of Sextupoles, l [m]	Initial Position, S[m]	Field Strength , $G_i[m^{-3}]$
Sextupole 1	0.2	5.3012	0.74776
Sextupole 2	0.2	24.3584	-0.67603
Sextupole 3	0.2	32.0060	0.74776
Sextupole 4	0.2	44.4084	-0.67603

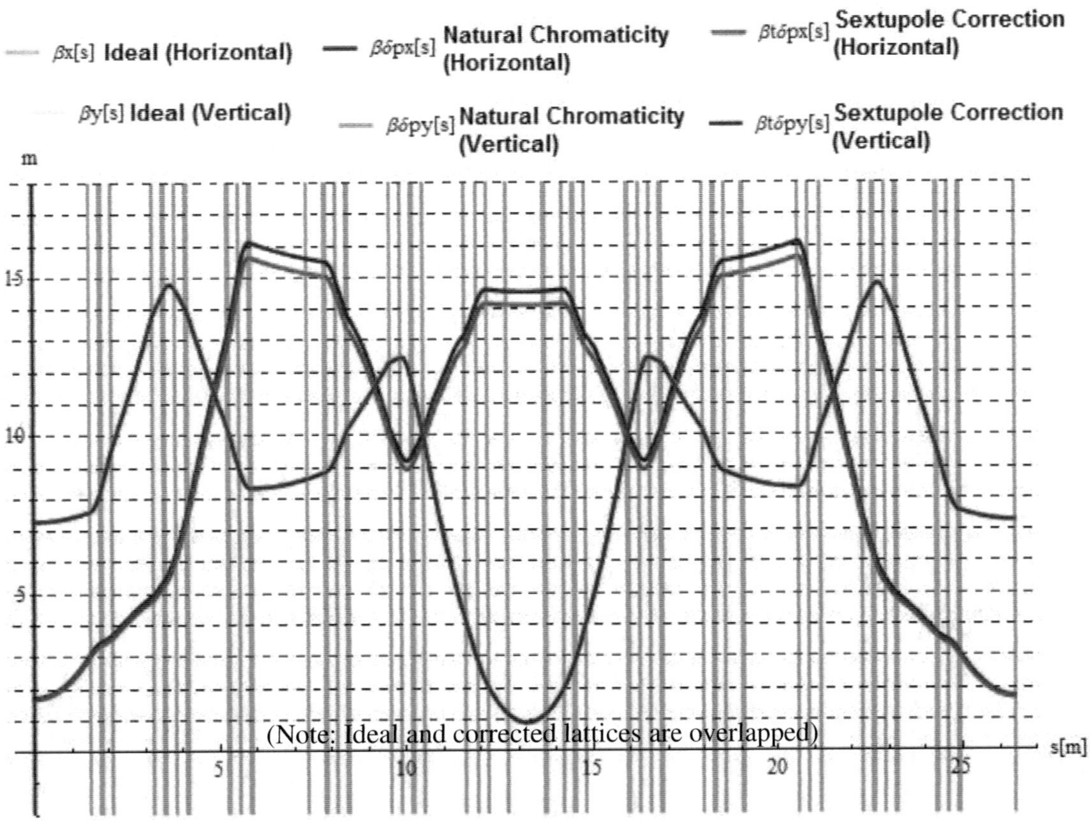

Figure 10. Horizontal and vertical lattice: Ideal, natural chromaticity and corrected lattice by four sextupoles with $\Delta p/p=0.01$

From Figure 10, the field strengths of four sextupoles with $G_{1,3}(s)$ as 0.74776 and $G_{2,4}(s)$ as -0.67603 can compensate the chromaticity perturbation ideally.

For three cases, we can summarize in Table 4. We can realized that more gentle correction uses two groups of sextupoles with individual magnets distributed more evenly around the circular accelerator and the total required sextupole strength is spread over all sextupoles and the strength could be reduced. These features are desired because the smaller strength will make the sextupole magnet much smaller.

Table 4. Comparison of Case A, B and C.

Case	**Types of Sextupoles**	**Number of Sextupoles**	**$G_1(s)[m^{-3}]$**	**$G_2(s)[m^{-3}]$**	**$\Delta\varepsilon_x$**	**$\Delta\varepsilon_y$**
A	2	2	28.1457	-38.367	0	0
B	2	4	14.0728	-19.1835	0	0
C	2	4	0.74776	-0.6760	0	0

According Table 4, we noticed that three cases able to compensate the natural chromaticites perfectly. In order to compensate the natural chromaticity effectively, two or more sextupoles with different strength and Courant-Snyder parameters are required. The number of sextupoles plays important role as well. With the increment of the number of sextupoles, it will decrease the required strength of the sextupoles. Meanwhile, case B and case C are used to evaluate the effectiveness of the sextupoles location. As obvious, the location is important and the larger Courant-Snyder parameters is a major compensator for the chromaticity correction. Hence, case C is an ideal configuration of natural chromaticity correction by using sextupoles.

4. Conclusion

In this paper, approach of natural chromaticity correction by sextupoles is introduced. The number of the sextupoles have been optmized and the strength is identified. Case C is an ideal case with 4 sextupoles and two field strengths and able to compensate the chromaticity perfectly. In near future, the nonlinear perturbation term of the Closed Orbit Distortion (COD) must be considered and evaluated.

5. References

[1] Leo K W, Monma T, Adachi T, Kawakubo T, Dixit T and Takayama K 2016 Compact hadron driver for cancer therapies using continuous energy sweep scanning *Phys. Rev. ST AB* **19** 042802.
[2] Leo K W, Adachi T, Kawakubo T, Monma T, Dixit T and Takayama K 2015 A compact hadron driver for cancer therapies using continuous energy sweep scanning *Proc. of the 13th Int. Conf. on Heavy Ion Accelerator Technol.* (Yokohama).
[3] Wiedemann H *Particle Accelerator Phys. 3rd ed.* (Berlin: Springer, New York: Heidelberg) ISBN-13 978-3-540-49043-2.
[4] Adachi T 2014 *KEK-Nuclear Malaysia-Sokendai Accelerator School in Nuclear Malaysia* (1st-4th December 2014).
[5] Peggs S and Satogata T 2017 *Intro. to Accelerator Dynamics 1st ed.* (Cambridge University Press)

Acknowledgements
We would like to express our gratitude to Ministry of Science, Technology and Innovation (MOSTI), Malaysia and High Energy Accelerator Research Organization (KEK), Japan for supporting the allocation to collaborate between Malaysian Nuclear Agency and KEK for this study.

A Malaysian Secondary Standard Dosimetry Laboratory participation in the IAEA/WHO Postal Dose Quality Audit: Data Analysis 2009 – 2019

N Abdullah[1,a] and M T Dolah[1]

[1]Radiation Metrology Group, Malaysian Nuclear Agency (Nuklear Malaysia), 43000 Kajang, Selangor, Malaysia

[a]hayatie@nuclearmalaysia.gov.my

Abstract. Participation of Secondary Standard Dosimetry Laboratory (SSDL) of Malaysia in the IAEA/WHO Postal Dose Quality Audit Service is very important in confirming the traceability of absorbed dose measurements provided to the radiotherapy centres. In the audit, the SSDL was supplied with a set of dosimeter for irradiation with absorbed dose to water of 200 cGy in reference conditions using ^{60}Co teletherapy or medical linear accelerator. In the period of 2009 to 2019, a total of 10 photon beams from ^{60}Co teletherapy unit, 6 MV X-ray and 10 MV X-ray produced by medical linear accelerator have been audited. The results of the audit demonstrate that the mean of percentage deviations between the SSDL and the IAEA measurements was -0.35% with standard deviation of 0.86%. The deviations vary between a minimum percentage relative deviation of -2.5% and a maximum of 0.6%. These results were considered satisfactory in which complied with the IAEA's acceptance limit of ± 3.5%.

1. Introduction

Radiation therapy is an important method to treat cancer with the aim to deliver as maximum doses as possible to the tumour while protecting the surrounding healthy tissue and organs at risk [1-2]. It is essential to confirm the accuracy of dose delivered to patient is within the tolerance limit of 5% at 95% confidence level [3-4]. Radiation underdose or overdose delivered clinically should be avoided in order to improve the effectiveness of the cancer treatment as well as to avoid radiation injuries to patient [5-11]. Therefore, a postal dosimetry audit often conducted as a part of Quality Assurance Programme (QAP) in radiotherapy where the dosimeters are mailed to the radiotherapy centres for irradiation to verify the accuracy of the dosimetry measurements [12-13]. In addition, the audit can be used as a tool for confirming the competency of medical physicist and ensuring the measurement traceability provided by national dosimetry laboratory [1][7-9].

The Secondary Standard Dosimetry Laboratory (SSDL) of Malaysian Nuclear Agency (Nuklear Malaysia) has been appointed by the National Metrology Institute of Malaysia (NMIM) as the designated institute for ionizing radiation measurement in Malaysia. The laboratory carries the responsibility for delivering precise, accurate and reliable dosimetry calibrating factors to the users in compliance to the Atomic Energy Licensing Act 1984 (ACT 304) as well as keeping abreast with the development of SSDLs network worldwide. Besides being a member of the International Atomic Energy Agency (IAEA) / World Health Organization (WHO) Network of SSDLs since 1979, the SSDL of Nuklear Malaysia is also a member of the Asia Pacific Metrology Programme (APMP) Technical Committee on Ionizing Radiation (TCRI) since 2005 to strengthen the international network in ionizing radiation standard measurement.

The SSDL of Nuklear Malaysia has been participating in the IAEA/WHO postal thermoluminescent dosimeter (TLD) and postal ionization chamber intercomparison programme to verify the accuracy of the secondary standard radiation source and ionization chambers used in the calibration service in Malaysia. Samat *et al.* [11] reported that the SSDL of Nuklear Malaysia has participated in this programme for 37 times since 1985 – 2008. Over a period of 28 years, the laboratory has ensured that therapy-level dose within acceptance limits of 1.5% and 3.5% for postal ionization chamber and TLD, respectively.

2. Materials and method

In the audit, the SSDL was provided with instruction and data sheets, a set of TLD (three capsules of TLD in one set where two capsules to be irradiated and a control TLD) and the IAEA standard TLD holder except in 2017 and 2019, the radiophotoluminescent dosimeter (RPLD) was provided. The control dosimeter labeled with a white sticker uses to monitor background radiation, undesirable accidental irradiation and unexpected fading.

Before irradiating the dosimeters, absorbed dose determination was performed according to the IAEA's TRS No. 398 code of practice with a calibrated 0.6 cm^3 Farmer ionization chamber, type NE 2571 (PTW-Freiburg, Freiburg, Germany) connected to a PTW Unidos electrometer, type 10005 and having a calibration in term of absorbed dose to water for ^{60}Co traceable to the IAEA Dosimetry Laboratory (Figure 1). The water tank and waterproof sleeve for the chamber used are made of Polymethylmethacrylate (PMMA) with density, $\rho = 1.19$ g/cm3. The water tank dimension is 30 cm x 30 cm x 30 cm. The thickness of the phantom wall is 1.5 cm. This water tank was designed with a 10 cm x 10 cm x 0.2 cm window for horizontal beam irradiation. The waterproof sleeve was fabricated with a thickness of 0.2 cm to allow the chamber to achieve thermal equilibrium with the water in less than 10 minutes and air gap of 0.2 cm to allow air pressure in the chamber to reach ambient air pressure quickly.

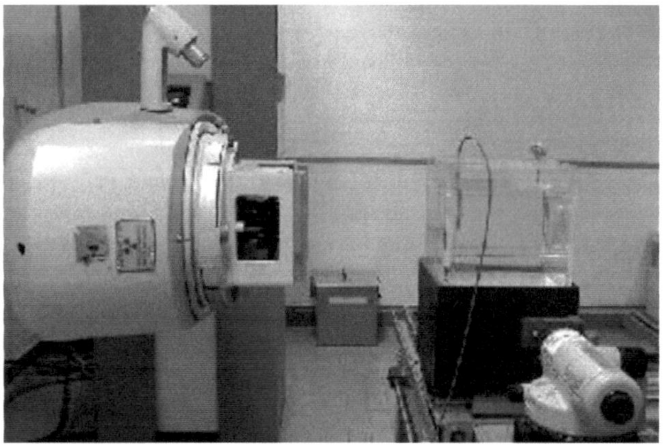

Figure 1. Measurement set-up for absorbed dose determination using the ^{60}Co. The ionization chamber is placed at 5 cm depth in water, 10 cm x 10 cm field size at the side wall of the water tank and 80 cm SSD.

The SSDL were asked to irradiate the dosimeters with an absorbed dose to water of 200 cGy in reference condition. Using a ^{60}Co teletherapy, the irradiation was carried out at 5 cm depth in water with field size of 10 cm x 10 cm and source-surface distance (SSD) of 80 cm. For medical linear accelerator, the irradiation was done at 10 cm depth in water with 10 cm x 10 cm field size at 100 cm SSD (Figure 2).

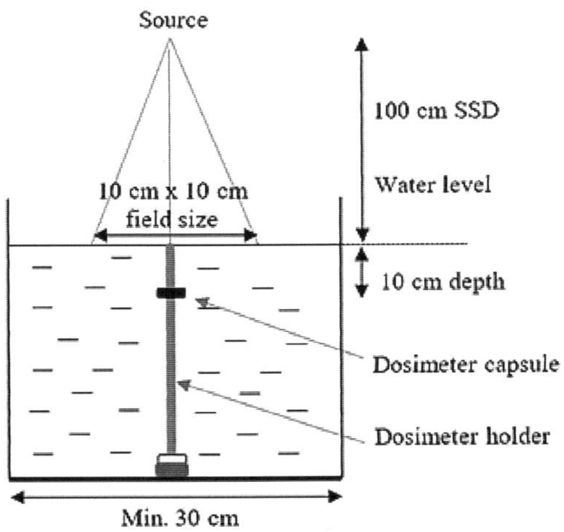

Figure 2. Measurement set-up for dosimeter irradiation using medical linear accelerator. The dosimeter capsule is placed at 10 cm depth, field size of 10 cm x 10 cm at water surface and 100 cm SSD.

Result of TLD audit was reported in the form of the percentage deviation between the SSDL stated dose and the IAEA measured dose where the acceptance limit of ± 3.5% should be complied. This value was calculated using equation 1.

$$Percentage\ deviation, \Delta\% = \left(\frac{D_{stat} - D_{TLD}}{D_{TLD}} \right) x100 \quad (1)$$

where the D_{TLD} is a mean measured dose by the IAEA from the readings of the TLDs irradiated by the SSDL and D_{stat} is the stated dose by the SSDL.

3. Results and discussion

A total number of 10 participations in the IAEA/WHO Postal Dose Quality Audit were recorded during 2009 to 2019 as presented in Table 1. Five participations using ^{60}Co teletherapy, three participations using 6 MV X-rays and 2 participations using 10 MV X-rays produced from medical linear accelerator. The SSDL were participated in the audit for one photon beam every year, except in 2018. The radiation unit used was a ^{60}Co teletherapy, type Eldorado 8 (Nordion, Ottawa, Canada) which belongs to Nuklear Malaysia and the medical linear accelerators were owned by the radiotherapy centres.

The results of the IAEA/WHO Postal Dose Audit from 2009 to 2019 performed by the SSDL are presented in Figure 3. The results demonstrate that the mean of percentage deviations between the SSDL stated dose and the IAEA mean measured dose was -0.35% with standard deviation of 0.86%. The percentage deviations vary between a minimum percentage deviation of -2.5% and a maximum of 0.6%. The mean of the distribution for ^{60}Co was -0.50% and the standard deviation was 1.17%. For 6 MV X-rays, the mean of the distribution is 0.07% and the standard deviation is 0.01%. The result of the mean of the distribution for 10 MV X-rays is -0.60% and the standard deviation is 0.01%. All results were considered satisfactory in which complied with the IAEA's acceptance limit of ± 3.5%. A percentage deviation with negative sign indicates that the SSDL estimates lower dose than what was measured by the IAEA. The uncertainties in the TLD and RPLD measurements of the dose were reported to be 1.8% and 1.5% (1 standard deviation), respectively.

Table 1. SSDL participations in the IAEA/WHO Postal Dose Audit from 2009 to 2019.

Year of participation	Radiation unit	Photon beam		
		1.25 MeV ^{60}Co	6 MV X-rays	10 MV X-rays
2009	Eldorado 8	√		
2010	Eldorado 8	√		
2011	Oncor Imp. +		√	
2012	Varian Trilogy			√
2013	Varian Trilogy		√	
2014	Eldorado 8	√		
2015	Varian Trilogy			√
2016	Siemens Oncor Impression		√	
2017	Eldorado 8	√		
2018	-			
2019	Eldorado 8	√		
	Total	5	3	2

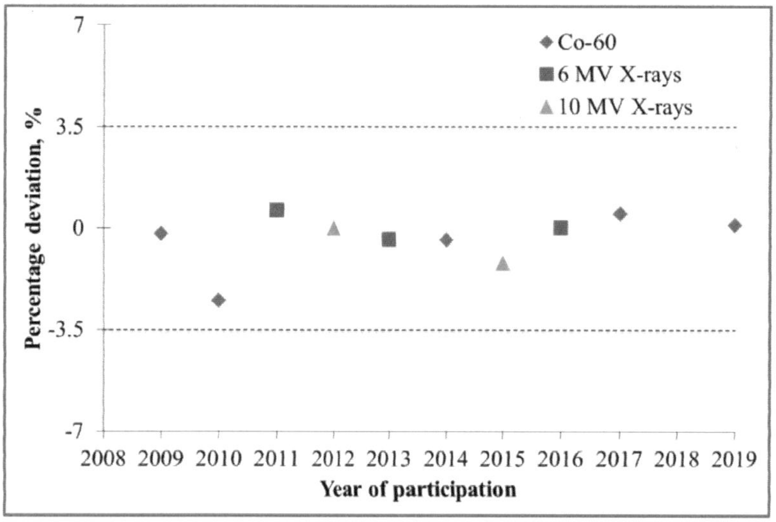

Figure 3. Summary of the results of the IAEA/WHO Postal Dose Quality Audit during 2009 to 2019. Data points represent the percentage deviation between SSDL stated dose and IAEA mean measured dose. The IAEA considers a result within ± 3.5% to be satisfactory.

4. Conclusion

The results of SSDL of Nuklear Malaysia participated in the IAEA/WHO Postal Dose Quality Audit using ^{60}Co teletherapy unit and medical linear accelerators have been presented. The results obtained from 2009 to 2019 demonstrate that all photon beams comply with the IAEA's acceptance limit of ± 3.5%. These audit results provide evidence that the SSDL of Nuklear Malaysia having a capability and competency in providing the calibration service of the radiotherapy dosemeter in compliance with the IAEA standard dosimetry protocols thus contributes to the improvement of quality of radiotherapy in Malaysia.

5. References

[1] Podgorsak E B 2005 *Radiation Oncology Physics* (Vienna: International Atomic Energy Agency).

[2] International Atomic Energy Agency 2000 Lessons learned from accidental exposures in radiotherapy *Safety Reports Series No. 17* (Vienna: International Atomic Energy Agency)

[3] Cunningham J, Cohen M and Dutreix A 1976 *ICRU Report 24: Determination of Absorbed Dose in A Patient Irradiated by Beams of X-Or Gamma-Rays in Radiotherapy Procedures* (Washington: International Commission on Radiation Units and Measurement).

[4] Van der Merwe D, Van Dyk J, Healy B, Zubizarreta E, Izewska J, Mijnheer B and Meghzifene A 2017 *Acta Oncologica* **56**(1) 1-6.

[5] Kry S F, Peterson C B, Howell R M, Izewska J, Lye J, Clark C H, Nakamura M, Hurkmans C, Alvarez P, Alves A and Bokulic T 2018 *Phys. and Imaging in Radiat. Oncol.* **7** 39-44.

[6] International Atomic Energy Agency 2008 *Setting up A Radiotherapy Programme: Clinical, Medical Physics, Radiation Protection and Safety Aspects* (Vienna: International Atomic Energy Agency).

[7] QUATRO I 2007 Comprehensive audits of radiotherapy practices: a tool for quality improvement (Vienna: International Atomic Energy Agency)

[8] International Atomic Energy Agency 2010 *Planning National Radiotherapy Services: A Practical Tool* (Vienna: International Atomic Energy Agency)

[9] Ortiz P, Oresegun M and Wheatley J 2002 *Safety* **21**(1) 11-230.

[10] Healy B J, Budanec M, Ourdane B, Peace T, Petrovic B, Sanz D E, Scanderbeg D J and Tuntipumiamorn L 2020 *Acta Oncologica* **59**(5) 503-510.

[11] Samat S B, Evans C J, Kadni T and Dolah M T 2009 *Radiat. Protec. Dosimetry* **133**(3) 186-191.

IOP Conf. Series: Materials Science and Engineering 1106 (2021) 012004 doi:10.1088/1757-899X/1106/1/012004

Power consumption comparison in three experimental circuits for radiation detection

Nor Arymaswati Abdullah[1,a)], Glam Hadzir Patai Mohamad[1], Nolida Yussup[1], Noor Farhana Husna[1], Nur Aira Abd Rahman[1 2], Khairul Salleh Mohamed Sahari[2] and Muhammad Fairuz Abdul Jalal[2]

[1]Malaysian Nuclear Agency, 43000 Kajang Selangor, Malaysia.
[2]Universiti Tenaga Nasional, Putrajaya Campus, Jalan IKRAM-UNITEN, 43000 Kajang Selangor, Malaysia.

[a]arymas@nuclearmalaysia.gov.my

Abstract. Geiger Mueller tube is a longstanding device to detect gamma ray in it range and it still preferred in many radiation detection devices due to robustness in rough environment. The high voltage supply to support Geiger Mueller activation needs a transformer at in order to operate. This paper will discuss the effectiveness of the transformer to get satisfactory and stable high voltage. The transformers were winded manually in few sets with limited size of the bobbin and ferrite core. The transformers were applied into three different circuits to analyse the high voltage. The result of the effectiveness of the transformer and the high voltage of the circuit were discussed.

1. Introduction

Safety is the most important issue when it comes to any kind of work place. The greatest issue in radiation exposure environment is the accuracy of the radiation detection device. In the past years, several studies have reported, including the study on oil insulating systems to investigate the high energy ionizing radiation phenomenon emitted by the electrical discharge, showed that the event occurred when the free electrons knocked out of the valence coating and positive ions that have lost these electrons and, also introducing argon bubbles [1]. Meanwhile, the research on high voltage transformer had been done with extremely low frequency electromagnetic radiation (160KVA and 250KVA) to investigate the effect of exposing it to selected students. The findings demonstrate that the increasing in heart pulse, arterial blood pressure and tympanic temperature but decreasing in blood oxygen saturation [2]. Besides that, in order to achieve an ideal scintillator design, the research work on scintillation and optical characteristics of ZnO:Sc based have been conducted to get the high time resolution. With 13.6 pA system noise current, this high time resolution is suitable for sub-nanosecond pulsed radiation detection [3]. As highlighted in a portable radiation detector application, a study concerning configuration of a sensor interface integrated circuit system has been performed. The study found that the high bias voltage of commercial silicon-photomultiplier detector could be driven by the proposed radiation detector system in mobile application [4]. In another study, by using 193 nm ArF excimer laser on nano-graphite electrical contacts on both diamond surfaces, the researchers created a radiation detector from a plate of high quality polycrystalline chemical vapour deposition diamond. These works proved that electrodes made by laser graphitisation are Ohmic electrically contacts [5]. Similarly in Geiger counter device, one of the oldest radiation detectors where numerous studies have been reported in a lot of papers till recent years. This common device is using Geiger Mueller (GM) tube, the oldest radiation detector which is simpler in design and robust in most environment. Compare to ion chamber, GM counters are more sensitive in monitoring low levels of contamination [6]. There are two points considered as an important topic in the latest invention of GM tube. According to [7], besides the improvements in manufacturing, the halogen-quenched GM tubes also of significant concern. There are large volume of published studies regarding the performance and effectiveness of the Geiger counter. Previous studies have reported that researchers have investigated GM counter with several methods including Monte Carlo [8][9] and correlation technique [10]. This paper will share the process of how to build and assemble all the components and devices to produce three circuits of high voltage and effectiveness of transformers ratio winding towards high voltage. Finally, the establishment of three type of circuits that are able to produce 450V-550V high voltage, where one of them to be chosen as part of the low cost survey meter.

2. Methodology

As shown in Figure 1, it is important to realize that certain requirement needs to be occupied for the purpose of GM counter to be functioning. The most important thing is high voltage supply so that the GM tube can be activated. Along with the product of stress and inter electrode distance, the entering ionizing radiation must be able to start the avalanche of a self-maintaining mechanism that leads to the break down [11]. Numerous studies have revealed that GM tube still on demand and a number of studies show the important of stable, low cost and efficiency of the high voltage supply [12].

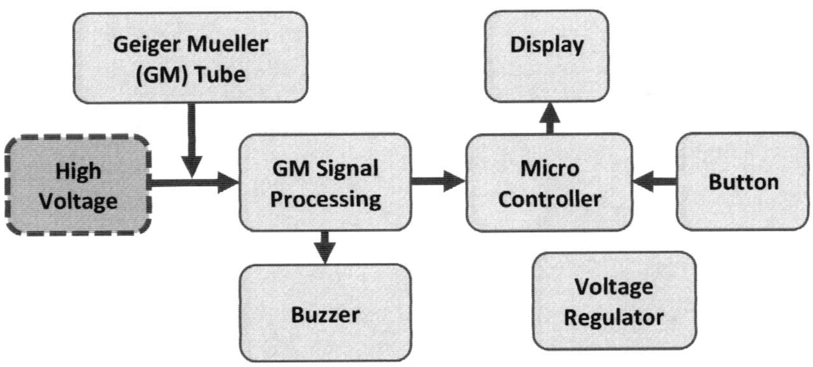

Figure 1. Block diagram of the radiation detection device.

The first process starts with the circuit design. There are three designs as shown in Figure 2, Figure 3 and Figure 4 that have been used in this study. The first one is a high voltage (HV) circuit using two coils transformer and Figure 3 is using three coils transformer. Meanwhile the third circuit not using transformer but inductance. First and foremost, all the components needed must be available. Normally it is quite difficult to get all the components to be there even a resistor.

Then, all components were assembled on breadboard before being transferred to strip board. At the same time, the transformers were winded using manual winder, as depicted below. Not to be forgotten, the size of transformer is limited into RM8 only since there is no other size obtainable and the ratio of the coils as shown in the Table 1 and Table 2. Once the result of HV can be accepted, the components were assembled into strip board. The acceptable HV, ranging from 450 volts to 550 volts, ever since the GM tube (LND7121) working at these point values. This process will be repeated if the result not achievable especially the transformer coils ratio. The flow of the process was depicted in Figure 5.

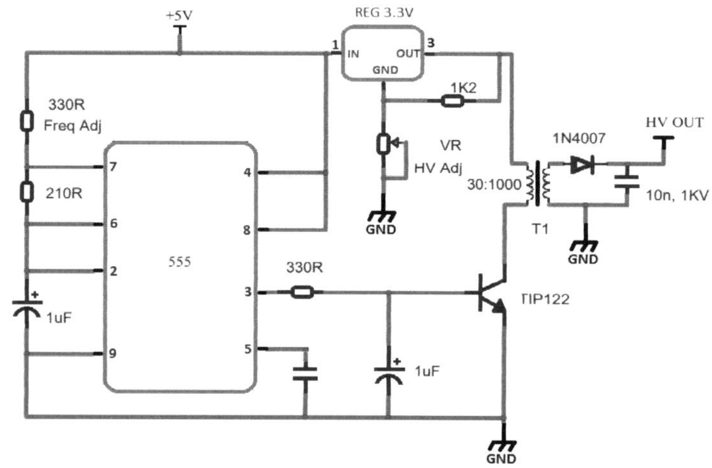

Figure 2. High voltage circuit with 2 coils ratio.

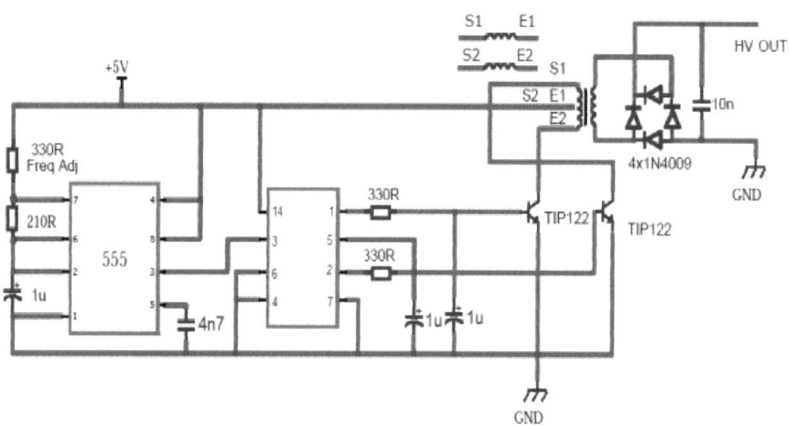

Figure 3. High voltage circuit with three coils ratio.

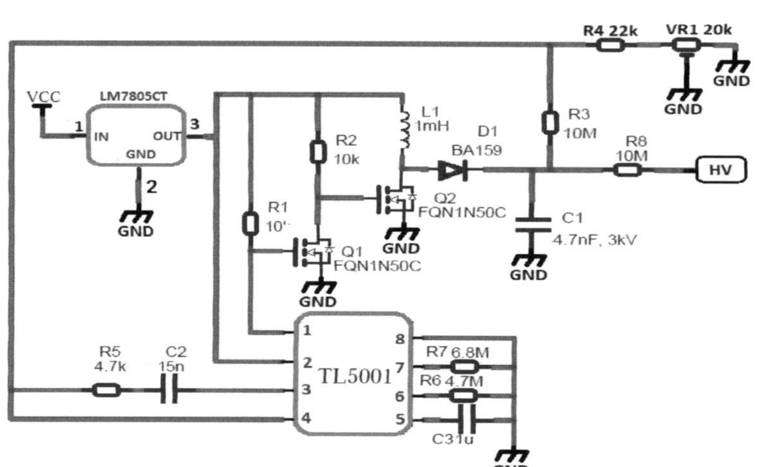

Figure 4. High voltage circuit with inductance without transformer.

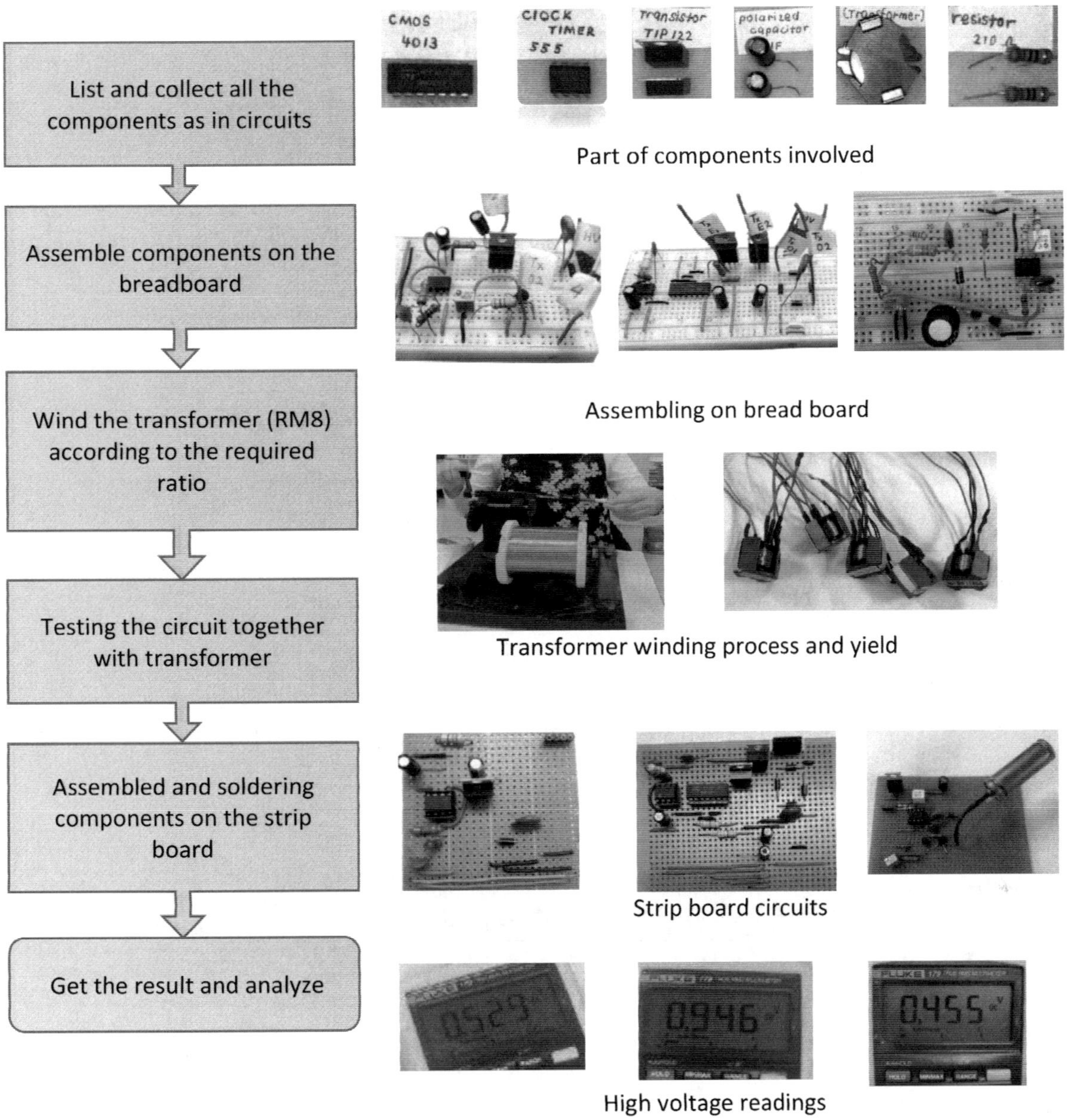

Figure 5. The flow of the assembling components from breadboard into strip board and the result.

3. Result and discussion

The resulting high voltage for each transformer configurations were shown in Table 1 to Table 3. Based on these results, it can be seen that the transformer application in the circuit needs more improvement. Both first and second circuits gave unpredicted results. Even the three coils ratio be able to produce 524 volts but this configuration is unstable and even drop between 0 to 290 volts. On the hand, it is obvious from the third table, that the HV circuit with inductance shows stable and good HV results. Even though the current is quite high compare to others, this circuit is the best solution at the moment. The plotted graph for voltage current and power for each circuit also depicted in Figure 6 to Figure 8.

Table 1. Current, high voltage and power for Circuit 1.

Transformer Configuration	Resistance (Ohm)	HV value (V)	Current (Amp)	Power (W)
10:1200	0.7:78.7	Drop	0	-
20:1200	1.2:72.2	300	0.207	62.100
30:1015	1.8:66.2	694	0.037	25.678
40:1100	2.2:65.6	600	0.034	20.400
50:1050	2.6:70.1	925	0.029	26.825
50:700	2.5:37.7	747	0.051	38.097
50:650	2.5:35.3	871	0.031	27.001
50:600	2.5:31.7	768	0.030	23.040

Table 2. Current, high voltage and power for Circuit 2.

Transformer Configuration	Resistance (Ohm)	HV value (V)	Current (Amp)	**Power (kW)
5:5:1100	0.5:0.5:63	704*	1.564	1.1011
10:10:1000	0.7:0.7:62	352*	1.562	0.5498
15:15:1020	1.3:1.3:64	334*	1.562	0.5217
20:20:1200	1.1:1.1:70	743*	1.563	1.1613
25:25:1050	1.6:1.6:60	524*	1.563	0.8190

*the value drop between 0-290V
**Starting power before drop

Table 3. Current, high voltage and power for Circuit 3.

Transformer Configuration	Resistance (Ohm)	HV value (V)	Current (Amp)	Power (W)
Nil	19.95k	241	0.08	19.28
	17.66k	255	0.09	22.95
	11.04k	305	0.11	33.55
	6.29k	355	0.14	49.70
	2.77k	405	0.18	72.90
	3.8	455	0.24	109.20

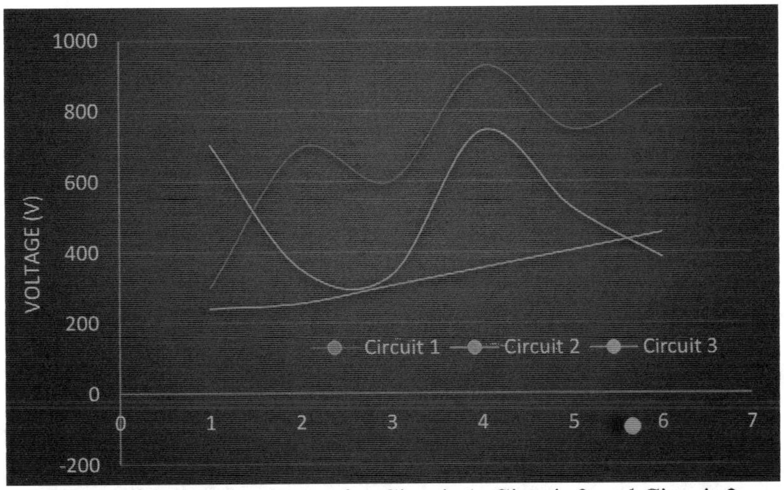

Figure 6. Plotted voltages for Circuit 1, Circuit 2 and Circuit 3.

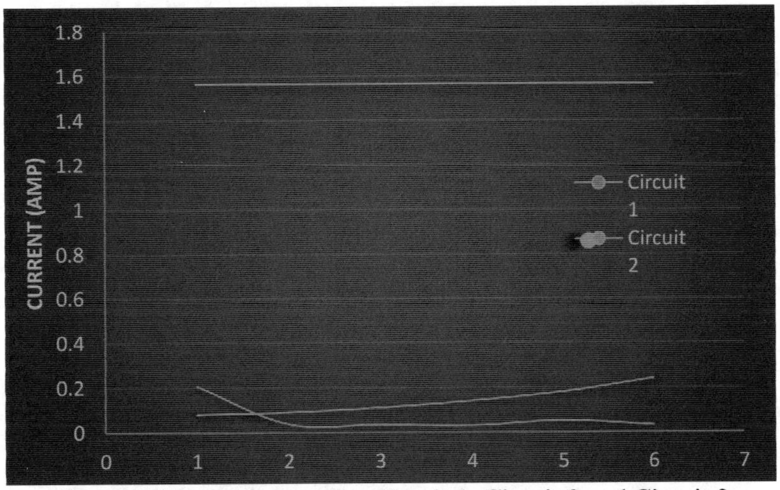

Figure 7. Plotted current for Circuit 1, Circuit 2 and Circuit 3.

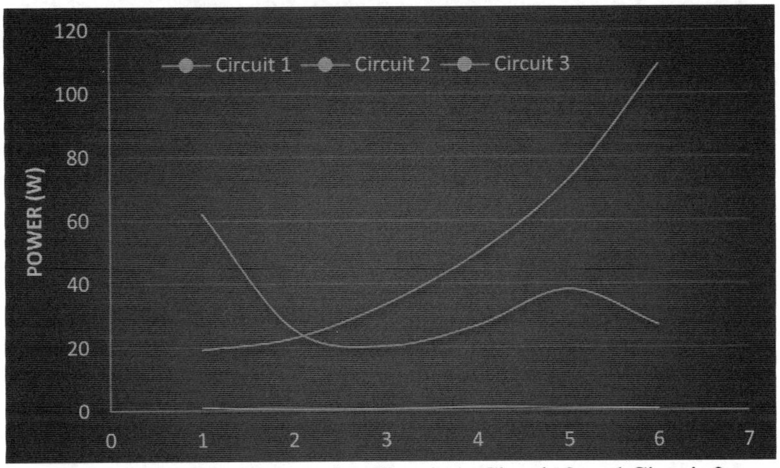

Figure 8. Plotted power for Circuit 1, Circuit 2 and Circuit 3.

In summary, large variations in high voltage values are exposed and the circuits with transformer configurations shown the unachievable high voltage value. The main findings suggest that in general the circuit with inductance element indicate that the circuit be able to produce the required voltage.

4. References

[1] Nagi Ł, Zmarzły D, Boczar T, and Frącz P 2016 Detection of high-energy ionizing radiation generated by electrical discharges oil 2036–2041.
[2] Abu-jafar M 2013 The effect of the electromagnetic radiation from high voltage transformers on students health in Hebron district An-Najah National University Faculty of Graduate Studies the effect of the electromagnetic radiation from high voltage transformers on student.
[3] Zhang K, Ouyang X, Song Z and Han H 2014 An ideal scintillator – ZnO :Sc for sub-nanosecond pulsed radiation detection *Nucl. Inst. Methods Phys. Res. A* **756** 14–18.
[4] Jeon H and Je M A 2019 Radiation Sensor Interface System for Mobile Application 4–26.
[5] Alemanno E *et al.* 2013 Laser induced nano-graphite electrical contacts on synthetic polycrystalline CVD diamond for nuclear radiation detection *Diam. Relat. Mater.* **38** 32–35.
[6] Theory of Operation and Laboratory Applications 2017 *RSSC Radiat. Detect. Surv. Instr.* 1–33.
[7] Company L 2016 (12) Patent Application Publication (10) Pub . No .: US 2016 / 0252627 A1 **1**(19) 11–14.
[8] Meric I, Johansen G A, Holstad M B and Gardner R P 2011 A Monte Carlo modelling of gamma-ray stopping efficiencies of Geiger – Muller counters *Nucl. Inst. Methods Phys. Res. A* **36**(1) 61–66.
[9] Meric I. *et al.* 2011 A single scatter electron Monte Carlo approach for simulating gamma-ray stopping efficiencies of Geiger-Muller counters *Nucl. Inst. Methods Phys. Res. A* **654**(1). 279–287.
[10] Almutairi B, Akyurek T and Usman S 2019 Voltage dependent pulse shape analysis of Geiger-Müller counter *Nucl. Eng. Technol.* **51**(4) 1081–1090.
[11] Tro I N and Tion D U C 2018 Statistical review of the insulation capacity of the Geiger-Mueller counter **33**(4) 369–374.
[12] Rahaman R and Kang S 2015 Design and development of low-cost, highly stable regulated high-voltage power supply for radiation detector 1–10.

Radionuclide analysis of the National Data Centre preparedness exercise 2019 (NPE2019): Malaysian National Data Centre (MY-NDC) findings

F I A Rashid[1, a)] and M Z Zolkaffly[1, b)]

[1]Planning and International Relations Division, Malaysian Nuclear Agency, 43000 Kajang, Selangor, Malaysia.

[a]faisal_izwan@nuclearmalaysia.gov.my
[b]zulfakar@nuclearmalaysia.gov.my

Abstract. The National Data Centre preparedness exercise (NPE) simulates fictitious violation of the Comprehensive Nuclear-Test-Ban Treaty (CTBT) and is regularly conducted by the National Data Centres (NDCs) for NDCs. The purpose of the exercise is to increase the NDCs' preparedness and capabilities to perform objective analysis of possible treaty violation events. In November 2019, the NPE2019 has officially launched as a combined radionuclide and waveform scenario. The scenario started as a fictitious state named RAETIA announced a reactor incident with a release of unspecified radionuclides into the atmosphere. The simulated concentration of particulate and noble gas isotopes at several radionuclide stations had been provided as a case study. Hence, the NDCsis tasked to check the consistency with the announcements as well as to identify potential source region of such release. This paper aims at presenting MY-NDC's radionuclide analysis findings of NPE2019. The analysis involved among others, multiple nuclide isotopic analysis, and atmospheric transport modelling (ATM). Consequently, the analysis had enabled MY-NDC to identify the potential source region of the radionuclide release as well as to discriminate the source of such release, whether coming from civil nuclear facilities or nuclear tests.

1. Introduction

The Comprehensive Nuclear-Test-Ban Treaty (CTBT) is a multilateral treaty that bans nuclear explosion anywhere on Earth whether for military or peaceful purposes. Malaysia is one of the contracting states to the treaty through its signature and ratification on 23 July 1998 and 17 January 2008, respectively [1]. Under the CTBT, a Malaysian CTBT National Data Centre (MY-NDC) has been established in December 2005 which provides technical information of CTBT related events to the Malaysian Nuclear Agency (Nuklear Malaysia) as the CTBT National Authority [2].

The National Data Centre Preparedness Exercise (NPE) is an exercise that simulates a fictitious violation of the CTBT. It aims to measure the readiness of the National Data Centres (NDCs) to fulfill their duties concerning the CTBT verification [3]. In the exercise, the NDCs are tasked to clarify the nature of the fictitious event based on independent technical judgment, whether the treaty incompliance has been triggered.

The NPE2019 had officially started on 22 November 2019. In the NPE2019, a hypothetical scenario describes that a TRIGA reactor accident had taken place on 30 July 2019 in a fictitious location of Pavia,

Raetia (45.18° N, 9.16° E) as shown in Figure 1 (a). Subsequently, the state of Raetia had announced a small release of radioactive isotopes to the environment but claimed the release was below the hazardous limit for human health. Upon the announcement, the neighboring state named Eastria (in real-world, Eastria is located in the territory of Austria with particulate and noble gas IMS station coded VIP00 and VIX00, respectively) has provided its national radionuclide measurements to the International Data Centre and requests IDC's assistance to identify potential sources. For the purpose of this exercise, the radionuclide measurements were also made available to the NDCs. The measurements consist of synthetic particulate and noble gas (radioxenon) data measured at selected International Monitoring System (IMS) radionuclide stations as shown in Figure 1(b). The particulate data consists of activity concentrations of Cs-134, Cs-137, Ba-140, La-140, and I-131 whilst the noble gas data comprising activity concentrations of Xe-131m, Xe-133m, Xe-133, and Xe-135. The chronology of particulate and noble gas detections after the Pavia' reactor accident is illustrated in Table 1. Hence, using the provided data, the NDCs are tasked to perform independent analysis in order to check the consistency with the announcements from the state of Raetia as well as to identify potential source regions of such release. Thus, this paper aims at presenting MY-NDC's radionuclide analysis findings of NPE2019.

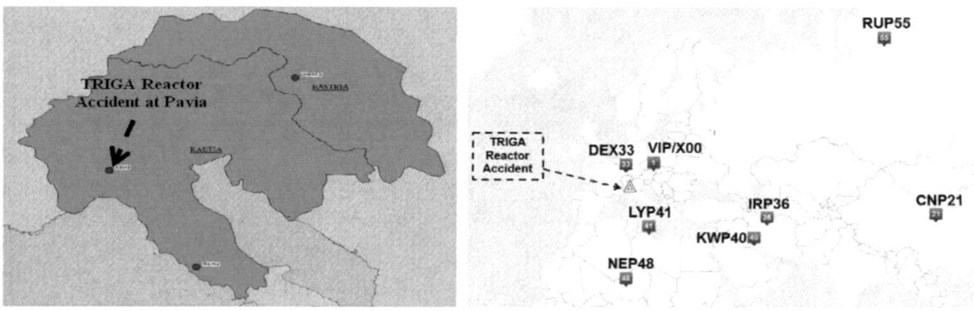

(a) (b)

Figure 1. (a) Location of the reactor accident in Pavia, Raetia and, (b) location of IMS stations.

Table 1. Chronology of detections.

IMS Station	Date of detection												
	30-Jul	**31-Jul**	**1-Aug**	**2-Aug**	**3-Aug**	**4-Aug**	**5-Aug**	**6-Aug**	**7-Aug**	**8-Aug**	**9-Aug**	**10-Aug**	**11-Aug**
Particulate Detection													
CNP21	**PAVIA's reactor accident occurred**												
KWP40													
LYP41													
NEP48													
IRP36													
RUP55													
VIP00													
Noble Gas (Radioxenon) Detection													
DEX33													
VIX00													

2. Methods

2.1. *Particulate data analysis*

The particulate data is analysed to identify nuclide ratios from relevant radionuclides. The ratios are important due to their usability to characterise the source in order to discriminate a civil reactor release or nuclear explosion. Moreover, the nuclide ratios are commonly used to acquire timing information by comparison with scenarios calculated from nuclear data [4]. In NPE2019, the provided measurements were in the form of nuclide activity concentrations. Thus, these measurements were converted into nuclide ratios using the following conversion equations proposed by Axelsson *et al.* [5]. Equation (1) and (2) were used to convert activity concentrations into nuclide ratios for independent decay nuclides and parent-daughter decay nuclides, respectively.

$$R = \frac{C_2\lambda_1 1}{C_1\lambda_2\xi_p\xi_c}(1) \qquad R = \frac{C_2^0\lambda_1 1}{C_1\lambda_2\xi_P\xi_C} - \eta_A \qquad (1)$$

where C is nuclide activity concentration, λ is the nuclide decay constant [s^{-1}],ξ is decay correction and η is correction term. The decay correction (ξ) is further divided into 3 types, namelyξ_c is decay correction during collection, ξ_p is decay correction during processing and ξ_A is decay correction during acquisition. The respective decay corrections and correction terms were calculated using the following equations.

$$\xi_C = \frac{(1-e^{-\lambda_1 t_C})}{(1-e^{-\lambda_2 t_C})}(3) \qquad \xi_P = e^{-(\lambda_1-\lambda_2)t_P}(4)\xi_A = \frac{(1-e^{-\lambda_1 t_A})}{(1-e^{-\lambda_2 t_A})} \qquad (2)$$

$$\eta_A = \frac{\lambda_2}{\lambda_2-\lambda_1}\left(\frac{\lambda_2}{\lambda_1}\xi_A - 1\right) \qquad (3)$$

where t_C, t_P and t_A is collection time, processing time, and acquisition time, respectively. Using these equations, the nuclide ratios of Cs-134/Cs-137 (independent decay nuclides) and La-140/Ba-140 (parent-daughter decay nuclide) were obtained. These ratios were then used to characterise the source as well as for the dating of a nuclear event.

2.2. *Noble gas data analysis*

By using the same conversion equations for particulate data analysis, the activity concentrations of radioxenon isotopes were converted into 2 main nuclide ratios, namely Xe-135/Xe-133 and Xe-133m/Xe-131m. These xenon nuclide ratios were then examined to distinguish nuclear explosion sources from civilian releases using multiple atmospheric xenon isotopic activity ratios as suggested by M B Kalinowski *et al.* [6].

2.3. *Atmospheric Transport Modelling (ATM)*

The ATM is used to assess the consistency between measurements and potential source locations. In the context of NPE2019, the ATM was simulated in two modes – backward and forward. For backward mode, the Linux-based software developed by the Comprehensive Nuclear Test-Ban Treaty Organization (CTBTO) called the Webconnected Graphics Engine (WEB-GRAPE) was utilised for the visualization of atmospheric transport models and identification of possible source regions of radionuclides detected by IMS stations [7]. The software computes trajectories of particles to determine the transport and dispersion of particles in the atmosphere, thus leading to the determination of possible source regions[8].Meanwhile, for the forward mode ATM, the Lagrangian Particle Dispersion Model HYSPLIT invented by the National Oceanic and Atmospheric Administration – Air Resources Laboratory was used. The software was utilised for predicting the potentially affected IMS radionuclide stations by simulating hypothetical release from possible source regions identified from backward mode ATM [9]. Table 2 lists the key parameters and data used to simulate both ATM modes.

Table 2. Key parameters and data used for simulating backward and forward mode ATM

Parameters	Backward ATM	Forward ATM
Meteorological data	European Centre for Medium-Range Weather Forecasts (ECMWF)	Global Data Assimilation System (GDAS), 10 degree
Total run time	14 days in reverse time	14 days in forward time
Number of particles release	Not applicable	200,000
Release time	Not applicable	30 July 2019, 0600 UTC
Duration of release	Not applicable	24 hours of continuous release
Rate of hypothetical release	Not applicable	1.04 E+8 Bq/hr
The initial point of projection	IMS radionuclide stations	Pavia's reactor site and selected radioisotope production facilities

3. Results and Discussion

3.1. Particulate data analysis

Figure 2 (a) and (b) depict the nuclide ratios of Cs-134/Cs-137 and La-140/Ba-140 from CNP21, KWP40 and LYP41. As can be seen clearly from the figures, the Cs-134/Cs-137 and La-140/Ba-140 ratios from the three IMS radionuclide stations were within 3.94 to 3.97 and 0.42 to 0.44, respectively. Since the variation of the ratios is very low, they can be considered constant over time. The smooth and constant Cs-134 /Cs-137 ratio indicates that radioactive material was predominantly released from the same type of source material. Moreover, the ratio of Cs-134/Cs-137 above 0.01 signified that the source was highly probable from a source other than the nuclear explosion [10].

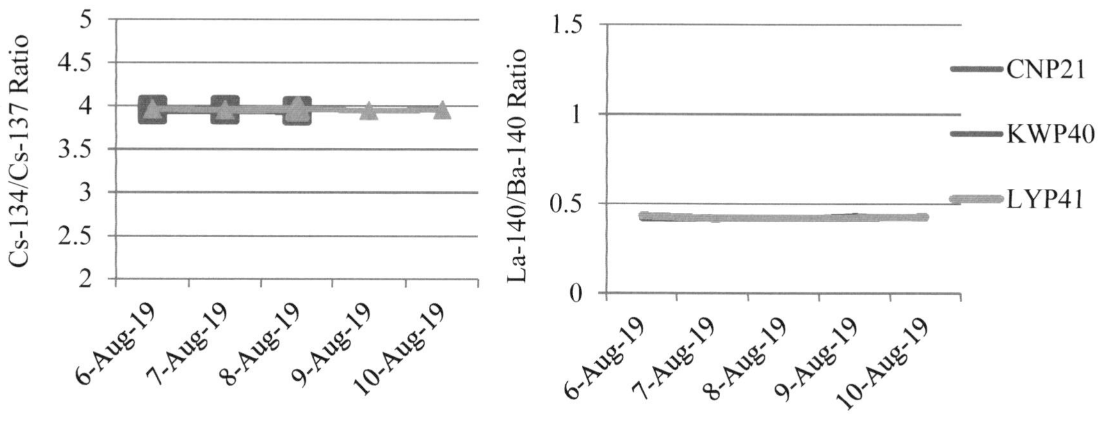

Figure 2. (a) Cs-134/Cs-137 ratio and (b) La-140/Ba-140 ratios from CNP21, KWP40 and LYP41.

3.2. Noble gas data analysis

Table 3 lists the nuclide ratios of Xe-135/Xe-133 and Xe-133m/Xe-131m measured at DEX33 and VIX00. These ratios were further examined using multiple atmospheric xenon isotopic activity ratios as illustrated in Figure 3. From Figure 3, it is evidently shown that radioxenon nuclide ratios measured at DEX33 for sample collection date of 2 August 2017 and VIX00 for sample collection date of 2, 3 and 7 August 2019 fall within the civil region. Consequently, it can be concluded that the xenon isotopes detected at these radionuclide stations were not a result of a nuclear test, but most likely from civilian applications that may include PAVIA's reactor, commercial nuclear power plants, and possibly medical radioisotope production facility.

Table 3. Xe-135/Xe-133 and Xe-133m/Xe-131m ratios at DEX33 and VIX00.

IMS	Collection date	Xe-135/Xe-133	133m/Xe-131m
DEX33	2-Aug-19	0.026	1.04
DEX33	3-Aug-19	0.01	-
DEX33	4-Aug-19	-	-
DEX33	5-Aug-19	-	-
DEX33	6-Aug-19	-	-
DEX33	9-Aug-19	-	-
VIX00	2-Aug-19	0.026	1.59
VIX00	3-Aug-19	0.005	4.5
VIX00	4-Aug-19	-	-
VIX00	5-Aug-19	-	-
VIX00	6-Aug-19	-	-
VIX00	7-Aug-19	0.001	1.07
VIX00	8-Aug-19	-	-

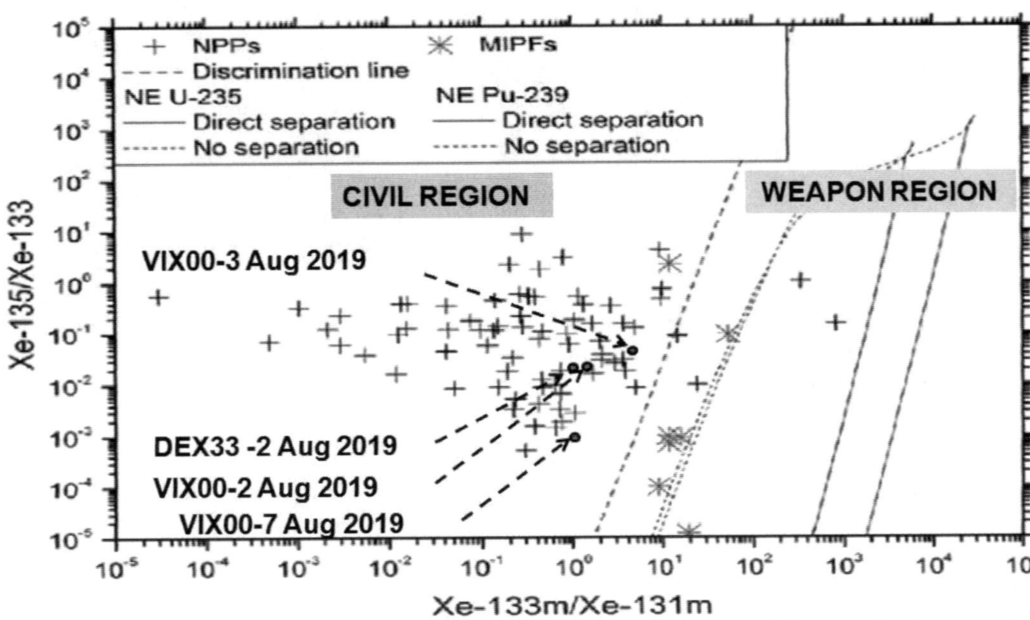

Figure 3. Multiple atmospheric xenon isotopic activity ratios for DEX33 and VIP00 samples

3.3. Backward and forward mode ATM

The backward mode ATM was simulated from the earliest sample collection date at the respective radionuclide station until the date of PAVIA's reactor accident in the morning of 30 July 2019. The results of backward mode ATM from CNP21, KWP40, LYP41, IRP36, RUP55, VIP/X00, and DEP/X33are shown in Figure 4. ATM simulation could not be performed for NEP48 due to unavailable SRS data. The plume is colour-coded according to its threshold sensitivity (m^{-3}) which varies from 10^{-19} (low sensitivity) to 10^{-13} (high sensitivity). Moreover, it should be noted from the figure that the region surrounding Raetia and Eastria has several operating nuclear power plants as well as 2 isotope production facilities (IPF), namely TYCO and IRE. Upon observation, the results concluded that the possible source regions for nuclides detected at all stations except DEP/X33 include PAVIA's reactor, nuclear power plants as well as TYCO and IRE. Interestingly, the backward ATM showed that DEP/X33 station will not be affected by the release from Pavia's reactor accident.

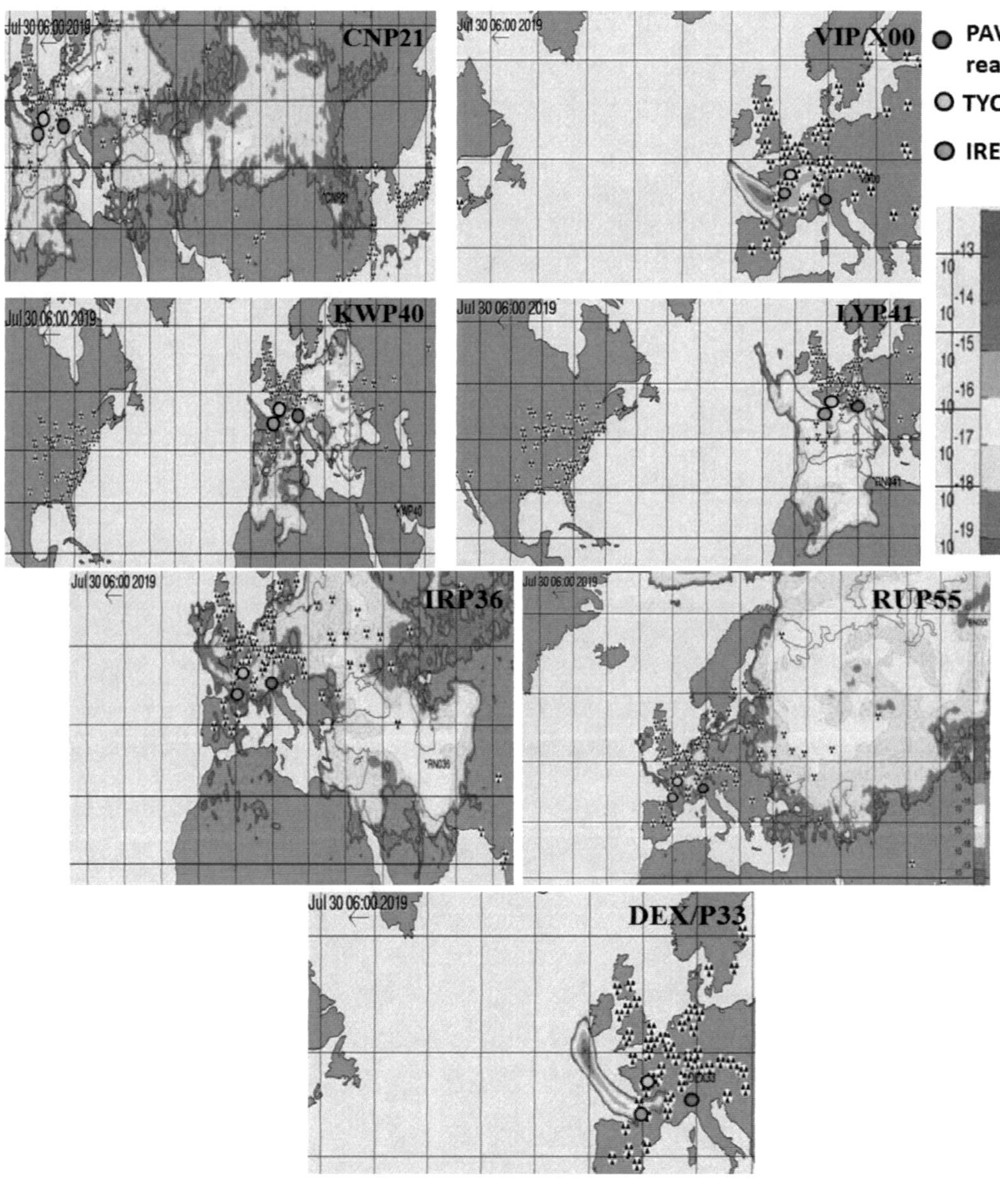

Figure 4. Results of backward mode ATM.

The forward mode ATM was performed for3 distinct locations, which are Pavia'a reactor site, TYCO, and IRE as illustrated in Figure 5 (upper left), Figure 5 (upper right), and Figure 5 (bottom left), respectively. In Figure 5 (upper left), the modeling showed that any release from Pavia's reactor on 30 July 2019 will affect all IMS radionuclide stations in the vicinity except DEP/X33. This finding concurred very well with the result from a backward mode ATM. Therefore, it can be concluded that the source of nuclides detected at DEP/X33 was originated from sources other than Pavia's reactor. Based on Figure 5 (upper right), the forward ATM demonstrated that any release from TYCO IPF on 30 July 2019 0600UTC will be measured at all IMS radionuclide stations excluding RUP55 and CNP21. Thus, the source of nuclides measured at RUP55 and CNP21 is most likely from sources other than TYCO IPF. Meanwhile, the release from IRE IPF on the same release data as TYCO IPF will result in detection at all IMS radionuclide stations as displayed in Figure 5 (bottom left). The results of these 3 release scenarios are summarised in Table 4.

Moreover, Figure 5 (bottom right) shows the comparison between measured and simulated concentrations of Xe-133 at VIP/X00.It can be seen that the pattern of detection is identical despite varying concentrations. This evidence shows the source of nuclides measured at VIP/X00 is most likely coming from Pavia's reactor site. Since the simulation presented in this paper is simplified modelling, thus, the varying concentrations between actual and simulated measurements are well expected. Moreover, the variation is also due to the fact that the air concentration results are very sensitive to the meteorology and processes that might affect the wet and dry scavenging, which were not taken into consideration in the simulation [11].

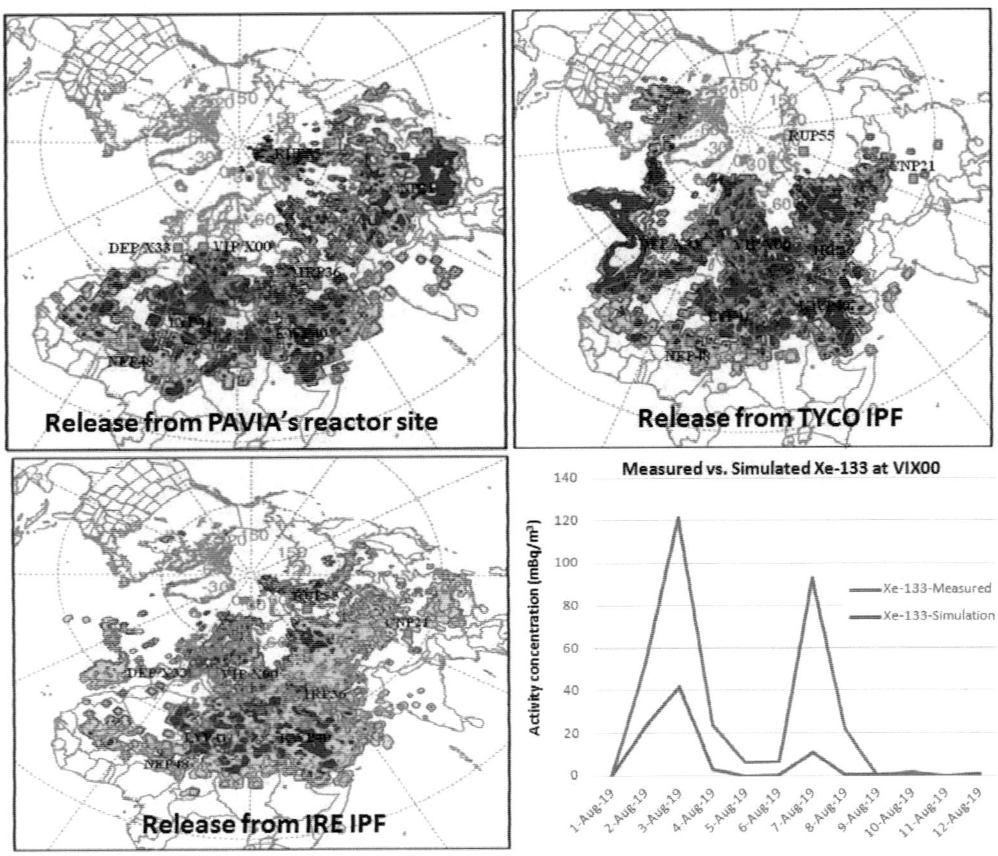

Figure 5. Forward ATM from Pavia's reactor site (top left), TYCO IPF (top right), IRE IPF (bottom left) and measured versus simulated concentrations of Xe-133at VIP/X00 for hypothetical release from Pavia's reactor site on 30 July 2019, 0600UTC (bottom right).

Table 4. Summary results of the forward ATM from 3 different sources of release.

Source of release	Affected IMS Stations	Non-Affected IMS Stations
Pavia's reactor	CNP21, KWP40, LYP41, IRP36, RUP55, NEP48, and VIP/X00	DEP/X33
TYCO IPF	KWP40, LYP41, IRP36, NEP48, VIP/X00, DEP/X33	CNP21 and RUP55
IRE IPF	CNP21, KWP40, LYP41, IRP36, RUP55, NEP48, VIP/X00, and DEP/X33	

Despite these IMS radionuclide stations detected the particulate and radioxenon isotopes, it is important to highlight that some of these stations were non-certified IMS stations. The non-certified stations were VIP/X00, RUP55, IRP36, LYP41. Thus, the only certified stations in NPE2019 were CNP21, KWP40, NEP48, and DEP/X33. Hence, the calibration parameters of the non-certified IMS radionuclide stations might be an issue.

The NPE2019 scenario is rather complex due to a challenging environment surrounding the area of the case study. In reality, Pavia's reactor accident is co-located in a region with a dense population of nuclear power plants as well as two global medical isotope production facilities, namely IRE and TYCO. In normal operation, such civil facilities emit related particulates and radioxenon gases into the atmosphere, limited to the release threshold as stipulated by their respective national regulations. As reported by A Pascal et al, the average release of radioxenon gases from IRE IPF and TYCO IPF is 2.7x 10^{12}Bq/day and 2x 10^{9}Bq/day, respectively [12]. Additionally, the same literature also reported that the continuous and pulsed releases from civil nuclear power plants are in the range of 1.3x 10^{12}Bq/year. Figure 6 shows the dispersion of radioxenon gases from the operation of medical radioisotope facilities and nuclear power plants around the globe [13]. The emission of radioxenon gases from these facilities could have a major effect, in terms of radioxenon measurements at the IMS radionuclide stations.

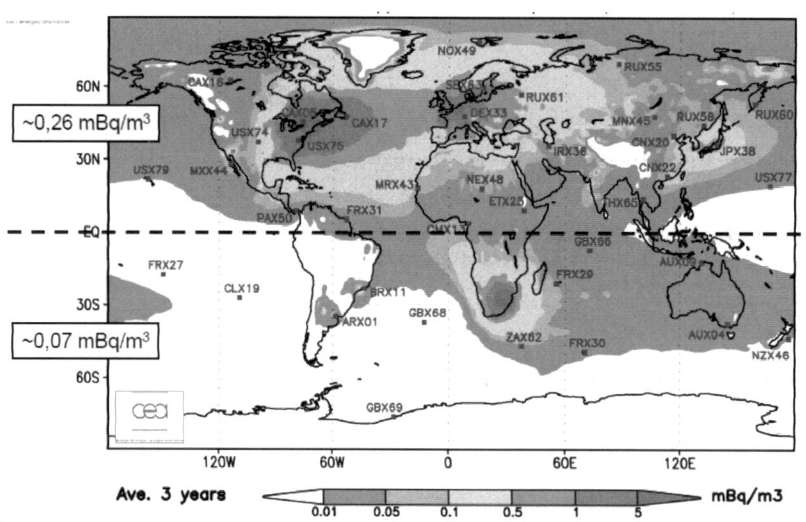

Figure 6. Dispersion of radioxenon gases from the operation of medical radioisotope facilities and nuclear power plants.

4. Conclusion

The radionuclide analysis for NPE2019 was classified into 3 types, namely particulate data analysis, noble gas data analysis, and ATM. The results from particulate data analysis confirmed that the source was predominantly released from the same type of source material and highly probable from a source other than a nuclear explosion, based on the ratio of Cs-134/Cs-137. On the other hand, the noble gas data analysis showed that using multiple atmospheric xenon isotopic activity ratios, the source of xenon isotopes detected at DEX33 and VIX00 were not a result of a nuclear test, but most likely from civil applications that may include PAVIA's reactor, commercial nuclear power plants, and medical isotope production facility. The results from backward mode ATM revealed that the possible source regions for nuclides detected at all radionuclide stations except DEP/X33 include PAVIA's reactor, nuclear power plants as well as TYCO and IRE. Moreover, the forward mode ATM concluded that any release from Pavia's reactor will affect the measurements at all radionuclide stations except DEP/X33. Taking all these scenarios into account, it can be concluded that the abnormal measurements of nuclide at DEP/X33 station were unlikely from Pavia's reactor. The particulates and radioxenon detected at VIP/X00 were most likely from Pavia's reactor site. Meanwhile, detection at other stations could also be originated from Pavia's reactor site. However, owing to the complex environment in the region, the nuclide emissions from IRE IPF, TYCO IPF, and civil nuclear power plants could not be totally ruled out as a possible source.

References

[1] Rashid F I A and Zolkaffly M Z 2018 Stakeholder engagement for promoting the Comprehensive Nuclear Test-Ban Treaty (CTBT): Malaysia's experience *IOP Conf. Ser. Mater. Sci. Eng* **298** 012054

[2] Rashid F I A and Zolkaffly M Z 2017 National data centre preparedness exercise 2015 (NPE2015): MY-NDC progress result and experience *AIP Conf. Proc.* **1799** 030013-1-9

[3] Gestermann N, Bönnemann C, Ceranna L and Wotawa G 2020 The National Data Center Preparedness Exercise 2009 - First Results *EGU General Assembly 2010, 2-7 May, 2010* (Vienna, Austria) 14008

[4] Pan P and Ungar R K 2012 Nuclear event zero time calculation and uncertainty evaluation *J Env Radioact* **106**(0) 65–72

[5] Axelsson A and Ringbom A 2014 On the calculation of activity concentrations and nuclide ratios from measurements of atmospheric radioactivity *Appl. Rad. and Isot.* **92** 12–17

[6] Kalinowski M B, Axelsson A, Bean M, Blanchard X, Bowyer T W, Brachet G, Hebel S, McIntyre J I, Peters J, Pistner C, Raith M, Ringbom A, Saey P R, Schlosser C, Stocki T J, Taffary T and Ungar R K 2010 Discrimination of nuclear explosions against civilian sources based on atmospheric xenon isotopic activity ratios *Pure Appl. Geophys.* **167** 517–39

[7] Preparatory Commission for the Comprehensive Nuclear Test Ban Treaty (CTBTO) 2018 WEB-GRAPE 1.8.4 Software User Manual (Austria: CTBTO).

[8] Kuśmierczyk-Michulec J and Bourgouin P 2018 Influence of mineral dust on changes of 7Be concentrations in air asmeasured by CTBTO global monitoring system. *J. Environ. Radioact.* **192** 454–66

[9] Stein A F, Draxler R R, Rolph G, Stunder B J B, Cohen M D and Ngan F 2015 NOAA's HYSPLIT atmospheric transport and dispersion modeling system *Bulletin of the American Meteorol. Soc.* **96** 2059-77

[10] Preparatory Commission for the Comprehensive Nuclear Test Ban Treaty (CTBTO) 2010 Radionuclide monitoring training course material for the technical training programme, (Vienna, Austria).

[11] Stein A F, Draxler R R, Rolph G, Stunder B J B, Cohen M D, and Ngan F 2015 NOAA's HYSPLIT atmospheric transport and dispersion modeling system tutorial.

[12] A Pascal, G Sylvia, M Mireille, G Philippe and L P Gilbert 2015, Simulation of worldwide Xe-133 atmospheric background WOSMIP V 12-14 May 2015 (Brussels Belgium).
[13] A Pascal and P Gross 2010 Contributions of isotope production facilities and nuclear power plants to the worldwide Xe-133 atmospheric background.

Development of Gaussian dispersion computer code enhanced with building downwash parameter

Nurlyana Omar[1] and Meng Hock-Koh[1,2]

[1]Department of Physics, Faculty of Science, Universiti Teknologi Malaysia, 81310 Skudai, Johor, Malaysia.
[2]UTM Centre for Industrial Applied Mathematics, 81310 Skudai, Johor, Malaysia

Nurlyana3@live.utm.my

Abstract. Atmospheric dispersion has been well known in environmental risk assessment. However, one parameter in atmospheric dispersion which is the building downwash was always neglected when calculating the dispersion of radioactive effluent using basic Gaussian dispersion model. In the Gaussian model, this parameter usually replaces by a value which is far from being accurate. The results of can differ almost to half of the actual result. Thus, the building downwash must be included in the calculation for more accurate dispersion results. This work focusses on the development of computer code to calculate the dispersion of radionuclide effluent which emphasis on the building downwash parameters. There are two parts of this work, the first is the basic Gaussian dispersion model computer code, named GD-Nx^3 and the second part is the Gaussian dispersion model enhanced with building downwash which called GDb-Nx^3. Both computer codes were build using Fortran computer language.

1. Introduction

During normal operating hours, radionuclide effluent was released from a nuclear power plant (NPP). However, these released complied to the standard and regulations from International Atomic Energy Agency (IAEA) [1][2][3] and it does not pose a danger to human since the concentration of radionuclide effluent is below the permissible limit.

However, when an accident happens, the radionuclide effluent released is beyond the permissible limit and can travel hundreds and thousands of kilometers away. This happened in early 2011 when the world was shocked by the serious multiple-reactor accident occurred in Japan. A huge amount of radioactive substance was released into the atmosphere from March 12 to the end of April and dispersed across the northern hemisphere due to the fast transport atmospheric system [4][5][6][7].

Many studies have been performed to simulate the dispersion of radionuclide effluents [8][9][10][11]. This is important to study the dispersion of the radionuclide effluent as it can be a potential health risk to the public. However, these simulation studies did not account for building downwash.

Measurement technique for atmospheric analysis has been improved in recent year. Gaussian atmospheric dispersion model was used to analyze the dispersion from a point source within 100 km radius. It is a simple model that needs few input data. However, it can only be used in an open space area and does not include building downwash. Thus, it is important to incorporate Gaussian atmospheric model with building downwash due to the possibility that radionuclide effluent can travel in the city area after release from the point source.

It is important to note that the simulations model need to have a high degree of accuracy and must be achieved faster than real-time detection for effective decision support tools. As stated by [12][13], atmospheric dispersion tools can contribute as good decision support tools in making a prevention action in an emergency.

According to [14] the concentration of radionuclide before and after a building can differ up to 90 per cent of initial reading. Most of the software in the market only focus on the open-air analysis and does not include building a washout parameter.

In this paper, we are working towards producing our very own dispersion computer code analysis by using Gaussian atmospheric dispersion model enhanced with building downwash parameter to assess the radioactive effluent dispersion released in an emergency such as in the Fukushima accident. There are two parts of this research, there is the basic Gaussian atmospheric model (GD-Nx3) and the second is enhanced with building downwash parameter (GDb-Nx3).

Our GD-Nx3 computer code is benchmarked to the HotSpot Health Physics code (hereafter referred as HotSpot) developed by Steven *et al.* in Lawrence Livemore National Laboratory's National Atmospheric Release Advisory Center (NARAC) [15]. Whereas our GDb-Nx3 is based on the technical paper by [16]. Using Fortran computer language, these analysis codes were developed.

2. Methodologies

2.1. Gaussian plume model

The transport of radionuclide effluent is given by the expression [17][18]:

$$\frac{\partial C}{\partial t} + \nabla . \vec{J} = S \qquad (1)$$

where C refers to the mass [kg/m^3], S the source function [kg/m^3s] and J being the mass flux [kg/m^2s] with the combinations of diffusion and advection effects. Following the Fick's law, the three-dimension advection-diffusion equation to be solved is in the form:

$$\frac{\partial C}{\partial t} + \nabla . (C\vec{u}) - \nabla . (-\boldsymbol{K}\nabla C) - S = 0 \qquad (2)$$

where:

K is the diffusion coefficients [m^2/s] and $\dot{u}$ is the wind velocity [m/s].

In our code, few Gaussian assumptions were made to solve the eq.2 namely:

i. The wind velocity in the y-direction or also known as the crosswind direction is negligible.
ii. The wind velocity in the z-direction is negligible.
iii. Radioactive effluent is released at a constant rate, Q from a single point source which located at a height, H from the ground. The source term may be written as

$$S(\vec{x}) = Q\,\delta(x)\,\delta(y)\,\delta(z-H) \qquad (3)$$

i. The solution takes place in a steady state condition.
ii. The diffusion due to advection in x-direction is overwhelming large compared to the diffusion because of turbulent in the x-direction.
iii. The radioactive effluent does not penetrate the ground.

Further explanation on Gaussian model that was used in GD-Nx^3 and GDb-Nx^3 can be referred to [19]. The equation to be computed is as per Eq.4:

$$C(x,y,z) = \frac{Q}{2\pi\sigma_y\sigma_z u} exp\left(\frac{-y^2}{2\sigma_y{}^2}\right)\left\{\exp\left(\frac{-(z-H)^2}{2\sigma_z{}^2}\right) + \exp\left(\frac{-(z+H)^2}{2\sigma_z{}^2}\right)\right\} \quad (4)$$

where x is the downwind distance, y is the crosswind distance from the centre line ofthe plume and z is the height above the ground for which the concentration is calculated (all in units of m). The parameters Q, u and h are the release rate [*Bq/s*], mean wind speed [*m/s*] and the effective release height [*m*], respectively.

Further explanation on GD-Nx^3 and GDb-Nx^3 can be referred to [19] as it explains more details on how the equation of Gaussian atmospheric dispersion model works which is the basic of our computer code.

2.2 Building downwash parameter

Radionuclide effluent example from Fukushima accident can travel far into the city area and encounter high rise building [8]. Therefore, it is crucial to incorporate the building downwash effect into the main atmospheric dispersion calculation as the concentration of radionuclide effluent can differ by 90% [20] from its real reading.

By understanding the building downwash effect, it can help in preventing the dangerous situation such as a high concentration of harmful effluent in one local area. For example, 'Good Practice Guide for Atmospheric Dispersion Modelling' prepared by the National Institute of Water and Atmospheric Research, AURORA, New Zealand was published to advise the best building dimension to avoid downwash effect as airflow around a building are often a complicated process [21].

Based on [16], there are four crucial part in incorporating building downwash with Gaussian atmospheric dispersion model (Figure 1). Segment A of the figure shows that the streamline slope is equal to zero which is the same as the Gaussian atmospheric dispersion model in an open space area. The segment, B, C and D will have a different calculation of the streamline slope. This calculation will be embedded into the Gaussian atmospheric dispersion model as part of the building downwash parameter.

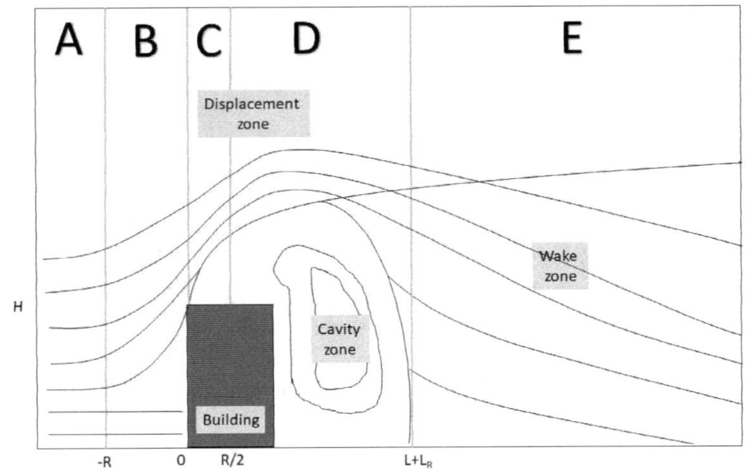

Figure 1. The building downwash effect due to the windward side.

Table 1. Table shows the streamline slope used to calculate the movement of the radionuclide effluent in building downwash.

• **Segment**	• **Streamline slope**	• **Downwind distance from the windward**
• A	• $\frac{\delta z}{\delta x} = 0$	• $(x < -R)$
• B	• $\frac{\delta z}{\delta x} = \frac{2(H_R - H)(x-R)}{R^2}$	• $(-R \leq x \leq 0)$
• C	• $\frac{\delta z}{\delta x} = \frac{(-4(H_R-H)(\frac{2x}{R}-1)}{R^2}$	• $(0 \leq x \leq 0.55R)$
• D	• $\frac{\delta z}{\delta x} = \frac{(H_R-H)(R-2x)}{\left(L+L_R-\frac{R}{2}\right)^2} * \left(\frac{z}{H}\right)^{0.3}$	• $(0.5 \leq x \leq L + L_R)$

The notation of H refer to the building height, HR is the maximum building height, R is the building length scale, L is the projected building length along with the flow and L_R is the downwind recirculation cavity. This streamline slope will be calculated depending on the building dimension and in segment B onwards, the assumption is that the effluent plume is consider moving as particle thus the streamline slope is to calculate the z-direction.

Table 2. Table shows the percentage of difference between GD-Nx^3 and HotSpot

Distance (km)	**GD-Nx3 (Bq-sec/m^3)**	**HotSpot (Bq-sec/m^3)**	**Differences value**
1.0	3.6×10^{10}	3.5×10^{10}	-0.1×10^{10}
5.0	3.0×10^{9}	3.0×10^{9}	-
10.0	8.8×10^{8}	8.8×10^{8}	-
15.0	4.4×10^{8}	4.4×10^{8}	-
20.0	2.7×10^{8}	2.7×10^{8}	-
25.0	1.8×10^{8}	1.8×10^{8}	-
30.0	1.4×10^{8}	1.4×10^{8}	-
35.0	1.1×10^{8}	1.1×10^{8}	-
40.0	8.8×10^{7}	8.8×10^{7}	-
45.0	7.3×10^{7}	7.2×10^{7}	-0.1×10^{7}
50.0	6.2×10^{7}	6.2×10^{7}	-
55.0	5.3×10^{7}	5.2×10^{7}	-0.1×10^{7}
60.0	4.7×10^{7}	4.6×10^{7}	-0.1×10^{7}
65.0	4.1×10^{7}	4.0×10^{7}	-0.1×10^{7}
70.0	3.6×10^{7}	3.6×10^{7}	-
75.0	3.2×10^{7}	3.2×10^{7}	-
80.0	2.9×10^{7}	2.9×10^{7}	-
85.0	2.7×10^{7}	2.7×10^{7}	-
90.0	2.4×10^{7}	2.4×10^{7}	-
95.0	2.2×10^{7}	2.2×10^{7}	-
75.0	3.2×10^{7}	3.2×10^{7}	-
80.0	2.9×10^{7}	2.9×10^{7}	-

3. Results and discussion

3.1. *Benchmarking GD-Nx³ computer code*

We benchmark our GD-Nx³ analysis computer code with the available software in the market which is HotSpot. We have utilized the data obtained from the Fukushima Dai-ichi accident in 2011 as the input data for this analysis. This analysis focus on the Cs-137 as it is one of the most emitted radionuclides during a nuclear accident. The half- life of Cs-137 is 30 years and assumed to exist longer in the air and can easily be used by selecting default settings in HotSpot software. **Table 2** shows the concentration of Cs-137 in the distance of x km from the point source.

It shows that our GD-Nx3 analysis computer code is in line with HotSpot software with the percentage different of 75 per cent which consider good.

3.2. *Complementing GDb-Nx³ analysis computer code*

The GD-Nx³ and GDb-Nx³ analysis computer code were developed using Fortran computer language. Fortran is the oldest computer language and does not need a Fortran Window, instead, it just needs a program to type it and can be saved using an editor which turned into an executable file by a Fortran compiler.

There are two stages in developing GDb-Nx³ analysis computer code. The first stage is to develop GD-Nx³ which is the open space dispersion analysis (Segment A) and then calculate the streamline slope for Segment B onwards for the GDb-Nx³. This part of the work is still in progress. Up to now, the GDb-Nx³ coding is still in Segment B, where the assumption needed such as the distance of the building from the point source and the dimension of the building is yet to finalized. Both parameters are crucial in developing the GDb-Nx³ code since it will affect the total result of the concentration.

4. Conclusion

In this paper, we discussed the current progress of our GD-Nx³ analysis computer code that has been benchmarked with the commercial HotSpot software for a hypothetical nuclear accident and also the current progress with GDb-Nx³. It shows that our GD-Nx³ analysis computer code in-line with HotSpot although there are some results underestimate the HotSpot results. We used the Fukushima Daiichi accident as the hypothetical nuclear accident for this study. From our analysis, it shows that if a nuclear accident to happen in Malaysia with the centre to be from Malaysia Nuclear Agency, the impact of radionuclide effluent can cover up to 95 km of radius. This prediction should cover an area such as Kuala Lumpur, Pahang, Negeri Sembilan and Melaka. However, the concentration would be different if the radionuclide to enter high-rise building area such as Kuala Lumpur and Selangor areas.

5. References

[1] International Atomic Energy Agency 2010 TECDOC-1638 Setting authorized limits for radioactive discharges: practical issues to consider (IAEA).
[2] International Atomic Energy Agency 2012 Generic models for use in assessing the impact of discharges of radioactive substances to the environment (IAEA).
[3] International Comission on Radiological Protection 2005 Draft For Consultation 2005 Recommendations Of The International Commission On Radiological Protection ICRP (ICRP).
[4] Masson O, Baeza A, Bieringer J, Brudecki K, Bucci S, Cappai M, Carvalho F P, Connan O, Cosma C, Dalheimer A, Depuydt G, De Geer L E, De Vismes A, Gini L, Groppi F,

Guonason K, Gurriaran R, Hainz D, Halldorsson O, Hammond D, Holy K, Homoki Z, Ioannidou R, Isahenko K, Jankovic M, Katzlberger C, Kettunen M, Kierepko R, Kontro R, Kwakman P J M, Lecomte M, Vitro L, L Leppane, A P, Lind, Lujaniene G, Mc Ginnity P, Mc Mahon C, Mala H, Menenti S, Manalopoulou M, Mattila A, Mauring A, Mietelski J W, Moller B, Nielsen S P, Nikoli, J, Overwater R M W, Palsson S E, Papastefanou C, Penev I, Pham M K, Povinec P P, Rameback H, Reis M C, Ringer W, Rodrigues A, Rulik P, Saey P R J, Samsonov V, Schlosser C, Sgorbati G, Silobritiene B V, Soderstrom C, Sogni R, Solier L, Sonck M, Steinhauser G, Steinkopff T, Steinmann P, Stroulos S, Sykora I, Todorovic D, Tooloutalaie N, Tositti L, Tschiersch J, Urgon A, Vagena E, Vergas A, Wershofen H, Zhukoya O 2011 Tracking of airborne radionuclides from the damage Fukushima Daiichi nuclear power plant *European Netw. Environ. Sci. and Technol.* **45** 7670-7677.

[5] Pittauerova D, Hettwig B, Fischer H W 2011 Fukushima Fallout In Northwest German Environmental Media Journal of Environmental Radioactivity **102** 877-880.

[6] Pham M K, Eriksson M, Levy I, Nies H, Osvath I, Betti M 2012 Detection of Fukushima Daiichi Nuclear Power Plant Accident Radioactive Traces in Monaco *J. of Environ. Radioact.* **114** 131-137.

[7] Ioannidou A, Manenti S, Gini L, Groppi F 2012 Fukushima fallout at Milan, Italy Journal of Environmental Radioactivity **114** 119-125.

[8] Man C K, Kwok Y H 2001 Assessment of risk to hong kong due to accidental release of radionuclides from a nearby nuclear power Plant *J. of Radioanalyt. and Nucl. Chemi.* 250(3): 485-490.

[9] Pirouzmand A, Dehghani P, Hadad K, Nematollahi M 2015 Dose assessment of radionuclides dispersion from Bushehr nuclear power plant stack under normal operation and accident conditions *Int. J. of Hydrogram Ener.* **40** 15198-15205.

[10] Sahin S, Ali M 2016 Emergency planning zone estimation for Karachi-2 and Karachi-3 Nuclear power plant using Gaussian puff model *Hindawi Pub. Corp. Sci. Technol. of Nucl. Installtion*.

[11] Aliyu A S, Ramli A T, Saleh M A 2014 Environmental impact assessment of a new nuclear power plant (NPP) based on atmospheric dispersion modeling. *Stoch Environ. Res Risk Assess.* **28** 1897-1911.

[12] Benamrane Y, Boustras G 2015 Atmospheric dispersion and impact modeling system: How are they perceived as support tools for nuclear crises management? *Safety Sci.* **71** 48-55.

[13] Benamrane Y, Wybo J L, Armand P 2013 Chernobyl And Fukushima nuclear accidents: What has changed in the use of atmospheric dispersion modeling? *J. of Environ. Radioact.* **126** 239-252.

[14] Capena E. 2004 An overview about the study of downwash effects on dispersion of airborne pollutants *Environ. Modelling* and *Softw.* **9** 1077-1087.

[15] Homann S G, Aluzzi F 2013 HotSpot health physics codes version 3.0 user's guide *Nat. Atmosph. Release Advi. Cent.* 3.

[16] Schulman L L, Strimaitis D G, Scire J S 2000 Development and evaluation of the PRIME Plume Rise and Building Downwash Model Air and Water Management Association **50** 378-390.

[17] Stockie J M The Mathematical of Atmospheric Dispersion Modelling. SIAM REVIEW **2011** **53**(2) 349-372.

[18] Veigele W J, Head J H 1978 Derivation of the Gaussian plume model J. of the Air Pollut. Contr. Assoc. **28**(11) 1139-1141.

[19] Shamsuddin S D, Omar N, Meng-Hock K 2017 Development of radionuclide dispersion modeling software based in Gaussian plume model *Matematika* **23**(2) 149-157.

[20] Schulze R H 1995 Balancing simplicity with accuracy in the use of dispersion modeling in the Unites-State International Journal of Environment Pollutan **5** 521-529.

[21] Bluett J and Gimson N, Fisher G, Heydenrych C, Freeman T and Gadfrey J 2004 Good

practise guide for atmospheric dispersion modelling *Minist. for the Environ.* (New Zealand).

Acknowledgement
This document is the results of the research project funded by the UTM Potential Academic Staff grant (Q.J130000.2726.02K70) UTM Research University Grant (Q.J130000.2626.15J74) and UTM ZAMALAH Scholarship.

IOP Conf. Series: Materials Science and Engineering 1106 (2021) 012007 doi:10.1088/1757-899X/1106/1/012007

Generation of Geiger-Muller detector dataset for simulation of radioactive source localization algorithm

Nur Aira Abd Rahman[1,2], Khairul Salleh Mohamed Sahari[1], Nor Arymaswati Abdullah[2], Muhammad Fairuz Abdul Jalal[1] and Nasri A Hamid[1]

[1]Universiti Tenaga Nasional (UNITEN), 43000 Kajang, Selangor, Malaysia
[2]Malaysian Nuclear Agency, 43000, Bangi, Selangor, Malaysia

nur_aira@nm.gov.my

Abstract. Detector dataset in digital data is required to simulate and evaluate a particle filter (PF) algorithm. The PF is used to estimate the position and intensity of unknown hotspots within a region of interest (ROI) from sparse sampling points around the ROI. The dataset will emulate GM measured data at these sampling points. GM radiation detection model, Poisson random variable generator, and Gaussian noise generator were implemented to generate the datasets. Next, using the datasets as input to the PF code, the corresponding output is evaluated for various configuration of hotspot locations and radiation intensities. The resulting detector datasets and the particle filter output were analyzed; and the results are presented in the paper.

1. Introduction

Radiation measurement module consists of radiation detector(s) as the front-end sensor to sense or detect the radiation. Interaction between the ionizing radiations (such as alpha, beta, gamma, and x-ray) with the detector materials is typically converted to voltage or current signal. For a Geiger-Muller detector (GM), a single radiation interaction event with the detector will produce one voltage pulse. The module will count the number of pulses within a period to measure the intensity of the radiation in count rate, such as count per second (CPS). Therefore, a high number of counts signifies higher radiation intensity and vice versa. Furthermore, calibration process could be performed to convert the count rate to dose rate into other units such as Sievert or Roentgen per hour[1].

Similarly, to calibrate or test the functionality of a system that involves radiation measurement, the detector is exposed to a radiation source such as a sealed radioactive source while the corresponding output is observed. Alternatively, a nuclear pulser could be connected to the system hardware to emulate the detector output signal [2]. This process enables the system output to be analyzed and compared against the desired result. However, these testing could only be performed when the system hardware or prototypes are ready. Meanwhile, evaluation of the system during algorithm or code development will require the detector datasets in digital data. These datasets will be used as input stimulus to run code simulation and analyze the flow of data. Hence, any error or bug could be captured earlier and corrected efficiently.

In this research, datasets were required to simulate the radioactive source localization algorithm [3][4]. This algorithm is used to estimate the position and intensity of unknown hotspots within a region of interest (ROI) from sparse sampling points by using Particle filter (PF) [5]. PF is a recursive Bayesian filter based on Monte-Carlo method that use a set of particles to represent the hypotheses of the hotspot's position and intensity. GM is used to measure radiation at designated sampling points around the ROI. Next, the data is utilized to compute each particle's probability, which subsequently determines the particle's importance weightage. Particles with higher weightages will be replicated while particles with smaller weightages will be eliminated. Iteratively with a new measurement, these particles will be updated and expected to converge to a final state where their means are taken as estimates of the true source intensity and position.

The aim of this paper is to establish a method to generate Geiger-Mueller (GM) radiation detector dataset that emulates measured data from an actual radiation counting event. The output is a set of data in the form of count rate (in count per second, CPS), proportional to the hotspot/radioactive source intensity. Next, the method is used to generate GM dataset as input for the PF algorithm/code simulation. Thus, the PF's accuracy to identify unknown hotspot can be verified and evaluated across various configurations of hotspot positions and intensities.

The paper is organized as follows; Section 2.0 describes the methodology to generate GM dataset. Section 3.0 presents the results and application of the dataset. Finally, Section 4.0 concludes the work and outlines future works.

2. Methodology

In this section, the methodology to generate GM dataset will be described. The section is divided into two parts; modelling of GM radiation detection and measurement, and generation of GM dataset for targeted sampling points.

2.1. Modelling of GM Detection Event

The hotspot is modelled as an isotropic point radioactive source, *S* that emits radiation radially with intensity I_s. Figure 1 shows the geometry of single-point radiation detection. The detector is located at a distance, *d* from the radioactive source. The intensity of the radiation I_j at the detector is given by:

$$I_j = \frac{I_S A}{4\pi d^2} e^{-\mu d} + I_{BG} \qquad (1)$$

where A is the solid angle of the cylindrical radiation detector, μ is the air attenuation coefficient, and I_{BG} is the mean of background radiation count rate for the target area.

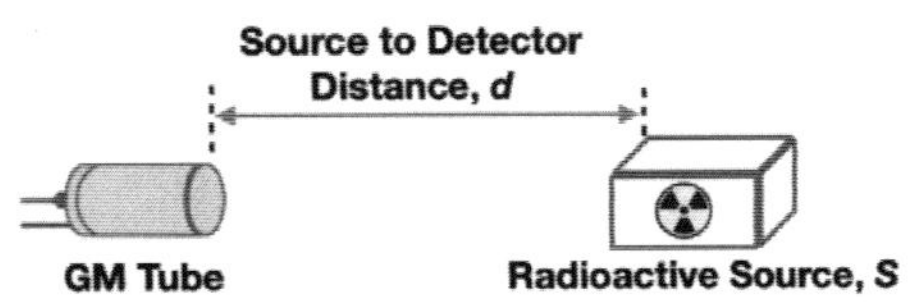

Figure 1. Single point measurement geometry with air as the attenuating medium.

Next, the radiation interaction events (as a product of radioactive source decay), x in a specified time interval, τ can be modelled by Poisson distribution:

$$P(x, \lambda) = \frac{\lambda^x e^{-x}}{x!} \qquad (2)$$

where λ is defined as the mean and the variance of the Poisson distribution. The Poisson random variable generator is implemented in Python to generate the detector count dataset [6]. For example, the detector count dataset in Figure 1 can be generated by defining the Poisson mean, λ as the expected count, I_j. Finally, Gaussian noise is added to the count data to simulate the detector signal noise, as shown in Figure 2.

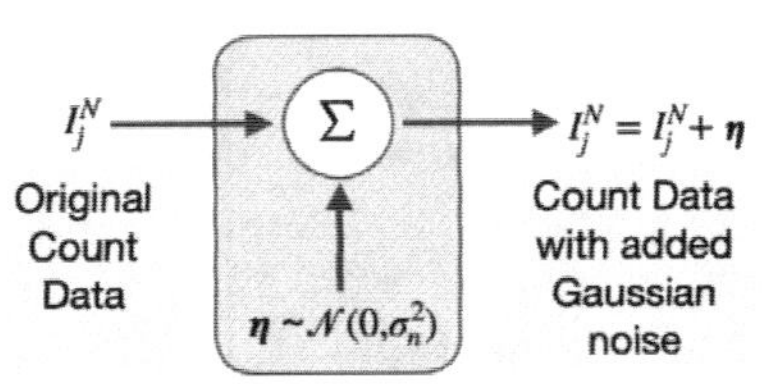

Figure 2. Signal noise modelled by Gaussian distribution.

2.2. Generation of GM Detector Dataset at Targeted Sampling Points

The complete flowchart to generate GM detector dataset is shown in Figure 3. The sampling geometry is essentially used to compute Euclidean distance between source to the detector, *d* for every sampling point within the geometry. The sampling parameters are source-to-detector Euclidean distance, attenuating material (air), detector solid angle, radioactive source intensity (in CPS), counting time, and the number of random variable/data, *N*. Based on these parameters, the expected intensity in the unit of CPS is calculated by using (1). Next, the set of count data is generated, and Gaussian noise is added to the data. Finally, the detector measurement dataset is generated.

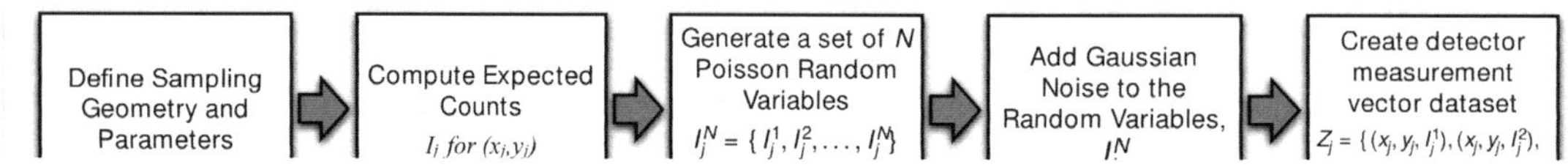

Figure 3. Flowchart for the generation of GM dataset.

2.3. Sampling Geometry for Simulation of Radioactive Source Localization Algorithm

The PF algorithm's objective is to identify the hotspot position and estimate the radiation intensity based on the measured detector data. In general, PF consists of a set of N particles $\{S^j\}_{j=1}^{N}$ that represents hypotheses of source parameters, $S_k^j = (x_k^j, y_k^j, I_k^j)$, where (x_k^j, y_k^j) refers to the hypothetical source position with intensity I_k^j. Data measured from the sampling points, Z_k is used to compute the weightage of each particle. Based on the measurements, particles with higher weightages will be replicated while particles with smaller weightage will be eliminated. The final particle set (after the last measurement) will be used to estimate the posterior density function, $P(S_k|Z_{1:k})$ of the source parameter.

Sampling geometry for the PF simulation is shown in Figure 4(a). This geometry is based on the real-world experimental setup, as shown in Figure 4(b). A radioactive source is placed at one of the designated hotspot locations; C1, C2, C3 or C4 as indicated in Figure 4(b). A detector is placed at sampling points, Sp_1 to Sp_8 (sequentially) to measure the radiation data. Four experimental datasets are collected, each corresponds to one of the hotspot locations and consists of measurements from Sp_1 to Sp_8. These datasets, Z_k will be used as input to the PF.

Similarly, to generate the detector dataset for PF simulation, the hotspot (x_s, y_s) position is fixed based on the C1 to C4 positions. In addition, the intensity (I_s) of the hotspot is adjustable. Thus, the generated dataset [illegible] wide range of hotspot intensities.

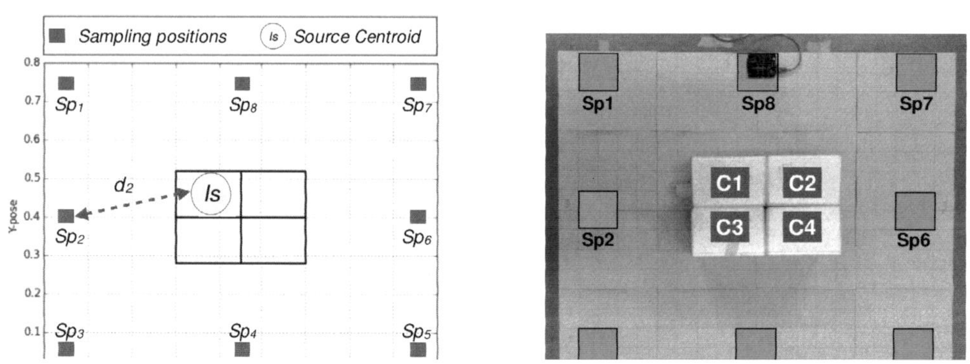

(a) (b)

Figure 4. Sampling geometries for radioactive source localization algorithm in (a) 2D grid visualization and (b) real-world experiment. The position of source centroid is varied based on the position of C1 to C4 in (b).

3. Results and discussion

In this section, the result and discussion for GM dataset generation and its implementation to simulate the PF algorithm will be presented.

3.1. GM dataset generation

GM expected intensity is generated for source intensity (expressed in count rate, i.e. count per second or CPS) ranging from 2000 to 12000 CPS using the radiation propagation model (1). The graph of expected intensity, I_j against source-to-detector distance, d is shown in Figure 5. In this graph, the model is configured based on the specification of a real GM detector component with a solid angle of 0.00061608 m^2. Meanwhile, the source-to-detector distance is varied between 0.2 m to 0.65 m. Notably, data trends in the graph obeyed the inverse-square law of radiation emission, where the expected intensity is inversely proportional to the squared of source-to-detector distance.

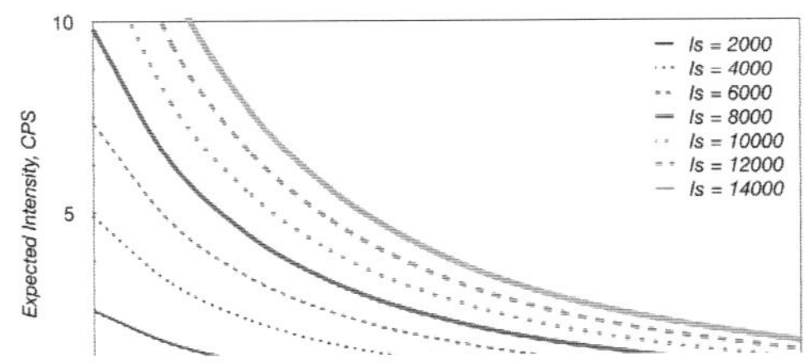

Figure 5. Graph of the GM expected intensity against the source-to-detector distance for source intensity, I_s range from 2000 CPS to 12000 CPS.

In the PF simulation experiment, the detector datasets are generated using the method described in Figure 3. The sampling geometry setup is shown in Figure 4(a) where detector datasets were generated for sampling points Sp_1 to Sp_8 for each source location C1 to C4. For example, with source intensity, I_s configured to 50000 CPS, the expected radiation intensity, I_j at the sampling point, Sp_1 is 13.62 CPS when the source is positioned at C1 (with XY coordinate (0.35,0.45)). Figure 6 shows the 100 datapoints generated by using a Poisson random generator, I_j^N and the datapoints with added Gaussian noise, $I_j^N + \eta$. The mean of generated Poisson datapoints, $\overline{I_j^N}$ is 13.68 CPS, while the mean of Poisson data with added Gaussian noise, $\overline{I_j^N + \eta}$ is 13.26 CPS.

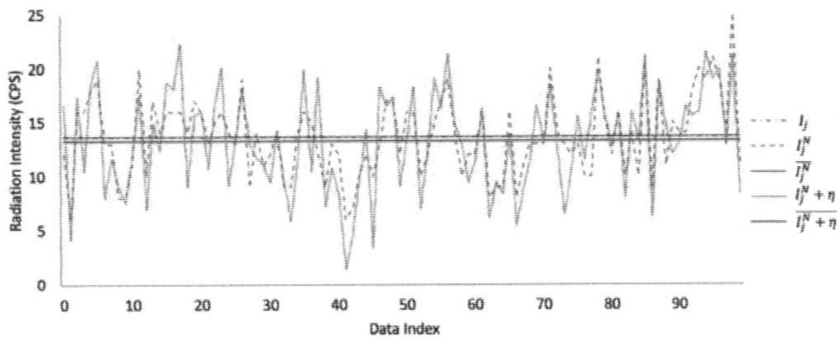

Figure 6. Plots of expected intensity, I_j, Poisson random data, I_j^N, mean of Poisson data, $\overline{I_j^N}$, Poisson data with added Gaussian noise, $I_j^N + \eta$, and mean of Poisson data with added Gaussian noise, $\overline{I_j^N + \eta}$ for Sp_1 with source intensity, I_s of 50000 CPS and positioned at C1.

Figure 7 shows the complete datasets generated for sampling points, Sp_1 to Sp_8 with source positioned at C1 and source intensity, I_s = 50000 CPS. The horizontal line represents the mean of Poisson data with added Gaussian noise, $\overline{I_j^N + \eta}$ that will be used as input to PF simulation. The plots showed that the generated datasets were randomly distributed with an independent and constant rate, which emulates radiation interaction with detector during radiation counting event. Furthermore, Figure 8 compares generated and measured datasets for source intensity, I_s of 5000 CPS. In both cases, data for sampling points Sp_2 and Sp_8 with smallest source-to-detector distances exhibited the highest mean of radiation intensities compared to other sampling points that are further away from the source. Figure 9(a) illustrates the generated data with respect to sampling geometry for I_s of 50000 CPS, and figure 9(b) compares the generated data against actual measured value for I_s of 5000 CPS.

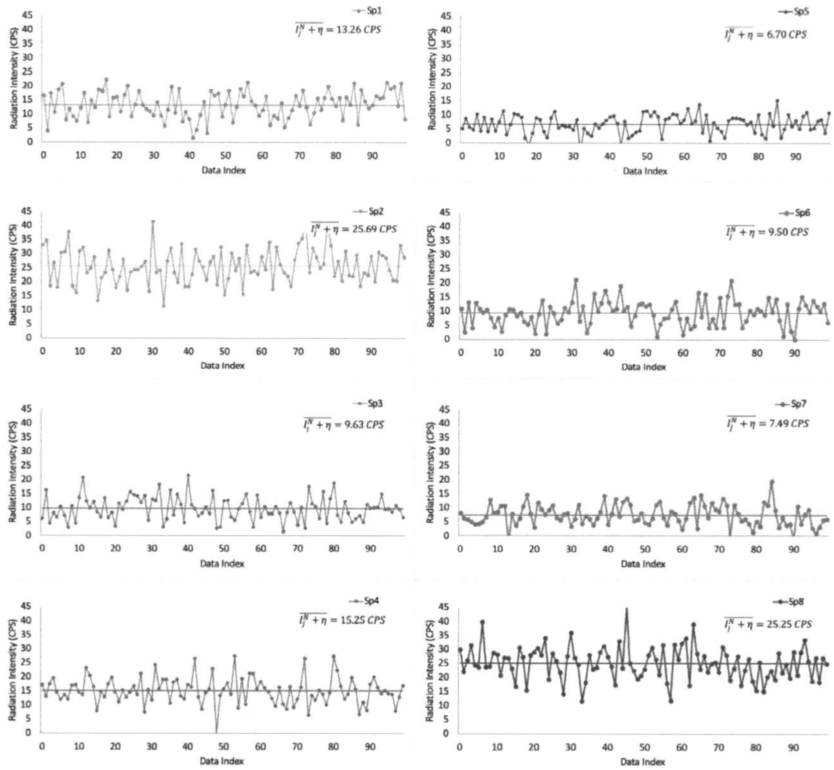

Figure 7. Datasets generated for detector positioned at sampling points Sp_1 to Sp_8 with source intensity, I_s of 50000 CPS and positioned at C1.

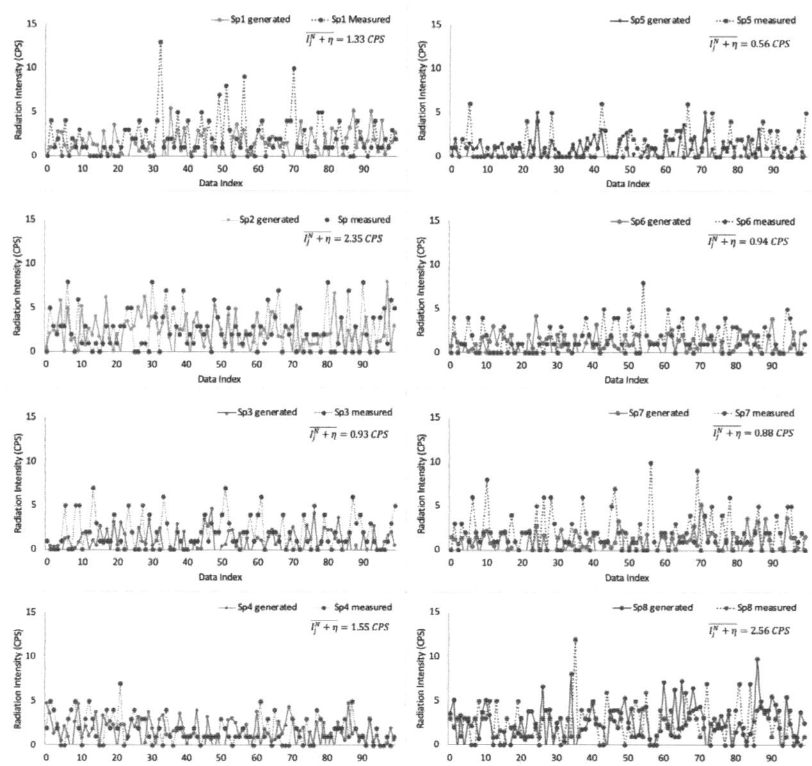

Figure 8. Comparison between generated and measured datasets for detector positioned at sampling points Sp_1 to Sp_8 with source intensity, I_s of 5000 CPS and positioned at C1.

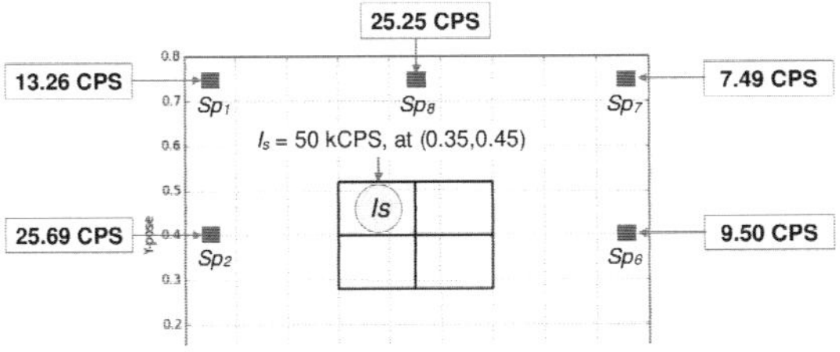

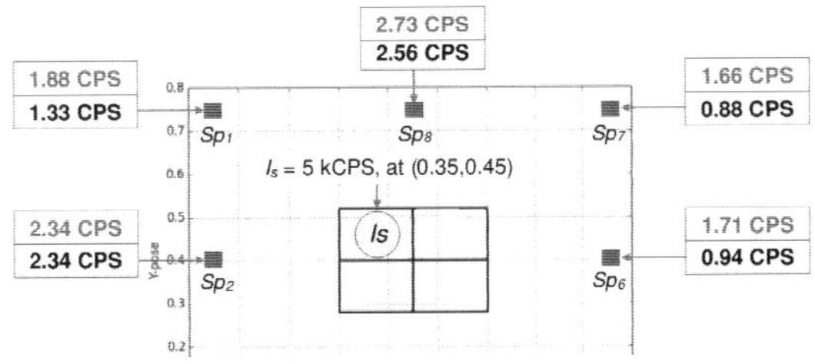

(a) (b)

Figure 9. Data generated with respect to sampling points for Is=50000 CPS in (a) and comparison between generated data (black) and actual measured data (red) for Is=5000 CPS in (b).

3.2. Application: evaluation of particle filter for hotspot localization and radiation intensity estimation

Figure 10 visualized the initial state of the PF with 50000 particles. In real applications, the hotspot, i.e. the source parameters are unknown; hence the particles are sampled from a uniform distribution and distributed across the whole ROI with intensities between 0 to 10000 CPS. Next, the PF is simulated with four testing conditions; where source locations varied between C1 and C4. The results had shown that with each measurement in the dataset, particles closer to source parameters were assigned with higher weightages and frequently duplicated. In contrast, particles with lower weightages were eliminated.

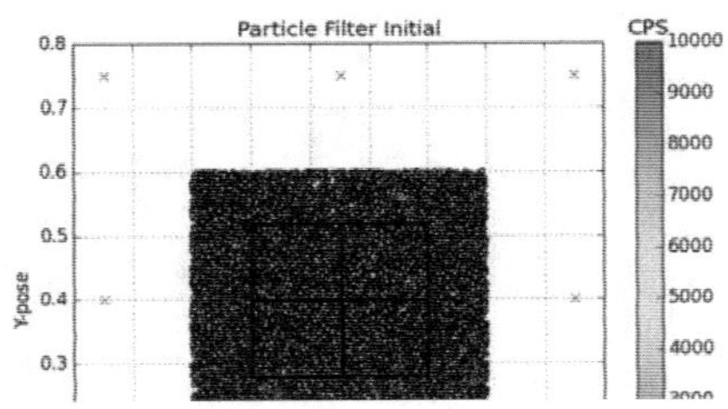

Figure 10. PF Initial state with 50000 particles representing the initial hypotheses of source position and intensity within the ROI. Each small circle visualized the hypothetical source position in XY coordinate while the colours reflect its respective radiation intensity.

Table 1 presents the PF results when tested with the generated dataset and measured dataset with the same sampling geometries and source parameter (true value). Gaussian curve fitting function is used to compute the means and the standard deviations for the final results. In addition, Figure. 11 visualized the PF results with generated datasets where the final state particles have converged to the true source positions and intensity. The results proved that the PF algorithm has successfully identified the correct location of hotspots, as indicated by the square red markers. The errors of hotspot intensity estimations were 1.98% to 12.58% for generated datasets and 4.74% to 35.06% for measured datasets. The noise from GM detector circuit (indicated as spikes of elevated counts in Figure 8) was identified as one of the potential factors contributing to the higher errors in the estimated intensity of the measured dataset.

Table 1. PF results with the generated dataset and real detector data.

Test Condition		*True Value (Hotspot Boundary)*	*Generated Dataset (Mean± Std Dev)*	*Measured Dataset (Mean± Std Dev)*
	x_s	0.3 – 0.45	0.339 ± 0.025	0.405 ± 0.048
C1	y_s	0.4 – 0.52	0.458 ± 0.028	0.438 ± 0.036
	I_s	5000	5099 ± 1502	5273 ± 927
	x_s	0.45 – 0.6	0.501 ± 0.027	0.489 ± 0.047
C2	y_s	0.3 – 0.52	0.455 ± 0.023	0.430 ± 0.033
	I_s	5000	5608 ± 1512	6251 ± 982
	x_s	0.3 - 0.45	0.395 ± 0.029	0.418 ± 0.045
C3	y_s	0.28 – 0.4	0.349 ± 0.022	0.370 ± 0.033
	I_s	5000	5083 ± 1482	6368 ± 915
	x_s	0.45 – 0.6	0.559 ± 0.022	0.500 ± 0.039
C4	y_s	0.28 – 0.4	0.344 ± 0.026	0.375 ± 0.033
	I_s	5000	5629 ± 1565	6753 ± 953

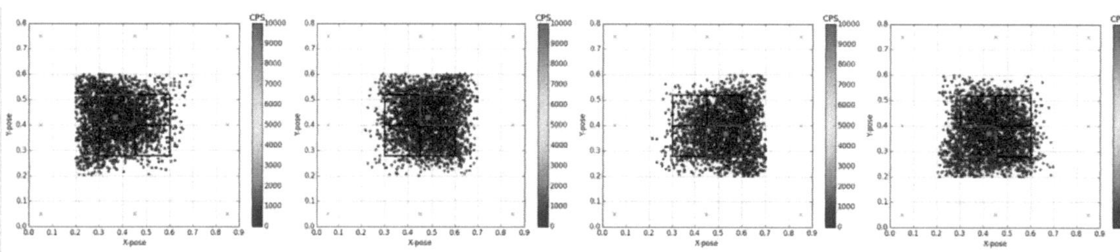

Figure 11. PF results with generated datasets where true hotspot locations are indicated with the red boxes (C1, C2, C3, and C4 from left to right) and PF final estimations are indicated by the red square markers.

4. Conclusion

In conclusion, a method to generate datasets for a GM detector was developed by implementing GM radiation detection model, Poisson random variable generator, and Gaussian noise generator. The results had shown that the datasets are randomly distributed with an independent and constant rate, which emulates the output from actual radiation counting event. Next, the dataset is used to simulate PF code for hotspot localization and intensity estimation. The method has enabled PF code evaluation on the various configuration of hotspot locations and radiation intensities. In the future, the work will be extended for other detector types to increase its flexibility and applications.

5. References

[1] IAEA 2000 Calibration of Radiation Protection Monitoring Instruments

[2] García-Durán Á *et al.* 2019 Nuclear pulse generator embedded in FPGA *Appl. Radiat. Isot.* **147** 129–135.

[3] Gao W, Wang W, Zhu H, Huang G, Wu D and Du Z 2018 Robust radiation sources localization based on the peak suppressed particle filter for mixed multi-modal environments *Sensors* (Switzerland) **18**(11) 1–20.

[4] Hellfeld D, Joshi T H Y, Bandstra M S, Cooper R J, Quiter B J and Vetter K 2019 Gamma-ray point-source localization and sparse image reconstruction using Poisson likelihood *IEEE Trans. Nucl. Sci.* **66**(90) 2088–2099.

[5] Cook J, Smith R C, Ramirez C and Rao N S V 2020 Particle filtering convergence results for radiation source detection 1–19.

[6] Community T S 20-Jul-2020 numpy.random.poisson (Online: https://numpy.org/doc/stable/reference/random/generated/numpy.random.poisson.html).

Acknowledgement
The work is supported by the Ministry of Education Malaysia through FRGS research grant FRGS/1/2019/TK04/UNITEN/01/2. The authors would like to acknowledge the Malaysian Nuclear Agency and CAMARO Research Group of Universiti Tenaga Nasional for the facilities and equipment.

A study on UVC irradiance emitted by various types of UVGI products

Wan Syazlin Wan Yunoh[1,a)], Roha Tukimin[1,b)], Naurah Mat Isa[1,c)], Mohd Yusof Hamzah[1,d)] and Khairul Azhar Abdul Halim[1,e)]

[1]Malaysian Nuclear Agency (Nuklear Malaysia), Bangi, 43000 Kajang, Malaysia.

[a]wan_syazlin@nm.gov.my
[b]roha@nm.gov.my
[c]naurah@nm.gov.my
[d]m_yusof@nm.gov.my
[e]azharhalim@nm.gov.my

Abstract. There are many types of Ultra Violet Germicidal Irradiation (UVGI) products available in the market used in many applications such as sterilization and disinfection. These products are popular during the spread of pandemic COVID-19 and become an option of disinfection method for surface and air such as in office, hospital, hotel room and car. The objective of the study is to investigate the UVC lamps of the UVGI products based on their wavelength. This is to validate whether the product samples are using UVC lamp as stated in the specification which normally has 253.7 to 254 nm wavelength. Besides that, the UVC intensity level has been measured to determine the irradiance emitted by the lamps. The measurement was carried out at various distance to see the effectiveness of the UVGI in disinfection process at different size of spaces. UVC is categorised as non-ionising radiation and has potential health effects due to excessive exposure specially to eye and skin. The equipment used for the irradiance measurement is IL 1700 radiometer with UVC sensor and the measurement data recorded in microwatt/cm^2. The irradiance level of the samples was compared with the permissible exposure limit to the members of public by International Commission on Non-Ionizing Radiation Protection (ICNIRP).

1. Introduction

Ultra Violet Radiation (UVR) referred as optical radiation which contain three type of UVR; UVA, UVB and UVC. UVC emits by germicidal lamps for disinfection purpose which is highly in demand during this pandemic of Covid-19 affecting globally. Non-Ionizing Radiation Group (NIR) had conducted test to the various type of UVGI products to determine the effectiveness of the UVGI and study on the related safety aspects by analyzing the irradiance and compared to the ICNIRP recommendation permissible exposure limit for UVR. This study involved of 6 products, with 3 types of UVGI with different lamp power. The relation between the irradiance with distance and the time for a complete disinfection with distance were studied.

The objective of this research was to validate the product samples are using UVC lamp as stated in the specification by investigating the products based on their wavelength and to analyse the intensity of UVC from the sample products used as Ultra-Violet Germicidal Irradiance (UVGI) in irradiance.

2. Methodology

Irradiance of UVC from UVGI products was measured by using Research Radiometer IL1700 attached with UVC Probe SED240. The measurement was conducted by adjusting the distance of UVC probe with the UVGI products by 10cm. Started with 10cm until 180cm and continue at 350cm until 400cm. The measurement process was repeated twice. Relation between the reading of irradiance in $\mu W/cm^2$ with distance in cm was recorded in the results. Time to reach 1 J/cm^2 was calculated by using the formula below:

$$t = \frac{1}{E} \tag{1}$$

where t is time in second, E is the measured irradiance in Wcm^{-2}. The calculated time converted to minutes and the relation between the time in minutes and distance in cm was plotted.

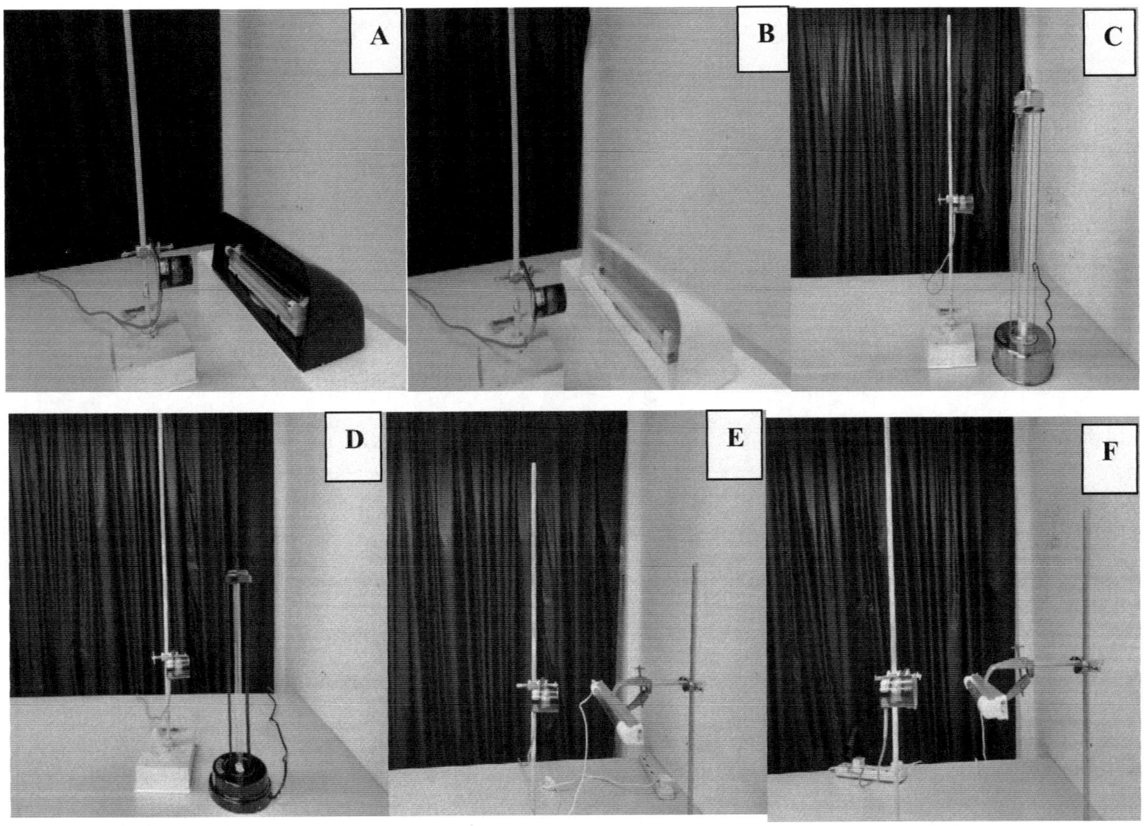

Figure 1. Set-up of equipment for the measurement of Irradiance from the UVGI products: (A) Wall Mounted UV Lamp 32W, (B) Wall Mounted UV Lamp 16W, (C) UV Mobile Lamp 75W, (D) UV Mobile Lamp 40W, (E) UVGI Lamp 16W, (F) UVGI Lamp 8W.

3. Results and discussion

The plot in Figure 2 shows the relation between the reading of irradiance ($\mu W/cm^2$) with distance (cm) for 6 products of UVGI.

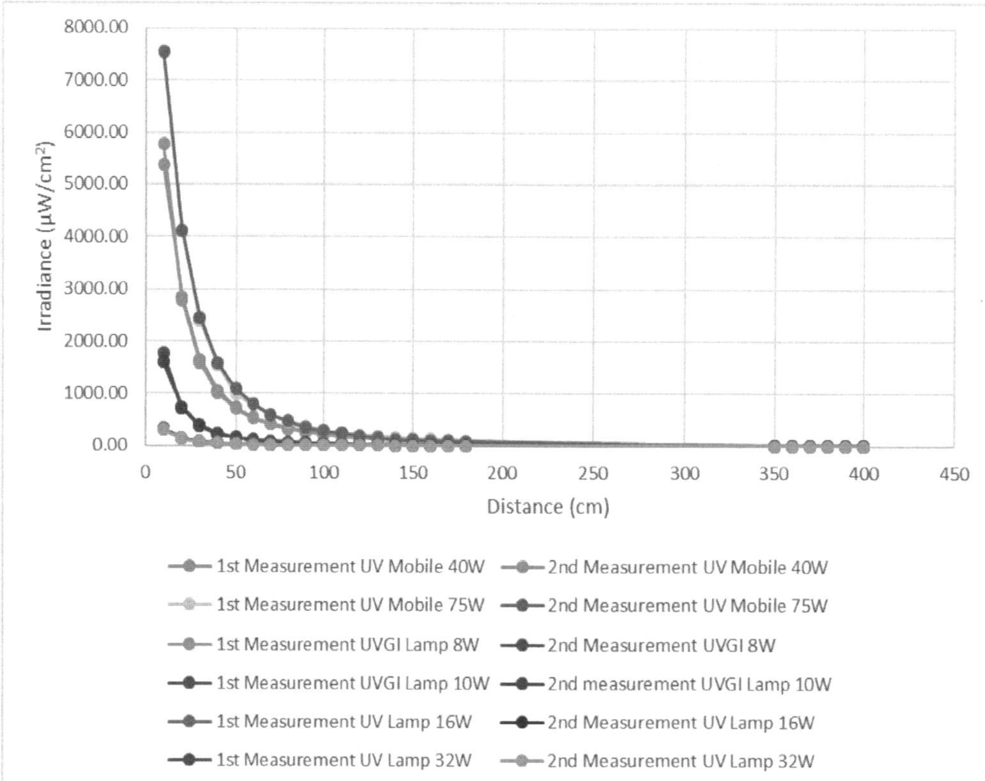

Figure 2. A plot of irradiance ($\mu W/cm^2$) against distance (cm) for 6 products of UVGI.

According to recent studies, 1 J/cm^2 is the sufficient dose for disinfection process of Covid-19. The relation between the time (minutes) to reach 1 J/cm^2 with the distance (cm) was plotted in Figure 3.

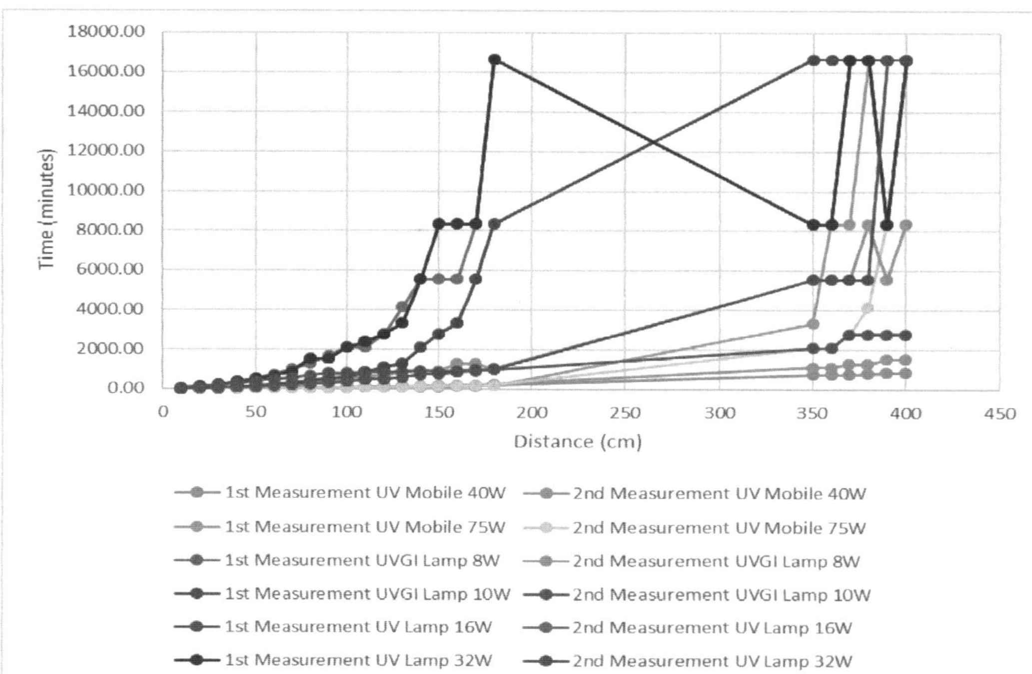

Figure 3. A plot of time to reach 1 J/cm^2 (minutes) against distance (cm) for 6 products of UVGI.

The measured irradiance was compare to the exposure limit for workers and public provided by International Commision on Non-Ionizing Radiation Protection (ICNIRP) for duration of 8 hours. The comparison plot is as indicate in FIgure 4.

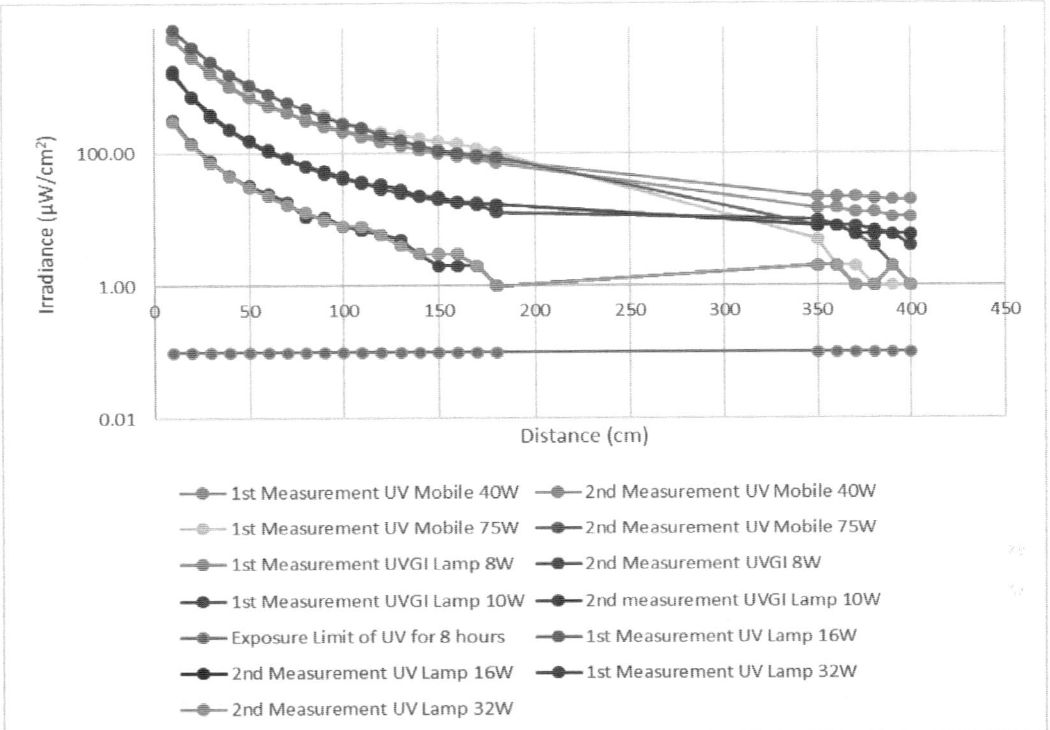

Figure 4. A plot of irradiance (µW/cm²) against distance (cm) for 6 products of UVGI and their comparison with ICNIRP exposure limit.

The maximum irradiance for all products was measured at the distance of 10 cm and the minimum irradiance was measured at the distance of 400 cm. The maximum and minimum of irradiance with the time to reach 1 J/cm² and the comparison with exposure limit for UVC was listed in Table 1.

Table 1. Maximum and minimum of irradiance, time to reach 1 J/cm² and the comparison with exposure limit for each product.

Product	Maximum Irradiance (μWcm^{-2})	Time to reach 1 J/cm² (minutes)	Minimum Irradiance (μWcm^{-2})	Time to reach 1 J/cm² (minutes)	Comparison of Maximum Irradiance with exposure limit for UVC (exposed for 8 hours)
UV Mobile Lamp 40W	5780.00	2.88	11.00	1515.15	1.7301×10^{-5} times higher
UV Mobile Lamp 75W	7550.00	2.21	1.00	16666.67	1.3245×10^{-5} times higher
UVGI Lamp 8W	1753.00	9.51	1.00	16666.67	5.7045×10^{-4} times higher
UVGI Lamp 16W	1758.00	9.48	4.00	4166.67	5.6883×10^{-4} times higher
UV Lamp 16W	324.00		1.00	16666.67	3.0864×10^{-4} times higher
UV Lamp 32W	343.00		1.00	16666.67	2.9155×10^{-4} times higher

4. Conclusion

Six sample UVGI products are validated for using UVC lamp as stated in the specification which has 253.7 to 254 nm wavelength. Based on the measurement and analysed data, the irradiance of UVC light decrease inversely proportional to the distance from the source. This means that the further the distance from the source, the lower the intensity of the UVC. The time to reach the dose of 1 J/cm^2 which indicate the time for a complete disinfection process is increases with the distance from the source, which implies that the further the source, the longer it takes to reach the dose of 1 J/cm^2. The reading of UVC from the products are exceed the permissible exposure limits by ICNIRP if it is exposed for duration of 8 hours at all distances. The hazard warning signs for UVC exposure, restriction of access and the need for personal protection while operating this device must be applied to avoid unnecessary exposure of UVC to personnel.

5. References

[1] International Commision on Non-Ionizing Radiatin Protection (ICNIRP) 2004 Guidelines on limits of exposure to ultraviolet radiation of wavelentghs between 180 nm and 400 nm (incoherent optical radiation) *Health Phys.* **87**(2):171-186.

[2] Protecting Workers from Ultraviolet Radiation, International Commission on Non-Ionizing Radiation Protection in Collaboration with International Labour Organization and World Health Organization, ICNIRP 14/2007, SBN 978-3-934994-07-2 (DCM, Meckenheim)

[3] Memarzadeh F, Olmsted R N, Bartley J M, Arbor B A and Hills B 2010 Applications of ultraviolet germicidal irradiation disinfection in health care facilities: Effective adjunct, but not stand-alone technology *Am. .J Infect Control.* **38**:S13-24.

[4] Kanaan M 2019 CFD optimization of return air ratio and use of upper room UVGI in combined HVAC and heat recovery *Case Studies in Thermal Engineering*.

[5] Yang J H, Wu U I, Tai H M, Sheng W-H 2019 Effectiveness of an Ultraviolet-C disinfection system for reduction of healthcare associated pathogens *J. of Microbiol. Immunol. and Infec.* **52** 487-493.

[6] Lee B, Bahnfleth W P 2013 Effects of installation location on performance and economics of in-duct ultraviolet germicidal irradiation systems for air disinfection *Building and Environ.* **67** 193-201.

[7] Xu P, Pecciaa J, Fabiana P, Martynyb J W, Fennellyc K P, Hernandeza M, Millerd S L 2003 Efficacy of ultraviolet germicidal irradia tion of upper-room air in inactivating airborne bacterial spores and mycobacteria in full-scale studies *Atmosph. Environ.* **37** 405–419

[8] Sung M, Kato S 2011 Estimating the germicidal effect of upper-room UVGI system on exhaled air of patients based on ventilation efficiency *Building and Environ.* **46** 2326-2332.

[9] Wang C, Lu S, Zhang Z 2019 Inactivation of airborne bacteria using different UV sources Performance modeling, energy utilization, and endotoxin degradation *Sci. of the Total Environ.* **655** 787–795.

[10] Scarpino P V, Jensen N J, Jensen P A and Ward R 1998 The use of ultraviolet germicidal irradiation (UVGI) in disinfection of airborne bacteria and rhinoviruses. *J. Aerosol Sci.* **29** (1) 778-799.

[11] Devin M B S, Harnish M S D A, Lawrence B S C, Sandoval-Powers B S M, Heimbuch M S B K 2018 Ultraviolet germicidal irradiation of influenza-contaminated N95 filtering facepiece respirators *Am. J. of Infect. Contr.* **46** e49-e55.

[12] Hamzavi I H, Lyons A B, Kohli I, Narla S, Parks-Miller A, Gelfand J M, Lim H W, Ozog D 2020 Ultraviolet germicidal irradiation possible method for respirator disinfection to facilitate reuse during COVID-19 pandemic *J. of the Am. Acade. of Dermatol.*

[13] Shu M D, Jun H, Sheng L, Li R 2013 Stability of SARS Coronavirus in Human Specimens and Environment and Its Sensitivity to Heating and UV Irradiation *Biomed. and Environ.Sci.* **16** 246-255.

[14] Kowalski W J and Bahnfleth W P 2000 UVGI Design basics for air and surface disinfection *HPAC Eng.*

[15] Wang B, Mortazavi R 2009 Evaluation of modelling and measurement techniques of ultraviolet germicidal irradiation effectiveness- towards the design of immune buildings; *Indoor and Built Environ.* **18** 101.

[16] Bahnfleth W, Lee B, Au J and Freihaut J 2009 Annual simulation of in-duct ultraviolet germicidal irradiation system performance *11th Int. Conf. Glasgow, Scotland July 27-30, 2009.*
[17] https://www.icnirp.org/en/activities/news/news-article/sars-cov-2-and-uvc-lamps.html.

Acknowledgments
The author would like to acknowledge the Non-Ionizing Group (NIR), Radiation Health & Safety Division (BKS), Malaysian Nuclear Agency for supporting this research. The author also thankful to all person involve and for the necessary help during the measurement activity conducted.

IOP Conf. Series: Materials Science and Engineering 1106 (2021) 012009 doi:10.1088/1757-899X/1106/1/012009

Study on mechanical and shielding properties of barite colemanite concrete

Safwan Shalbi[1], N Sazali[1,a)], Muhammad Aliff Ashraff Rosdi[1], Faridah Mohamad Idris[2]

[1]Faculty of Mechanical Engineering, Universiti Malaysia Pahang, 26600 Pekan, Pahang, Malaysia.
[2]Malaysian Nuclear Agency, 43000 Kajang, Selangor, Malaysia.

[a]azlianie@ump.edu.my

Abstract. Natural and artificial mineral additives have been frequently used for improving the physical, mechanical and radiation shielding properties of concrete in recent years. However, it was found that the mechanical properties, including hardness, tensile strength and elongation at break, of these radiation shielding blocks were reduced with increasing of the amount of radiation shielding substance mixed in the radiation shielding In this study, the mechanical and shielding properties of concrete mix with barite and colemanite was use in order to produce optimize shielding for Boron Neutron Capture Therapy at the thermal column of Malaysia research reactor. The barite colemanite concrete formulations are based on ASTM standard and each sample was tested with both mechanical and radiation testing. In mechanical testing, the slump test, ultrasonic pulse velocity test and compressive strength test. For instance, the shielding properties of barite colemanite concrete was tested using neutron test using small angle neutron microscopy (SANS) and gamma test using ^{60}Co. The significance of mechanical and shielding test for barite colemanite concrete beneficial for the production of optimize use of barite and colemanite in concrete for high mechanical and high shielding properties for BNCT shielding purpose.

1. Introduction

In the application of nuclear technology, the protection of biological organisms from the adverse effects of radiation exposure is a basic requirement [1]. Many areas are now using nuclear technology in their application, such as nuclear research reactors, nuclear power plants, industrial inspections, and research facilities, which have put people and the atmosphere around the facilities at significant exposure from ionizing radiation. To ensure that people and the atmosphere surrounding nuclear facilities are protected from the adverse effects of ionizing radiation, sufficient safety is therefore necessary. Methods involving time, distance, and shielding [2] will prevent radiation exposure. Compared to others, shielding is the most effective form [3]. Commonly, concrete is widely used and known as radiation shielding for and nuclear facility [4]. Concrete is made up of cement, water, and aggregate (sand and rocks). Concrete is good for shielding gamma and thermalizing fast neutrons because of its density and content of hydrogenous material, which is water [5]. The concrete use in nuclear facility must be from Grade 40 and above and, their limitation is the concrete that is required has higher thickness to ensure that the nuclear facility is safe to use, as the cost of concrete is much less than lead. Lead proved to have greater photon absorption but less effective in term of weight, cost and mobility. In this research, the BNCT research facility is developed by having minor modification at thermal column and with portable shielding. Therefore, the shielding must be designed with effective absorption cross section, cost and weight. In conjunction, the idea of this research to improve the mix concrete capability by using the barite (good for gamma shielding) and colemanite (good for neutron shielding) with high mechanical strength. Previous research shows that the barite colemanite concrete mix can be used for X-ray facility bunker but not nuclear facility due to mechanical strength problem [6]. Thus, this research will improve and fulfilled this gap to ensure the barite colemanite concrete can be used as a shielding in nuclear facility.

2. Materials and methods

In this research, various design mixtures will be prepared based on the Department of Environmental (DOE) method of barite colemanite concrete. The mixture will be designed to achieve at least grade 40 concrete and has a density of more than 3.5 g/cm^3. The procedure for conducting the test cube samples is the same as other tests of concrete mixing. All of the materials, which are mixed into the concrete, namely Portland cement, water, sand, plasticizer, barite, and colemanite, will firstly be weighed. Next, all of the materials are mixed inside the rotary cement concrete mixer. To determine whether the target slump is reached, a slump test will be performed as per [7]. Then, the concrete mixture will be poured into several cube molds of 15 cm in dimensions. The concrete will be allowed to rest for 24 hours. After 24 hours, the concrete will be taken out of the mold and cured inside a water tank for 28 days. During the preparation of the samples, the concrete mixture will be going through the slump test to assess the consistency of the fresh concrete. Next, after 28 days of curing, the barite colemanite concrete samples will be tested in terms of its mechanical and radiation shielding capability. The concrete samples will be going through various tests, namely for mechanical properties testing and shielding properties testing.

2.1 Mechanical properties testing

2.1.1 Slump test. The test will be carried out following the ASTM C143. The steel slump cone will be placed on a solid, impermeable, and level base surface. Next, fresh concrete will be filled into the slump cone in three equal layers. Each layer is rodded twenty-five (25) times to ensure compaction. The third layer will be finished off level with the top of the cone. Then, the cone will be carefully lifted up, leaving a heap of concrete that slump slightly. The slump cone will be placed on the base to act as a reference, and the difference in the level between the top of the cone and the top of the concrete is measured and recorded to the nearest 10 mm to analyze the slump of the concrete. Illustration of the concrete slump test is shown in Figure1[7].

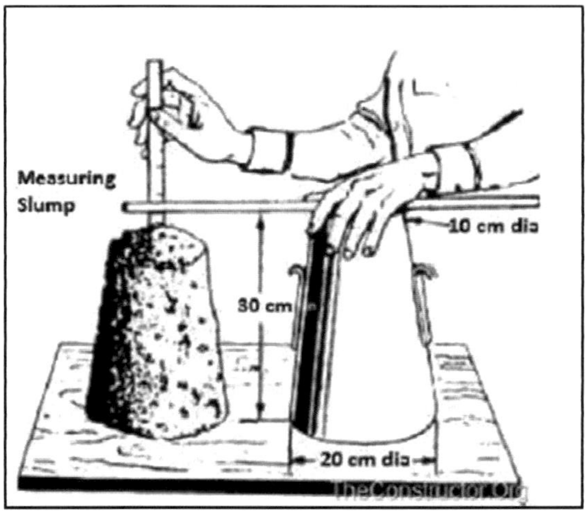

Figure 1. Slump test illustration[7].

2.1.2 Ultrasonic pulse velocity test. Ultrasonic pulse velocity test is an in-situ NDT method to evaluate the durability of the concrete. The evaluation will be carried out based on the correlation between time of propagation or velocity longitudinal ultrasonic waves in the concrete. Measurements are performed by the surface scanning with a fixed base. Ultrasonic pulse velocity testing equipment included a pulse generation circuit, consisting of an electronic circuit for generating pulses and a transducer for transforming electronic pulse into a mechanical pulse having an oscillation frequency in a range of 40 kHz to 50 kHz, and a pulse reception circuit that received the signal. The procedure is employed to perform an ultrasonic velocity test on the concrete [8]. The quality of concrete in terms of uniformity, incidence or absence of internal flaws, cracks, and segregation, etc. can be assessed using the guidelines shown in Table 1, which have been evolved for characterizing the quality of concrete in structures in term of the ultrasonic pulse velocity.

Table 1. Guidelines for characterizing the quality of concrete structure in term of ultrasonic pulse velocity [8].

Pulse Velocity (km/s)	Concrete Quality (Grading)
Above 4.5	Excellent
3.5 to 4.5	Good
3.0 to 3.5	Medium
Below 3.0	Doubtful

2.1.3. Compressive strength test. The compressive strength test is important to determine whether the concrete mixture meet the target strength or not. The compressive strength test will be conducted on the hardened concrete samples [9]. This test also provides an idea about all the characteristics of the concrete. By this test, one may judge whether concreting had been done properly or not. Concrete compressive strength for general construction varies from 15 MPa (2200 psi) to 30 MPa (4400 psi) and higher in commercial and industrial structures. The compressive strength of concrete depends on several factors such as water-cement ratio, cement strength, quality of concrete material, and quality control during the production of the concrete, etc. The test of compressive strength can be carried out either on cube or cylinder concrete samples as display on Figure 2. In this research, the cube concrete samples will be used to test its compressive strength. These specimens will be tested by a compression testing machine after 28 days of curing [9]. The load is applied gradually at the rate of 140 kg/cm^2 per minute until the specimens fail while the load at the failure is divided by area of the specimens to get the compressive strength of the concrete.

$$Compressive\ strength\ (grade) = \frac{Compressive\ force(kN)}{Surface\ area(mm^2)} \quad (1)$$

Figure 2. Compressive strength test.

2.2. *Shielding properties*

2.2.1. Neutron scattering test. Neutron scattering test is performed after the sample hardened in 28 days from mixing. The concrete samples are placed in the front of the neutron beam at the small angle neutron scattering (SANS) facility at Reactor RTP and aligned at a 180° with the collimator. A neutron survey meter as shown in Figure 3 below is placed in after the concrete sample to obtain radiation readings transmitted after the sample. Each sample from A to E is shot with the beam and the reading of the survey meter was recorded. Readings of the survey meter without the sample in placed is also recorded to get the bare readings. In this test, the reactor power was 750kW with background of 692.57 µSv/hr.

Figure 3. Neutron scattering test.

2.2.2. Gamma transmission testing. The radioactive source used in this experiment was ^{60}Co with an activity of 111 MBq. Geiger Mueller (GM) detector will be used to measure the intensities of gamma radiation as show on Figure 4. The attenuation coefficient for gamma radiation of ^{60}Co is calculated using Equation 2. The total mass attenuation coefficients, μm is shown in Equation 2 [10]:

$$\frac{\mu m}{\rho} = \frac{\frac{1}{t}\ln\left(\frac{I_0}{I_d}\right)}{\rho} cm^2 g^{-1} \quad (2)$$

where ρ is the density of the specimen. In this test, the counts of I_0 and I_d will be measured on 15 cm before and after cubic barite colemanite concrete samples respectively. All of the samples will be tested with the same experimental set up and environmental conditions to avoid any inconsistency. A digital counter is used to measure the count of the incident gamma-rays, I_0 applied perpendicularly without the barite colemanite concrete sample. Next, the intensities of gamma-rays, I_d passed through the barite colemanite concrete samples will be measured [10].

Figure 4. Gamma emission test.

3. Results and discussion

Table 2 shows the mixture percentage of the barite colemanite concrete based on the calculation from DOE. The calculation shows the five promising mixture formula for barite colemanite concrete based on Grade 40.

Table 2 The mixture percentage of the barite colemanite concrete samples.

	Mixture (%)				
Material	A	B	C	D	E
Cement	12.87	12.39	18.74	18.74	20.18
Water	9.94	8.17	7.5	6.88	4.65
Sand	18.71	20.28	25	25	24.71
Colemanite	11.7	11.27	1.25	1.88	1.85
Barite	46.78	45.07	46.5	46.5	46.95
Plasticizer	0	2.82	1	1	1.65

3.1. *Shielding Properties*

3.1.2. Effects of barite and colemanite in mix concrete on neutron. Table 3 and Figure 5 show the percentage of neutron absorbed and average equivalent neutron radiation on the concrete samples. The results showed that the highest amount of absorption of neutron radiation is from sample that consists of 46.95% barite and 1.85% of colemanite (Sample E), while the sample that absorbed the least amount of neutron is from sample which consists of 46.5% of barite and 1.25% of colemanite (Sample C). Sample E also shows lowest average equivalent neutron radiation detected show a good absorbent for neutron compared to other samples. This explains that the higher the percentage of neutron radiation being absorbed, the lower the average equivalent neutron detected. The results also concluded that the 46.95% barite and 1.85% of colemanite is the optimum amount to be added in the concrete mixture to achieve highest neutron absorption.

Table 3. Average data of neutron scattered percentage by barite colemanite samples.

Sample	Neutron Scattered (%)
A	95.85
B	96.22
C	89.91
D	96.41
E	98.88

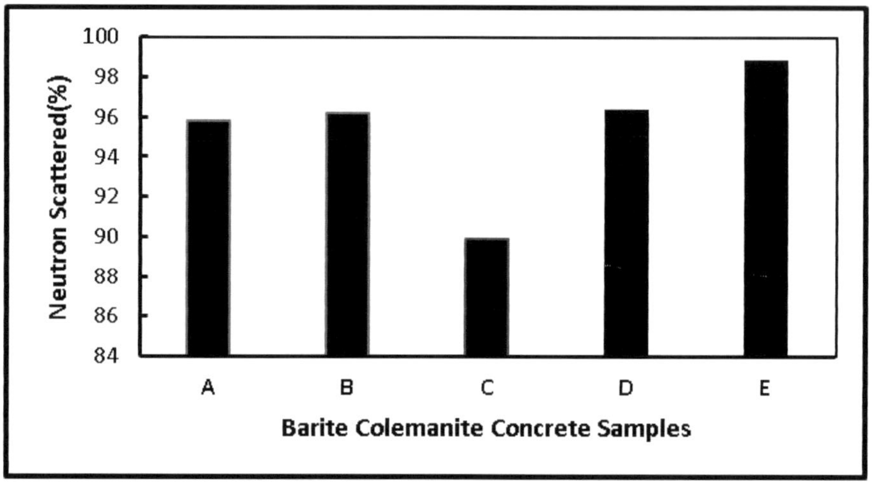

Figure 5. The neutron scattered percentage by barite colemanite samples.

3.1.3. Effects of barite and colemanite in mix concrete on gamma. Table 4 and Figure 6 shows the amount of gamma absorbed and gamma count rate of detector. Concrete sample D which consisted of 46.5% of barite and 1.88% of colemanite addition shows highest amount of gamma adsorbed with 75.18%. This explains that the more the amount of gamma radiation absorbed in the concrete mixture, the lower amount of gamma radiation pass through the concrete. The gamma count rate detected strengthens the finding, in which concrete Sample A shows the minimum amount of count rate detected. The addition of colemanite does not really improve the gamma radiation absorption as the Sample A have more than 10% colemanite on it.

Table 4. Average data of gamma absorption percentage by barite colemanite samples.

Sample	Gamma Absorbed (%)
A	71.8
B	73.31
C	75.03
D	75.18
E	75.15

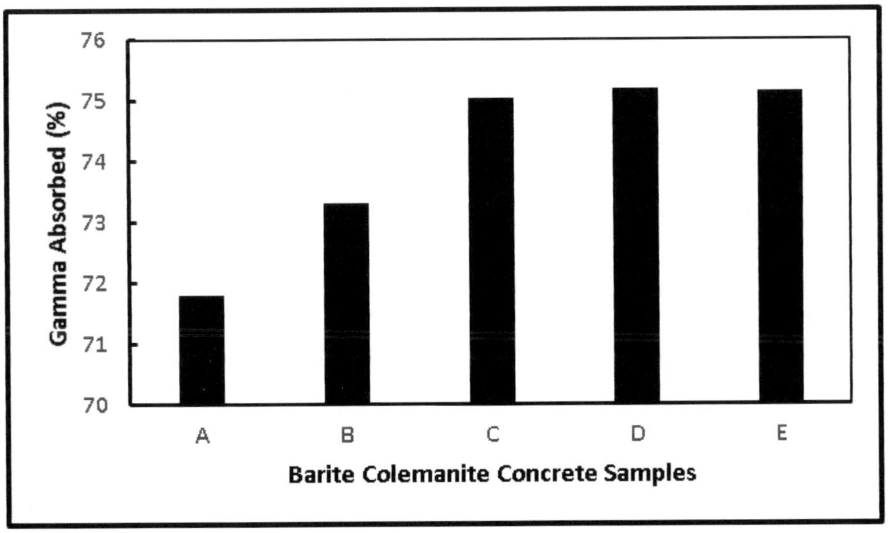

Figure 6. The gamma absorption percentage by barite colemanite samples

3.2. Mechanical properties

The mechanical properties of the barite colemanite concrete samples were tested using the three standard testing methods, namely the ultrasonic pulse velocity test, slump test, and compressive strength test. Table 5 shows the mechanical properties of barite colemanite concrete design. Based on the results shown in Table 5, the average pulse velocity for the design mix barite colemanite was 3.77 km/s. It shows that the concrete quality for barite colemanite was excellent based on the standard guidelines. However, the slump test for barite colemanite shows that Sample A has a high degree of workability compared to other samples. In order to achieve Grade 40, the standard slump test of the concrete produce must be between 30mm-60mm. In terms of compressive strength test for concrete samples of barite, colemanite shows that only one sample manage to achieve higher than Grade 40 which is Samples E that consist of 46.95% barite and 1.85% of colemanite. This mechanical property testing proved that to produce a good concrete mix higher than Grade 40, the design must pass all those testing and the standard guideline. Based on this research, the performance of the Sample E barite colemanite concrete for both shielding and mechanical properties has met the minimum requirements such as concrete quality, concrete density, and concrete strength.

Table 5. Mechanical properties testing result for barite colemanite samples.

Sample	Slump Test (mm)	Pulse velocity(km/s)	Compressive Strength Test (N/mm^2)
A	75	3.48	0
B	33	3.62	33
C	63	4..11	32
D	40	3.89	32
E	36	3.76	42

4. Conclusion

This study has demonstrated that varying the ratios of colemanite in the barite concrete mixture has different reaction towards neutron and gamma radiation absorption. In addition, the varying ratio of colemanite shows higher impact on the mechanical properties of the concrete mix design. Based on this research, only one samples succeeded to meet the minimum requirements such as concrete quality, concrete density, and concrete strength and good shielding properties for both neutron and gamma. 46.95% barite and 1.85% of colemanite are recommended to be considered to add on concrete for nuclear facility development.

5. References

[1] Murray R L and Holbert K E 2015 Chapter 11 - Radiation protection in nuclear energy (7th ed.) (Boston: Butterworth-Heinemann) 153-176.
[2] Martin J E 2008 Radiation Shielding in physics for radiation protection (Wiley-VCH Verlag GmbH) 367-423.
[3] Özavcı S and Çetin B 2017 Radiation shielding properties of concretes including quiclime (CaO) *AIP Conf. Proc* **1815**(1) 130009.
[4] El-Khayatt A 2010 Radiation shielding of concretes containing different lime/silica ratios *An. of Nucl. Ener.* **37**(7) 991-995.
[5] Stanković S, *et al.* 2010 Gamma radiation absorption characteristics of concrete with components of different type materials *Acta Physica Polonica A* **117**(5) 812-816.
[6] Subramanian D V *et al.* 2015 Comparison of neutron attenuation properties of ferro boron slabs containing 5% natural boron with other high-density materials *Radiat. Protec. and Environ.* **38**(3) 109.
[7] American Society for Testing and Materials (1999) Standard test method for slump of hydraulic-cement concrete *ASTM Int.*, (West Conshohocken, PA) C143 / C143M – 12.
[8] Komlos K, Popovics S, Nürnbergerová T, Babal B, Popovics J S 1996 Ultrasonic pulse velocity test of concrete properties as specified in various standards *Cement and Concrete Compos.* **18**(5), 357-364.
[9] American Society for Testing and Materials 2016 Standard test method for compressive strength of hydraulic cement mortars (using 2-in. or [50-mm] cube specimens) *ASTM Int.*, (West Conshohocken, PA) C109 / C109M-16a.
[10] John A J B, Lamarsh R, 2001 *Intro. to Nucl. Eng.* 3rd ed. (United States of America: Prentice-Hall).

Acknowledgement

Authors would like to extend their gratitude to Ministry of Higher Education Malaysia and Universiti Malaysia Pahang (UMP) with grant number RDU1902615 and PGRS200304.

IOP Conf. Series: Materials Science and Engineering 1106 (2021) 012010 doi:10.1088/1757-899X/1106/1/012010

Gadolinium-doped polymeric as a shielding material for X-ray

S M Tajudin[1], A H Aminordin Sabri[1], M Z Abdul Aziz[2], F Tabbakh[3]

[1] Faculty of Health Sciences, Universiti Sultan Zainal Abidin (UniSZA), Terengganu, Malaysia.
[2] Oncological and Radiological Science Cluster, Advance Medical and Dental Institute, Universiti Sains Mala[1]ysia, Penang, Malaysia.
[3] Institute of Nuclear Science and Technology, Plasma and Nuclear Fusion Research School, Tehran, Iran

[1]suffian@unisza.edu.my

Abstract. A relatively new polymeric base compounds (C_nH_{2n}) had been proposed as a shielding material for 150 keV photon for an X-ray laboratory. When doped with 8 % gadolinium (Gd), a thickness of 2 cm shielding compound (6 g cm^{-3}) could attenuate more than 90 % of the incident photons with an average dose rate reduction of more than 88 % for 150 keV incident photon. While considering the transmitted photons after the shielding, it is necessary to account for reflected photons and its dose contribution. The reflected photons and ambient dose equivalent were calculated for single-photon energy of 150 keV and an X-ray of 150 kVp. The shielding compound was successfully attenuated most of the incident photon energies particularly below 90 keV. In both cases of photon sources, two significant reflected photon peaks at ~40 and 50 keV were observed as a result of Gd characteristic X-rays. A Compton scattered photon at energy of 95 keV appeared as a result of large scattering angle within 130° -140° for 150 keV incident photons. Thus it is necessary to add a thin inner layer at the source-facing side of the shielding compound to shield the radiation workers and patient inside the X-ray room during the X-ray procedure. An iron layer with thickness of 0.5 cm was adequate to shield almost completely the reflected photons and ambient dose for X-ray 150 kVp source

1. Introduction

Lead (Pb-82) material in the order of millimeters thickness had been widely used since 1970 to attenuate the photons either to shield the emission of photons from radioactive source or used in a large scale to shield the room [1,2], such as at UniSZA X-ray laboratory (150 kVp). While attenuate the photons, the lead material itself is poisonous and and expensive [3][4][5]. Many scientists had proposed other materials to replace the lead element as a gamma or X-ray shielding such as clay [6][7][8][9][10][11][12][13][14]. As an example the authors had successfully demonstrated the polymeric compounds doped with heavy element other than lead as an alternative shielding materials for 150 keV photons with an af[2]fordable thickness [15]. The gadolinium (Gd-64)-doped polymeric compounds had been calculated by the authors to have ~90 % photon attenuation with less than 2 cm thickness for 150 keV incident photon, which will be adequate for shielding an X-ray room.

While Gd-doped the polymeric compounds is successfully attenuated the transmitted photons significantly, the impact of reflected photons due to the polymeric compounds itself need to be evaluated to ensure the radiation safety area either outside or inside of the irradiation facility. The medical radiation workers may need to attend to the patients inside the room during an X-ray procedure such as fluoroscopy, thus the effect of reflected photons and its dose must be evaluated. As an example, the author had concluded the contribution of reflected photons and dose from the clay material is negligible for low energy photon [16].

In this study, the reflected photons and ambient dose equivalent (Sv/photon) were evaluated by Monte Carlo simulation (EGS5 code, [17]) for the previous recommended thickness of Gd (8 %) doped-polymeric compounds (6 g cm^{-3}). The selection photon energy of 150 keV is based on our X-ray machine at UniSZA which have maximum setting of 150 kVp. The energy used is closed to the one who used Co-57 gamma source (0.0144 keV, 0.122 keV and 0.136 keV photons). We also include the calculation for X-ray of 150 kVp. Consequently, an additional layer of light element at the inner side (source-facing side) of the polymeric compounds is important to avoid unnecessary dose due to reflected photons. This outcome enabled us to shield properly the photons that transmitted and reflected by the Gd doped-polymeric compound in our facility.

2. Calculation of transmitted and reflected photon energy spectra of 150 keV photon

It was assumed Gd element might have kind of 'reflective effect' to the photons. As example, Gd had been used as a reflective material coating for X-ray mirror in some applications [18]. Thus we investigated the photons energy and dose that reflected from the shielding compound. As recommended in previous study, for subsequent calculations, a thickness of 2 cm shielding Gd-doped polymer compound (10 cm x 10 cm x 2 cm) was chosen for 150 keV parallel photon beams. Surface crossing was used to calculate the reflected dose and photon spectra on the surface of 2 x 2 cm^2 scoring region.

Figure 1 shows the intensity of transmitted and reflected photon spectra when a parallel beam of 150 keV incident photon pass through a 2 cm thickness of gadolinium (Gd). The detector response was not considered in the calculated photon spectra. The dotted and full lines of the Figure 1 represent transmitted and reflected photons, respectively. The transmitted and reflected average energy is 145.5 keV and 73.4 keV, respectively. A very small reflected peak of 150 keV photons appeared as a result of backscattered photons as represent by the full line.

In both cases, two peaks at 40 and 50 keV photons are evidence due to the emission of K-L X-rays and K-edge X-rays of Gd element doped in the polymer. The tendency of characteristic X-rays to be reflected after undergoing Compton scattering interactions in the polymer is higher than the transmitted photons by a factor of 10. It should be noted that the y- axis of the graph was plotted in log scale. The dominant of reflected photon energies for 40 keV, 50 keV and 95 keV values could be easily absorbed by the human body and contribute to an additional whole body dose particularly for the radiation workers who are working inside the shielding environment. Thus, it is necessary to have an additional layer of low Z element at the inside surface of the polymer (on the face side of photons beam) to attenuated such effect.

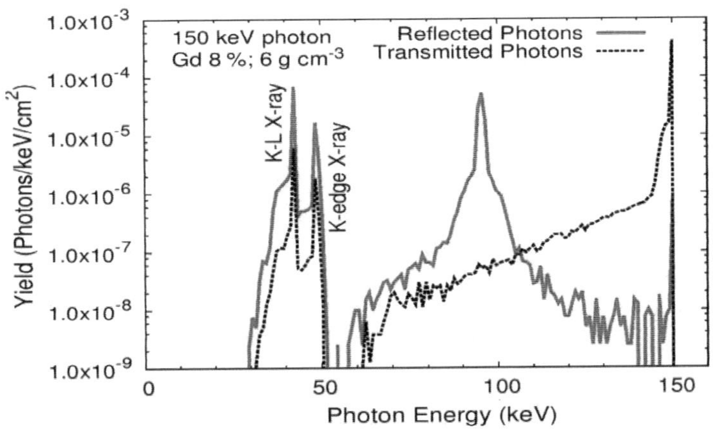

Figure 1. The transmitted and reflected photon spectra for a Gd-doped polymer. A very small peak for backscattered photon of 150 keV photon was observed. A dominant reflected photon peaks were observed at ~95 keV as a result of Compton scattered with large angle (130 – 140°). The peak at ~ 40 and 50 keV due to the K characteristic X-rays of Gd doped in the polymer.

3. The transmitted and reflected photon energy spectra of 150 kVp X-ray

Report 78 Spectrum Processor of IPEM 1997 software [19] was used to obtain the unfiltered X-ray spectra as a source input for EGS5 code as shown in Figure 2. Figure 3 (a) and (b) shows the intensity of transmitted and reflected photon spectra of X-ray 150 kVp. In Figure (a), 2 cm of Gd-doped polymer significantly attenuated the photons below 90 keV of an X-ray spectra. The reflected photon spectra show a significant contribution of the K characteristic X-rays in comparison to the transmitted one as previous spectra. The second peak of Compton scattered photon at ~95 keV not evidence as the average energy X-ray of 150 kVp is ~ 50 keV. The reflected peak at ~95 keV (as a result of 120-130° scattering angle) from the Compton scattered photons will be appear, as an example when utilized a monoenergetic source such as Co-57. The energy of Compton scattered photon at the similar scattering angle will be higher up to ~ 200 keV when a higher energy gamma sources used such as Cs-137 and Co-60 as per Compton scattering equation.

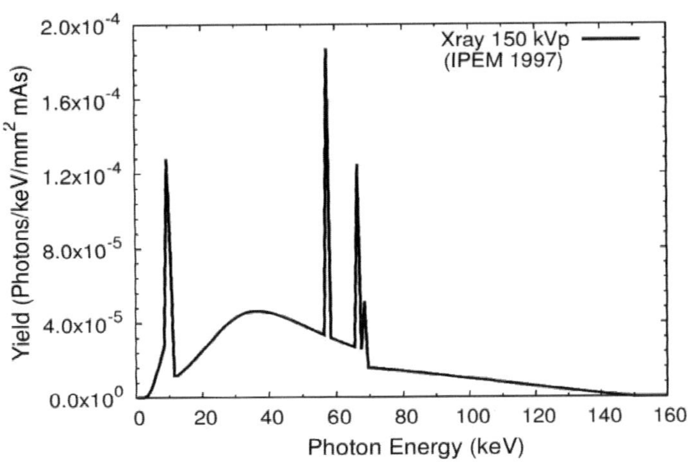

Figure 2. X-ray photon spectra of 150 kVp as a source input in the calculation. The average energy is 49.8 keV. The X-ray spectra was adopted from Report No. 78 [19].

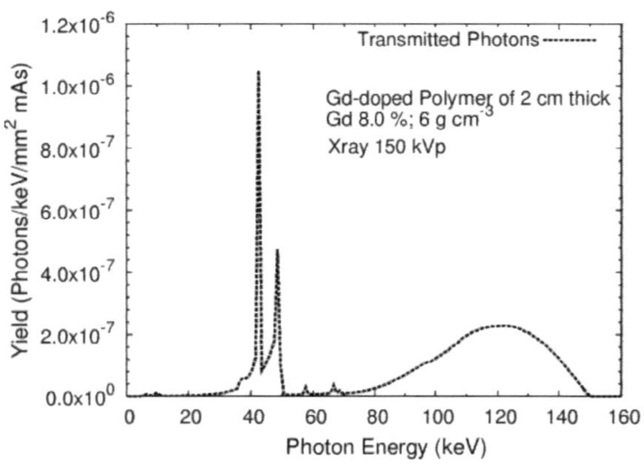

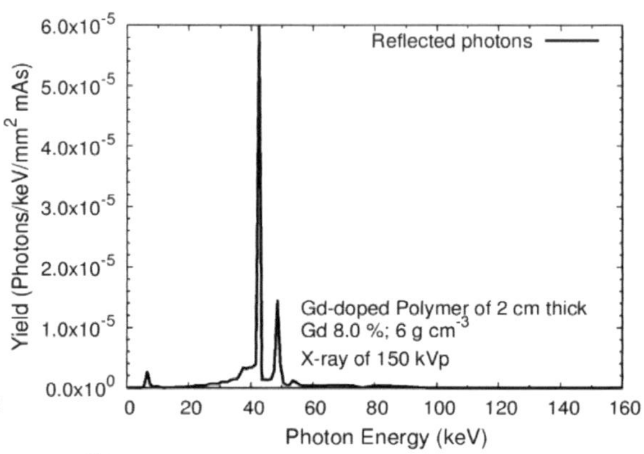

Figure 3. Figure 3 (a) show the transmitted photon spectra while (b) is the reflected photon spectra for the X-ray source of 150 kVp, respectively. The amount of K characteristic X-rays were significant in the reflected photon spectra (b), thus an additional layer is necessary at the inner side to reduce the effect.

4. Designing an inner layer of the Gd-doped shielding compound

Having confirmed through simulation that the Gd-doped polymer compound has significant reflected photons at lower energies photons due to K characteristic X-rays and Compton scattered photons at large scattering angle, the next step is to configure the material as inner layer and apply it to the current shielding problem. The layer should be applied at the inner side of the wall shielding (Gd-doped polymer compound) of an X-ray facility. As example, a thin layer of Fe-26 with thickness of 0.5 cm was sufficient to attenuate the reflected photons at 40 and 50 keV as shown in Figure 4. The reflected peak of K-L X-rays at 40 keV photon almost completely disappeared while the reflected K-edge peak of 50 keV photon was reduced almost by a factor of 10. The reflected peak at 7 keV is due to the emission of Fe-26 K-edge and will be attenuated easily as a function of distance in air.

Our finding supported McCaffrey's study [20] when the author had suggested a bilayer non-Pb shielding in the order of low Z and followed by high Z element to have a good gamma attenuation while not mention the reflected one.

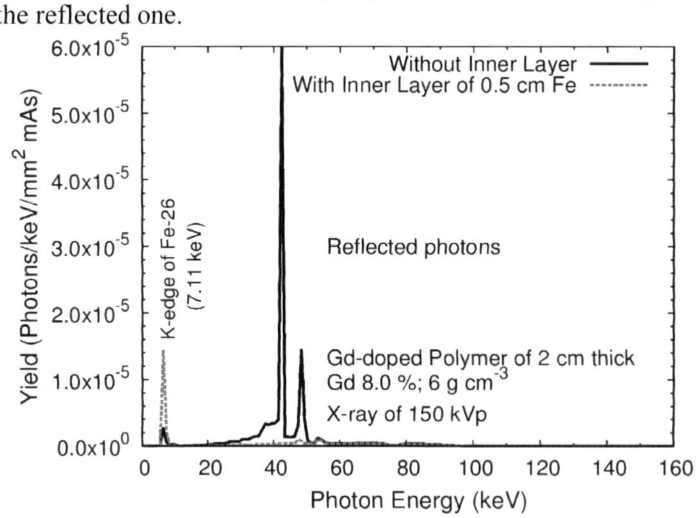

Figure 4. Figure show the reflected photon spectra with (dotted line) and without (full line) the inner layer added to the Gd-doped shielding compound. Both reflected photons at ~40 and 50 keV photons were successfully attenuated with 0.5 cm iron (Fe-26).

5. Conclusion

An approach of photon calculations either outside or inside of an X-ray room must be considered when designing a radiation shielding to ensure radiation safety area. The shielding materials must be studied both for its transmitted and reflected photons particularly if the shielding compound has a higher Z element or the source is higher energy photons. The Gd-doped polymeric compound has a significant reflected K characteristic X-rays at ~40 and 50 keV photons. If the source is a monoenergetic source such as Co-57 with the energy more than 120 keV, another significant Compton scattered peak will be reflected to have an energy ~ 95 keV due to large scattering angle within 130 – 140°. Thus an additional layer is necessary for the inner side of the attenuating materials. The shielding compound density and thickness have no effect on the intensity and energy of reflected photons. A thin layer of iron with 0.5 cm thickness is adequate to attenuate such effect significantly.

References

[1] Stukenbroeker G L; Bonilla Charles F; Peterson Ruben W 1970 The use of lead as a shielding material *Nucl. Eng. and Design* **13** 3-145

[2] Hyun S J, Kim K J, Jahng T A, Kim H Jib 2016 Efficiency of lead aprons in blocking radiation – how protective are they? *Heliyon* **2**(5).

[3] Wani A, Ara A, Usmani J 2015 Lead toxicity: A review *Interdiscipl. Toxicol.* **8**(2) 55-64.

[4] Billen P, Leccisi E, Dastidar S, Li S, Lobaton L, Spatari, S, Baxter J B 2019 Comparative evaluation of lead emissions and toxicity potential in the life cycle of lead halide perovskite photovoltaics. *Energy* **166** 1089–1096 (http://doi.org/10.1016/j.energy.2018.10.141).

[5] Flora G, Gupta D, Tiwari A 2012 Toxicity of lead: A review with recent updates *Interdiscipl. Toxicol.* **5**(2) 47-58.

[6] Olukotun S F, Gbenu S T, Oladejo O F, Sayyed M I, Tajudin, S. M., Amosun A A, Fasasi M K 2020 Investigation of gamma ray shielding capability of fabricated clay-polyethylene composites using EGS5, XCOM and Phy-X/PSD. Radiation Physics and Chemistry **7** 109079.

[7] Akkurt H Canakci 2011 Radiation attenuation of boron doped clay for 662, 1173 and 1332 keV gamma rays *Iran. J. Radiat.* Res. **9** 37-40.

[8] Yue K, Luo W, Dong X, Wang C, Wu G, Jiang M, Zha Y 2009 A new lead-free radiation shielding material for radiotherapy, Radiat. Prot. Dosim., **133** 256–260.

[9] Sabri A, Aziz A H A M Z, Olukotun S F, Tabbakh F, Tajudin S M 2020 Study on the shielding materials for low-energy gamma sources. *IOP Conf. Ser.: Mater. Sci. and Engineering* **785** 012007.

[10] Agar O, Sayyed M I, Akman F, Tekin, H O, Kaçal M R 2018 An extensive investigation on gamma ray shielding features of Pd/Ag-based alloys *Nucl. Eng. and Technol.* **51**(3) 853–859. (http://doi.org/10.1016/j.net.2018.12.014).

[11] Sabri A H A, Tajudin S M 2019 Simulation of Photon Attenuation through Aluminium Using E-Radiation Simulator (EGS5 Code) *J Fundam. Appl. Sci.* **1** 173–183.

[12] Tajudin S M, Hanim A, Sabri A 2019 SIMU-RAD programme: A learning tool for radiation (photons and charged particles) interaction. *The J. of Polish Soc. of Med. Phys.* **25**(3) 189–192.

[13] Bashter I I 1997 Calculation of radiation attenuation coefficients for shielding concretes. Annals of Nuclear Energy **24**(17) 1389–1401 (http://doi.org/10.1016/S0306-4549 (97)00003-0).

[14] Aziz A, Yani M Z, Haryanto S, Ali F Y, Tajudin N K, Iwase S M, Musarudin 2020 Monte Carlo simulation of X-ray room shielding in diagnostic radiology using PHITS code. *J. of Radiat. Res. and Appl. Sci.* (https://doi.org/10.1080/16878507.2020.1828020).

[15] Tajudin S M, Tabbakh F 2019 Biological polymeric shielding design for an X-ray laboratory using Monte Carlo codes. *Radiol. Phys. and Technol.* **12**(3) 299–304. (https://doi.org/10.1007/s12194-019-00522-w).

[16] Tajudin S M, Sabri A H A, Aziz M Z Abdul, Olukotun S F, B.M. Ojo, Fasasi M K 2019 Feasibility of clay-shielding material for low-energy photons (Gamma/X), *Nucl. Eng. and Technol.* ISSN 1738-5733.
[17] Hirayama H, Namito Y, Bielajew A F, Wilderman S J, Nelson W R 2005 EGS5 Code System. SLAC Report SLAC-R-730 and KEK Rep. 8(**2010**).
[18] https://universe.gsfc.nasa.gov/xrays/MirrorLab/xoptics.html 24-Jul-2012
[19] Cranley K, Gilmore B J, Fogarty G W A, Desponds L 1997 Catalogue of Spectral Data for Diagnostic X-rays *IPEM report 78*.
[20] McCaffrey J P, Mainegra-Hing E, Shen H 2009 Optimizing non-Pb radiation shielding materials using bilayers. *Med. Phys.* 36.

IOP Conf. Series: Materials Science and Engineering 1106 (2021) 012011 doi:10.1088/1757-899X/1106/1/012011

Effect of lead oxide on concrete density for radiation shielding purposes

Siti Amira Othman*

Department of Physics and Chemistry, Faculty of Applied Sciences and Technology, Universiti Tun Hussein Onn Malaysia

sitiamira@uthm.edu.my*

Abstract. Selection of suitable shielding materials depends on the type, intensity and energy of radiation source, as well as the shielding properties such as mechanical strength, economic preparation and resistance to radiation damage. When compared to ordinary concrete, the addition of some materials with various fractions in the composition of concrete has provided concrete forms more efficient for gamma ray shielding. Concrete has proven to be an tremendous and adaptable shielding material with a well-established linear attenuation coefficient for gamma rays and neutrons. The objectives of this study is to determine the exact composition of concrete constituents and its density. Concrete samples were prepared by mixing cement, water, aggregate and additive with different composition and densities. Density was increases by decreasing the aggregates percent and increasing the additives such as lead oxide. Sample will be compacting to forming a solid material through pelleting and sintering process. Finally, it will undergo for radiation attenuation test. It was found that, substitution of lead oxide by aggregate causes increasing in the specific density and compressive strength of the concrete.

1. Introduction

Concrete is a composite material composed of coarse aggregate bonded together with a fluid cement which hardens over time. Most concretes used are lime-based concretes such as Portland cement concrete or concretes made with other hydraulic cements, such as cement fondu. However, road surfaces are also a type of concrete, asphalt concrete, where the cement material is bitumen, and polymer concretes are sometimes used where the cementing material is a polymer. Increasing the amount of cement in the concrete mixture increases of the yield stress[1]. Good quality aggregate should consist of particles of adequate strength and desirable engineering properties as well as resistance to exposure conditions. The aggregate properties affect the strength, stiffness, and long-term deformation of hardened concrete [2]. For example, some aggregates may react negatively with cement or, in contrast, they may interact beneficially with the cement paste, enhancing the concrete's strength or stiffness. As a result, concrete properties such as elastic modulus, creep, or shrinkage can vary as much as 100% depending on the aggregate type.

2. Methodology

The materials used for construction of concrete are Portland cement (type I) with a density of 400kg/ m^3. To increase work ability of concrete, about 2.5wt% of cement was used. Distilled water for dilution process. The water to cement ratio is 0.45 by weight. An aggregate with a density 2400kg/cm^3 was used in concrete production. Aggregates were graded in accordance to grain size classification ASTM C150, Type I. Lead oxide was used as an additive in forms of powder.

2.1. Sample preparation

Cement, water, aggregate (sand and gravel) and additive will be mix with percentage value of aggregate is vary. Ordinary concrete (without additive) is prepared as a sample reference. By using Ball Mill machine, samples will be grind and blends with speed 250 rpm for 2 hours. Samples will be add with water. By using Hydraulic Press machine, the samples will be press 2 ton about 1 minute. Samples were undergo water curing process for about 7 days by immersed in water at about 25°C. After 7 days of water curing, the samples were exposed to heat at temperature of 180°C for about 2 hours. Sample will measure including dose of radiation, intensity of radiation, thickness of concrete and concrete strength.

3. Results and Discussion

3.1DensityTest

In order to produce a high quality concrete, various trial of water to cement ratio of 0.4, 0.8, 1.2, 1.4, 1.6 and 2.0 were prepared. Fig. 1(a) shows the graph of density of concrete and water to cement ratio. The graph shows a decreasing density from 0.4 to 2.0 water to cement ratio. The results obtained were related to the complete hydration in concrete. Each of the compounds in Portland cement will undergoes hydration and contributes to the final concrete product when the water is added. Complete hydration of Portland cement will occur at approximately w/c = 0.4. For the water to cement ratio below than 0.4, the hydration process might be incomplete as the water should fill all the gel pores in order to accomplish the hydration process.

As for water to cement ratio equal to 0.46, Abram's Law stated that higher water to cement ratio will lower the strength of concrete [3]. This is because as water was added over and over, it will be evaporated which will leave cavities in concrete and indirectly reduce the density of concrete. In other words, the more water was use to mix the concrete, the weaker the concrete mix. The less water use to mix the concrete, the stronger the concrete mix [4]. A mix with little water and more concrete mix will be dryer and less workable but stronger. Table 1 shows the density results. The samples were polished before conducting the test to remove all entire unwanted particles from the surface of concrete. The next phase in producing concrete was to find the best ratio between fine aggregates and cement. Table 1 shows the results obtained from the mix design of concrete using cement, water and sand.

Besides that, table 1 shows the results obtained from the density test of plain concrete samples. From figure 1(b) and 1(c) shows density of concrete is decrease when weight of fine and coarse aggregate is increase due to bond characteristics [5]. The use of larger maximum size of aggregate affects the concrete in several ways. First, since larger aggregates have less specific surface area and the aggregate–paste bond strength is less, aggregate fails along surfaces of aggregates resulting in reduced compressive strength of concrete. Secondly, for a given volume of concrete, using larger aggregate results in a smaller volume of paste, thereby providing more restraint to volume changes of the paste [6].

IOP Conf. Series: Materials Science and Engineering 1106 (2021) 012011 doi:10.1088/1757-899X/1106/1/012011

Table 1. Ratio of concrete sample

Density of concrete to determine water to cement ratio						
Samples	Water (g)	Cement (g)	Water : Cement	Density (g/cm^3)		
A	2.0	5.0	0.40	2.5117		
B	4.0	5.0	0.80	2.3605		
C	6.0	5.0	1.20	2.1561		
D	8.0	5.0	1.60	1.7218		
E	10.0	5.0	2.00	1.5086		
Density of concrete to determine fine aggregate to cement ratio						
Samples	Water (g)	Cement (g)	Fine aggregate (g)	Fine aggregate : Cement	Density (g/cm^3)	
A	2.0	5.0	2.0	0.4	2.610	
B	2.0	5.0	2.5	0.5	2.551	
C	2.0	5.0	3.0	0.6	2.502	
D	2.0	5.0	3.5	0.7	2.440	
E	2.0	5.0	4.0	0.8	2.411	
Density of concrete to determine coarse aggregate to cement ratio						
Samples	Water (g)	Cement (g)	Fine aggregate (g)	Coarse aggregate (g)	Coarse aggregate : Cement	Density (g/cm^3)
A	2.0	5.0	2.0	2.0	0.4	2.732
B	2.0	5.0	2.0	2.5	0.5	2.705
C	2.0	5.0	2.0	3.0	0.6	2.621
D	2.0	5.0	2.0	3.5	0.7	2.543
E	2.0	5.0	2.0	4.0	0.8	2.487
Density of High Density Concrete						
Samples	Water (g)	Cement (g)	Fine aggregate (g)	Coarse aggregate (g)	Lead (g)	Density (g/cm^3)
A	2.0	5.0	2.0	2.0	2.0	2.811
B	2.0	5.0	2.0	2.0	4.0	2.925
C	2.0	5.0	2.0	2.0	6.0	3.207
D	2.0	5.0	2.0	2.0	8.0	3.296
E	2.0	5.0	2.0	2.0	10.0	3.433

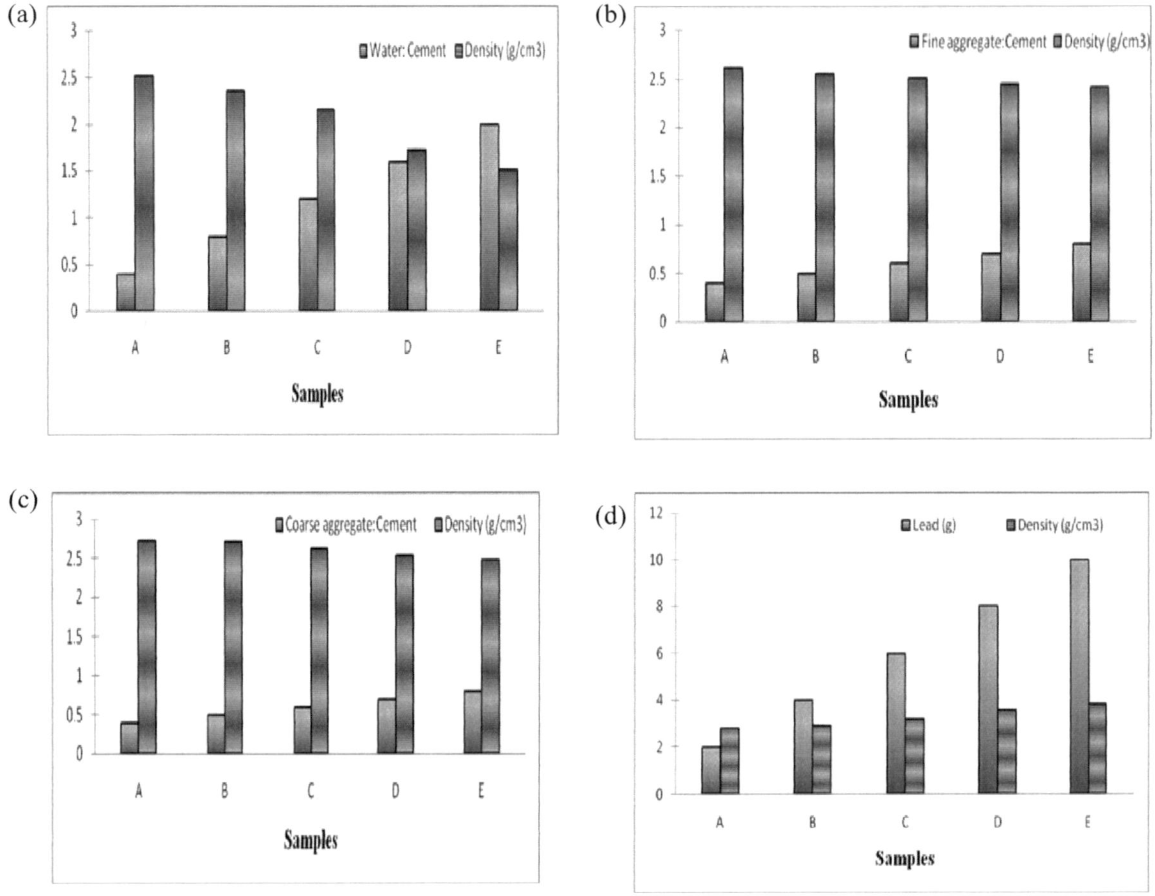

Figure 1. (a) Graph density of concrete and water to cement ratio (b) Graph density of concrete and fine aggregate to cement ratio (c) Graph Density of Concrete and coarse aggregate to cement ratio (d) Graph density of concrete and weight of lead in concrete (all in gram unit).

Density of concrete is determined by the aggregate density as well as soft with porous concrete produce weak concrete with lower wear resistance [7]. Bulk density of aggregates is the mass of aggregates required to fill the container of a unit volume after aggregates are batched based on volume. A new concrete mix design from different weight of lead was done based on the previous density test and the density was determined using the same equipment. The samples were polished before conducting the test as it will remove unwanted particle from the surface of concrete. The result obtained was shown in Table 1. The mix design of concrete actually really important as it was the key to the production of the high quality concrete [8]. In Fig. 1 above, it can be observed that the pattern of density of concrete is increasing with the higher weight of lead. Instead of the suitable mix design, the pattern of increasing in density of concrete was the result from the lead powder which helps a lot in increasing the density. Lead is a high density material. It also displays a marked tendency toward covalent bonding. Because of that, it is kind of suitable to produce a high dense concrete using this admixture. High density concrete is mainly used for the purpose of radiation shielding, for counterweights and other uses where high density is required. The high density concrete has a better shielding property, so that it can protect harmful radiation [9].

According to the recent study made by previous researchers[4][10], the highest density of concrete that they can produce using lead is 3.542 g/cm^{3}which is higher compared to the concrete made in this study which the highest density is 3.433 g/cm^3. This might be due to the difference in the period of water curing which water curing process in this research was made for 7 days. As mentioned earlier, water curing plays an important role in concrete making as the process will improve the workability and the structure thus effect the density of concrete.

3.2 Analysis for radiation attenuation test
Radiation attenuation test was done using NaI Scintillation Detector with Cs-137 as the source of radiation. The count time was set as 150 seconds. Table 2 below shows the results obtained from the radiation test. The background reading was repeated three times to reduce errors and the average counts rate was calculated.

Table 2. Result of radiation count rate

Background reading				
Background reading (Counts/s)	Average Counts rate (Counts/s)	Corrected count rate (Counts/s)		
28476	30020	30020/150 = 200.13		
30118				
31467				
Radiation counts rate for concrete samples				
Sample	Thickness of sample (g/cm^2)	Counts rate (Counts/s)	Average Counts rate (Counts/s)	Corrected Counts rate (Counts/s)
	0.0000	46523 49512 50013	48682	124.42
A	2.4160	41864 42817 43582	42754	84.89
B	2.5891	41929 41511 41766	41735	78.10
C	2.7161	41507 41473 42096	41692	77.81
D	2.8350	41316 41604 41372	41430	76.07
E	2.9726	41379 41198 41589	41389	75.79

From the graph plotted in Fig. 2, it can be observed that the pattern of the counts rate is decreasing as the density of the concrete increase. The decline of the count rate shows that γ-ray had been absorbed by the concrete. Gamma ray will be interact with atomic particles of concrete and make them lose the energy. The particles in the concrete samples were packed as the density is increase which will contribute to the loss of gamma energy as many interactions occurred between concrete particles [11]. In conclusion, gamma-ray absorption will increase as the density of materials increased. In this radiation attenuation test, the count rate represented the intensity that passes through the concrete samples. It can be said that the intensity of the gamma ray decreased as the density of samples increases. The thickness relate to the density proportionally [12].

(a)

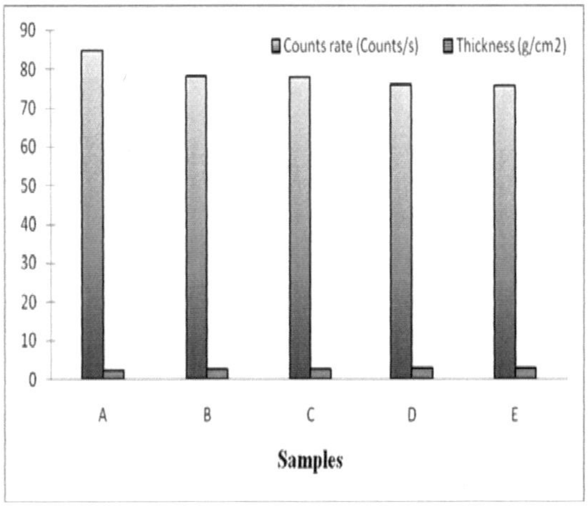

(b)

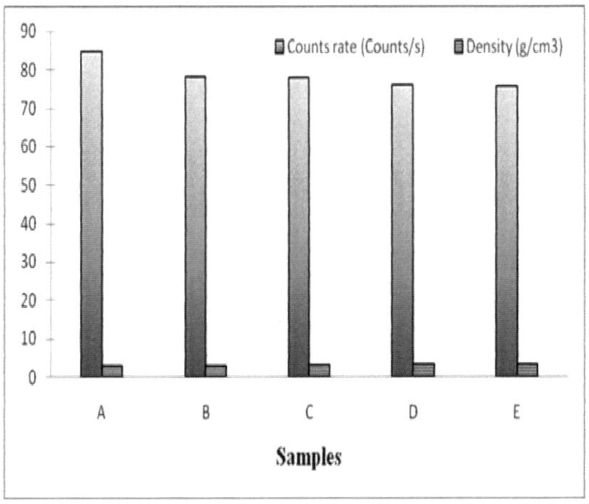

Figure 2. (a) Graph counts rate and concrete thickness (b) Graph counts rate and concrete density

3.3 Analysis for linear attenuation coefficient and half value layer

The determination of linear attenuation coefficient can present the performance of concrete towards radiation shielding [13]. In simple explanation, the higher the attenuation of gamma ray, the performance of concrete will be good. In this part, concrete samples were radiated with Cs-137 with the energy of 0.662 MeV. The value of linear attenuation coefficient for every samples were calculated using Eq. 1.

$$\mu = (1/x) \ln (I / I_o) \quad (1)$$

where μ is the linear attenuation coefficient in cm^{-1} and x is the thickness of the samples in cm. In this study, the thicknesses, x of the samples were fixed as 1 cm. I_o is the gamma ray counts without an absorber and I is the gamma ray count with absorber. Half-value layer was defined as the thickness of the material that reduces the intensity of the beam into half of its original value[14]. It was related to the linear attenuation coefficient by Eq. 2 below.

$$HVL = \ln 2 / \mu \quad (2)$$

where μ (cm^{-1}) is the linear attenuation coefficient

Table 3. Half value layer for concrete samples

Sample	Thickness (g/cm^2)	Density (g/cm^3)	I	I_o	μ (cm^{-1})	HVL (cm)
A	2.4160	2.811	84.89	124.42	0.382	1.814
B	2.5891	2.925	78.10	124.42	0.465	1.491
C	2.7161	3.207	77.81	124.42	0.469	1.477
D	2.8350	3.296	76.07	124.42	0.492	1.408
E	2.9726	3.433	75.79	124.42	0.495	1.400

From table 3, it can be conclude that HVL was decreased as the thickness is increases. HVL is an indirect measure of photon energy or beam hardness. HVL is an important quality control test as it is used to measure whether or not there is sufficient filtration in the gamma ray beam to remove energy radiation, which can be damaging [15]. It also helps to determine the type and thickness of shielding required. When a gamma photon passes through a concrete, it does not deposit gradually its energy as an alpha particle on its way : either it interacts and disappears as such, or passes all the way through without interacting. In the case of a large number of gamma, a beam or a number of gamma emerge unscathed. The beam is attenuated. The stronger the 'probability of interaction', the faster the attenuation.

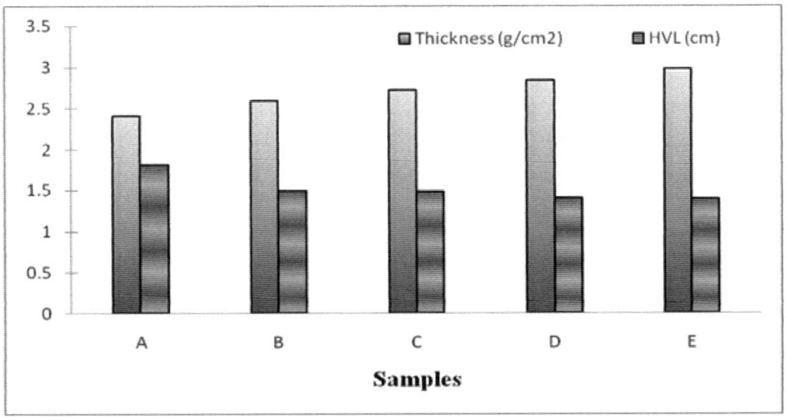

Figure 3. Graph thickness and HVL for concrete

3.4 Analysis for Hardness Test

In this research, hardness test was done using Vickers Hardness Method. This test was done in order to relate the hardness and the structure of concrete. The table 4 below shows the results obtained from the hardness test. The value of test force that has been used in this hardness test is quite high which is 4.903 N. In Vickers Hardness test, the value of test force is quite important as it can indicate the degree of hardness of material.

Table 4. Hardness of concrete

Lead (g)	Hardness (GPa)			Average hardness (GPa)
2	2.275	2.325	2.312	2.304
4	4.374	4.549	4.599	4.507
6	5.884	5.355	5.433	5.557
8	7.257	7.591	7.217	7.355
10	9.414	9.581	9.752	9.582

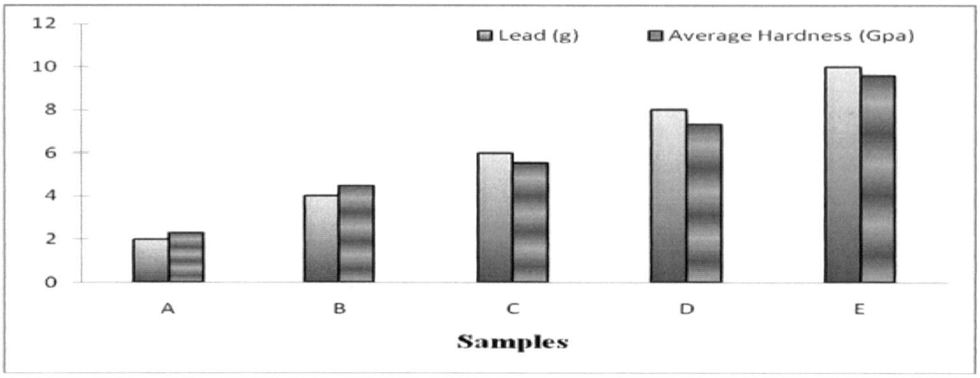

Figure 4. Graph hardness and weight of lead

Fig 4. shows a graph of hardness and weight of lead. Hardness was increase as weight of lead increase. Many factors influence the rate at which the hardness of concrete after mixing. Strength of concrete is directly related to the structure of the hydrated cement paste. Air in concrete produces voids [16]. Excess of water in concrete evaporate leave the voids in the concrete. Consequently, as the w/c ratio increases, the porosity of the cement paste in the concrete also increases. As the porosity increases, the compressive strength of the concrete decreases. Air also is created during improper mixing, consolidating and placement of the concrete. Air pockets, or irregularly sized air voids, are spread throughout the concrete and have negative effects on product appearance, strength and durability.

Conclusion

It was found that, use lead oxide as additives in concrete causes increasing in specific density and compressive strength of the concrete. Besides that, level of radiation intensity towards concrete composition can be identified. As a result enhances the shielding material efficiency against gamma ray. Based on the result and the discussion obtained from the research, the following conclusion can be drawn:

1. After a few trials, water to cement ratio was decided as 0.40 according to the density test.
2. The highest density of the concrete produced in this research was 3.433 g/cm^3.
3. The intensity of gamma ray was reduced with the increasing in the density of concrete.
4. The hardness of concrete increase as the composition of lead increase.

References

[1] Khaled Marar, Ozgur Eren. Effect of cement content and water/cement ratio on fresh concrete properties without admixtures, International Journal of the Physical Sciences 24 (2011) 5752-5765.

[2] Ihab Adam, Kenji Sakata, Toshiki Ayano. Influence of coarse aggregate on the shrinkage of normal and high- strength concretes, J. Faculty of Environmental Science and Technology 6 (2001) 41- 45.

[3] Sandor Popovics, Janos Ujhelyi. Contribution to the concrete strength versus water-cement ratio relationship, J. Materials in Civil Engineering 7 (2008) 459.

[4] Saleh H. Alsayed, Mohammad A. Amjad. Strength, water absorption and porosity of concrete incorporating natural and crushed aggregate, J. King Saud University- Engineering Sciences 1 (1996) 109-119.

[5] Shohana Iffat. Relation between density and compressive strength of hardened concrete, Concrete Research Letters 4 (2015) 182- 189.

[6] S. Kolias, C. Georgiou. The effect of paste volume and of water content on the strength and water absorption of concrete, Cement and Concrete Composites 2 (2005) 211-216.
[7] Nwokoye S.O, Ezeagu C.A. Optimization of the strength of concrete made from Nigerian processed aggregate, Internatiol Refereed J. of Engineering and Science 4 (2016) 45-61.
[8] Mohd El Khatieb. Trial concrete mix design comparison using local materials under local environment, International Journal of Science and Technology 10 (2017) 1-11.
[9] Suleyman Ozen, Cengiz Sengul, Teoman Erenoglu, Unercolak, Iskander Atilla Reyhancan, Mehmet Ali Tasdemir. Properties of heavyweight concrete for structural and radiation shielding purposes, Arabian J. for Science and Engineering 4 (2016) 1573- 1584.
[10] El-Sayed A. Waly, Mohamed A. Bourham. Comparative study of different concrete composition as gamma-ray shielding material. Annals of nuclear energy. 85(2015) 306-310.
[11] Emna Bouali, Abdelhak Ayadi, El- Hadj Kadri, Abdelhak Kaci, Hamza Soualhi, Amjad Kallel. Rheological and mechanical properties of heavy density concrete including barite powder, Research Article- Civil Engineering 45 (2020) 3999- 4011.
[12] O. Agar, M.I.Sayyed, F. Akman, H.O. Tekin, M.R. Kacal. An extensive investigation on gamma ray shielding features of Pd/Ag- based alloys, Nuclear Engineering and Technology 3 (2019) 853- 859.
[13] Almahdi Mousa, Kus Kusminarto and Gede Bayu Suparta. A new simple method to measure the X-ray linear attenuation coefficients of materials using Micro- Digital Radiography Machine, International Journal of Applied Engineering Research 12 (2017) 10589- 10594.
[14] Akkas. Determination of the tenth and half-value layer thickness of concretes with different densities, Acta Physica Polonica A. 129 (2016) 770- 72.
[15] M. Begum, A.S. Mollah, M.A. Zaman, A.K.M.M. Rahman. Quality control tests in some diagnostic X-Ray Units in Bangladesh, Bangladesh Journal of Medical Physics 4 (2011) 59-66.
[16] P. Shawnim, F. Mohammad. Porosity, permeability and microstructure of foamed concrete through SEM images, J. Civil Engineering, Science and Technology 10 (2019) 22-23.

IOP Conf. Series: Materials Science and Engineering 1106 (2021) 012012 doi:10.1088/1757-899X/1106/1/012012

Potential use of polyvinyl alcohol (PVAL) gel materials as mammography phantoms for the detection of calcifications

Franca Oyiwoja Okoh[1,3], Mohd Fahmi Mohd Yusof[2,a)], Siti Nor Azizah Abdullah[1]

[1]School of Physics, Universiti Sains Malaysia, 11800 Penang, Malaysia
[2]School of Health Sciences, Universiti Sains Malaysia, 16150 Kelantan, Malaysia
[3]Federal University Wukari, 234 Wukari-Taraba State, Nigeria

[a]mfahmi@usm.my

Abstract. The study focused on the construction of phantoms made of the polyvinyl alcohol (PVAL) gel materials for mammography based on the dimensions of available RMI phantom for quality control in mammography. The PVAL gel phantoms were fabricated at 20% percentages of PVAL to water to provide target density of 1.0 g/cm^3 and physical strength similar to that of human soft tissues. The calcifications were simulated by using calcium based powders at three different depths. Mammography images were obtained based on the automatic exposure control (AEC) at different compression forces. The results showed That all samples showed good visualisation of calcifications in the mammography images of all fabricated phantoms. The calcifications were well defined with no significant differences of the visualisation of calcifications in all mammography images of all fabricated PVAL gel phantoms. The results indicated the potential of PVAL gel materials to be developed and used as phantoms for the quality control works in mammography.

1. Introduction

Mammography became the most common method for the detections of breast cancer. Mammography can be classified into two namely screening and diagnostic mammography. The screening mammography is usually conducted in healthy persons for the early detection of breast cancer whilst the diagnostic mammography is more specific to provide better localizations of the breast cancer. Calcification is often used as indicator for the detection of breast cancer using mammography imaging. Calcification is the hardening of the breast tissues due to the deposition of calcium compounds. Breast calcifications is categorized as benign, probably benign and malignancies [1][2]. The pattern of calcifications would determine between benign and the risk of malignancies [3][4]. In the quality assurance of mammography imaging, the calcification detection became one of the important criteria to be complied. The quality assurance for mammography is commonly conducted by the acrylic based phantoms such as the Mammographic Accreditation Phantom model 18-220 (Fluke Biomedical, Cleveland, USA). This phantom provides the Al_2O_3 specks to simulate the calcifications. The acrylic phantoms however did not simulate the compressing features that would change the shape and patterns of the calcifications during mammography imaging.

IOP Conf. Series: Materials Science and Engineering 1106 (2021) 012012 doi:10.1088/1757-899X/1106/1/012012

Several studies had been conducted to fabricate a new type of phantoms for the quality assurance in mammography that focused on the determination of glandular dose [5-7]. PVAL is a water-soluble synthetic polymer that would form gel-structure materials through the process of freezing and defrosting [8]. It has approximate mass density of 1.19 g/cm^3 which is near to the value of water made it suitable to be developed as water and tissue equivalent bolus materials. A previous work suggested the suitability of PVA gel to be used as phantom for the magnetic resonance imaging and bolus materials for external beam therapy due to its close physical and radiological parameters to the human soft [9][10]. Another study also indicated the suitability of PVAL phantom for visualisation of calcifications by using dual energy technique [11].

The phantoms, made of alcohol (PVAL) in the current study, were fabricated in application for mammography. Low mammography range x-rays were used to visualise the calcifications in the fabricated PVAL gel phantoms at different depths from the surface of the phantoms.

2. Methodology

2.1. Preparation of the PVAL gel phantom materials

The polyvinyl alcohol (PVAL) material was purchased from the Sigma-Aldrich Inc. (Sigma-Aldrich, Darmstadt, Germany). It has the monomer unit of $(g)_n$ and was prepared in a crystalline powder form by the manufacturer. PVAL crystals were mixed into an amount of water and heated to enhance its solubility. The PVAL-water solutions were prepared at 20% based o the weight of the water to provide sufficient physical strength of the PVAL gels [10]. The mixtures were filled in a container with dimensions of 5 × 7 × 5 cm. The PVAL-water solutions were cooled down into room temperature to eliminate the air bubbles from the solutions before being frozen by using a refrigerator for approximately 12 hrs to from solid PVAL-water materials. The solid PVAL-water materials were then being defrosted at room temperature until the translucent gel PVAL bolus materials were achieved. Calcifications were simulated by using the $CaCO_3$ powders and being added onto the formed PVAL gels at different depths between of 1.5, 2.5 and 3.5 cm from the surface before another layer of PVAL gel was added as shown in Figure 1. The completed fabricated PVAL gel phantoms are shown in Figure 2.

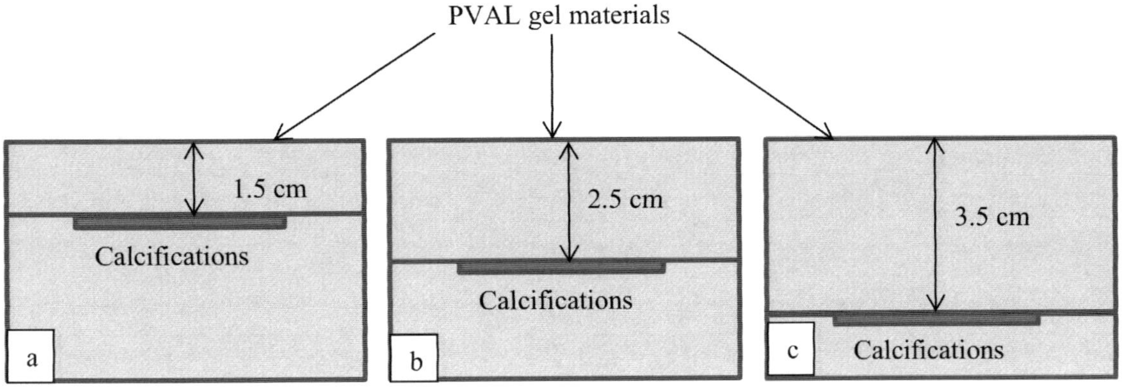

Figure 1. The schematic diagrams of fabricated PVAL gel phantoms for mammography with calcification features at (a) 1.5 (b) 2.5 and (c) 3.5 cm depths .

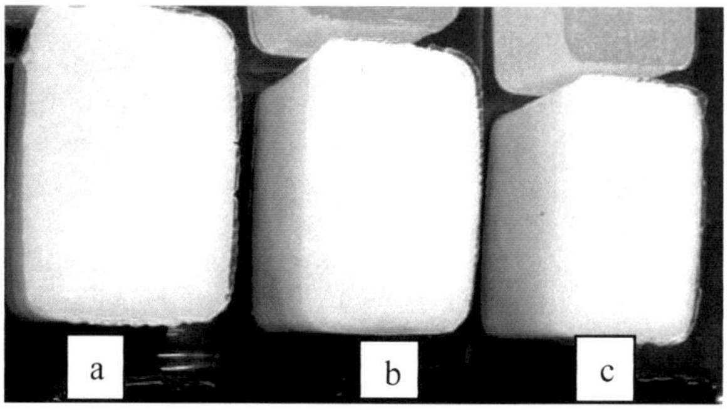

Figure 2. The completed fabricated PVAL gel phantoms for mammography with calcification features at (a) 1.5 (b) 2.5 and (c) 3.5 cm depths.

2.2. Mammography imaging of the fabricated PVAL gel phantoms

The mammographic images of the fabricated PVAL gel phantoms were obtained by using a computed mammography unit in the Radiology Department, Hospital Universiti Sains Malaysia (HUSM). The mammography imaging was carried out by using the automatic exposure control (AEC) based on the thickness and the compression loads of the mammography unit towards the fabricated PVAL gel phantoms. The mammography images were analysed by using the ImageJ software to evaluate the visual of the calcifications at various exposure factors and depths in the fabricated PVAL gel phantoms.

3. Results and discussion

The exposure factors by the AEC setups on the fabricated PVAL gel phantoms are shown in Table 1. The thickness of the fabricated PVAL phantoms reduced when higher compression forces were applied. The exposure factors of the mammography by the AEC were generally decreased when the thickness of the fabricated PVAL gel phantoms decreased. This is due to the exposure factors by the AEC that were based on the size and thickness of the tissue being imaged [12][13]. A few increment of mAs were also seen when the compression forces were increased due to variation of entrance surface air kerma (ESAK) when AEC is used [14].

Table 1. The force of compression, resulting thickness and exposure factors (kVp and mAs) on the fabricated PVAL gel phantom A, B and C during the mammography imaging.

Depth of calcifications (cm)	Force (N)	Thickness (cm)	kVp	mAs
1.5	0	4.4	28	333
	22.72	4.0	28	286
	68.18	3.7	27	304
	113.64	3.2	26	284
	159.10	2.9	26	259
2.5	0	4.3	28	334
	22.72	4.2	28	304
	68.18	3.9	27	304
	113.64	3.4	26	312
	159.10	3.1	26	272
3.5	0	4.7	28	345
	22.72	4.2	28	312
	68.18	3.8	27	300
	113.64	3.3	26	282
	159.10	2.9	26	255

The compression in mammography would provide uniform density in the breast tissues during imaging procedure. This would give better uniform attenuation by the x-ray giving better image quality as the absorption and scattering is uniform throughout the tissue being imaged.

The mammography images of the fabricated PVAL gel phantoms at all experimented compression forces are shown in Figure 3. The results showed that all the calcifications in all fabricated PVAL gel phantoms were clearly visualized at all mammography images. The calcification patterns at higher compression forces are appeared to be slightly dispersed and easier to be distinguished from each cluster. The compression forces and the resulted thickness of the fabricated PVAL gel phantoms did not significantly affect the mammography images at all experimented compression forces [15-17].

Table 2. The mammography images of the fabricated PVAL gel phantoms at different exposure factors (kVp and mAs) resulted from the compression forces and phantom thicknesses.

Force (N)	**Fabricated PVAL phantom images**		
	A	**B**	**C**
0	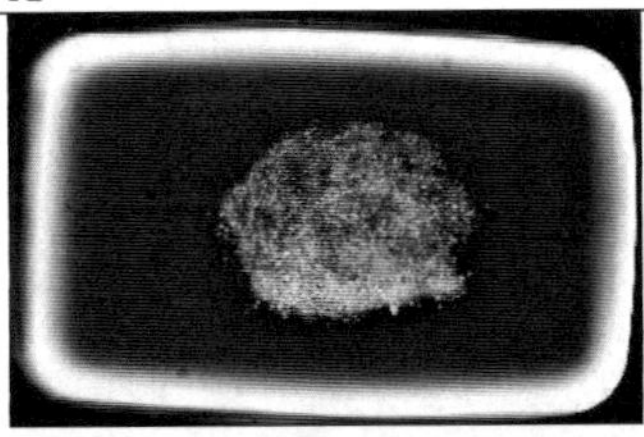	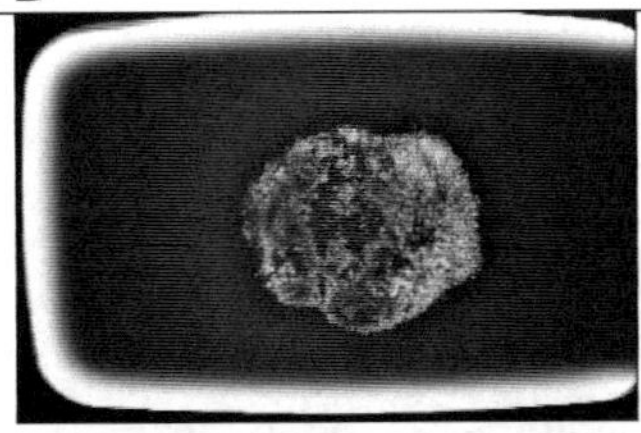	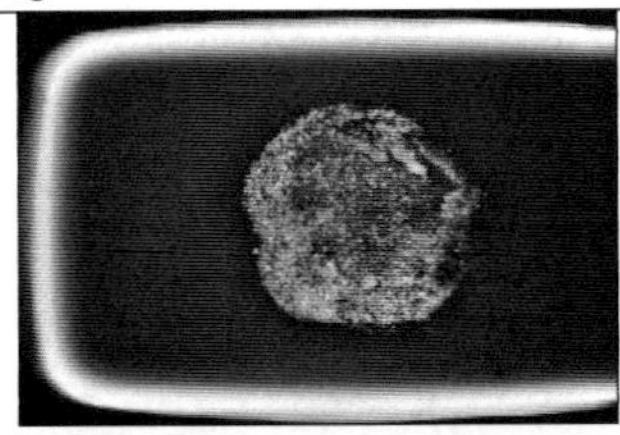
22.72	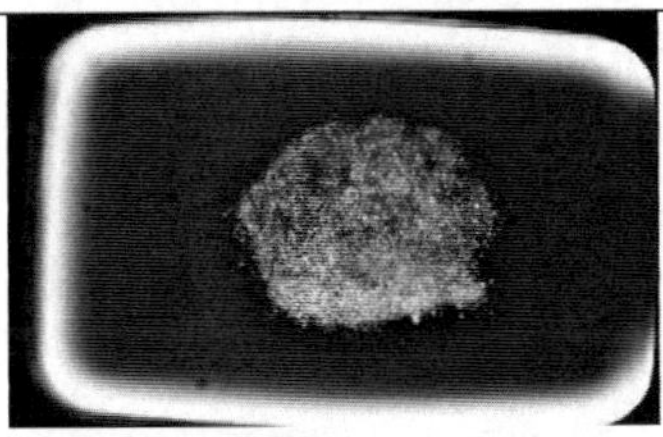	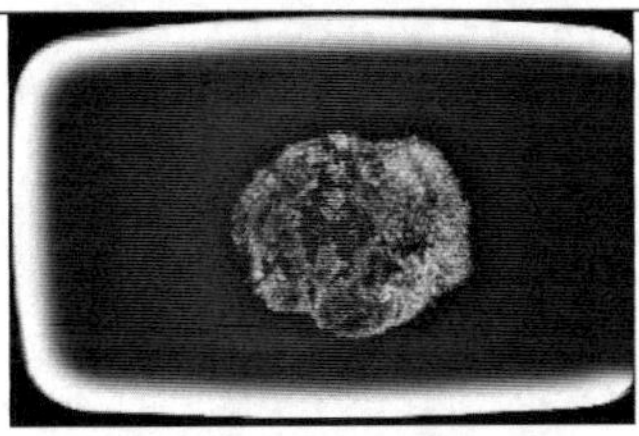	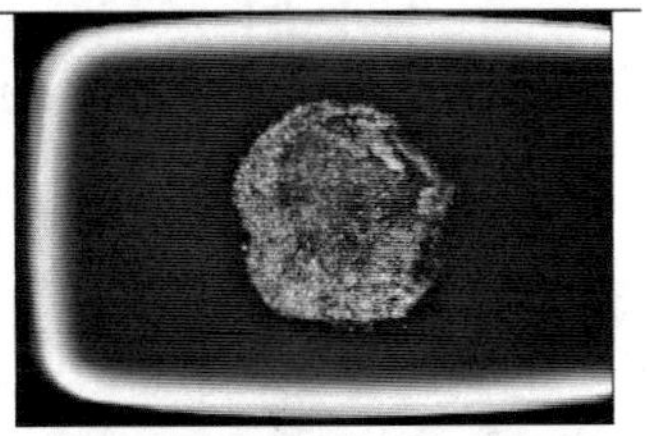
68.18	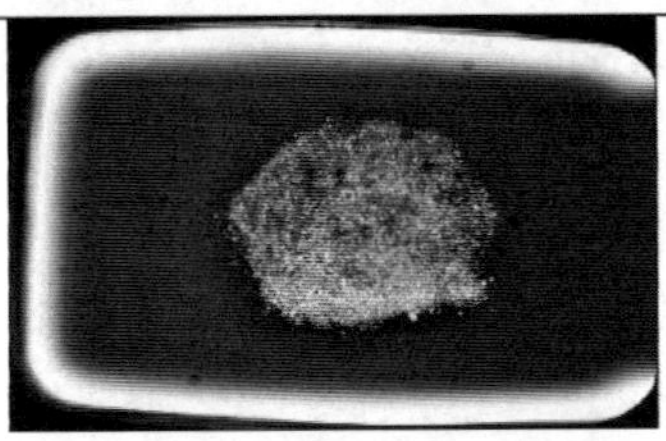	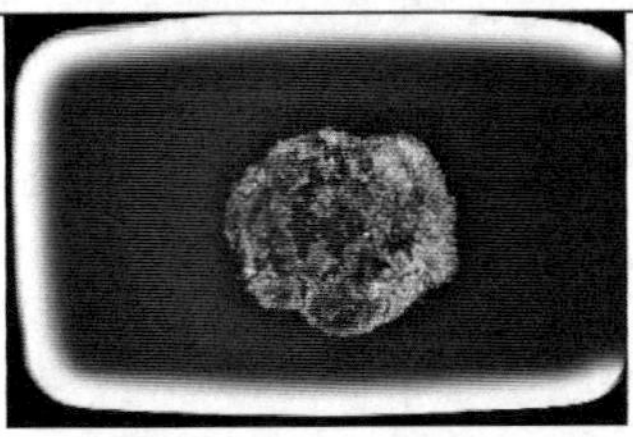	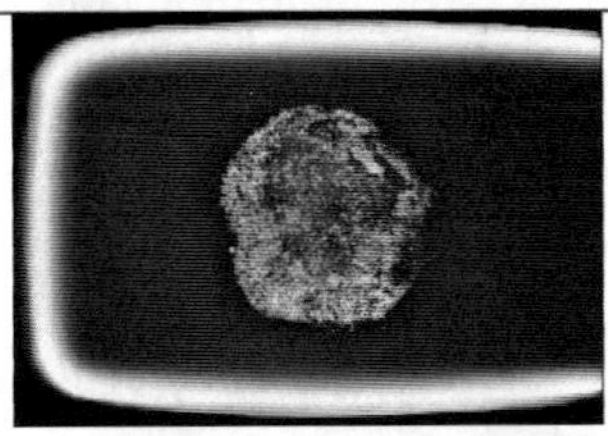
113.64	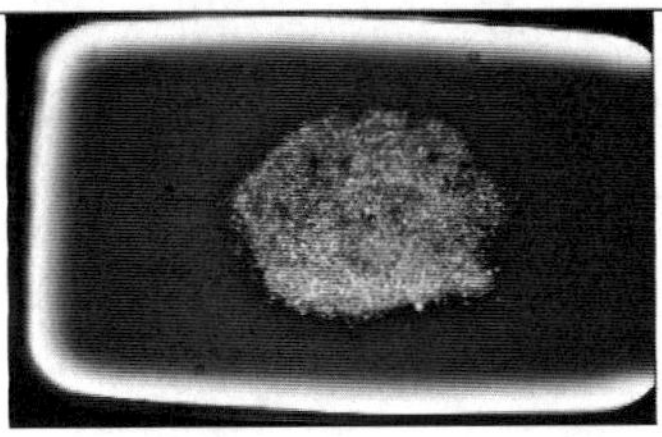	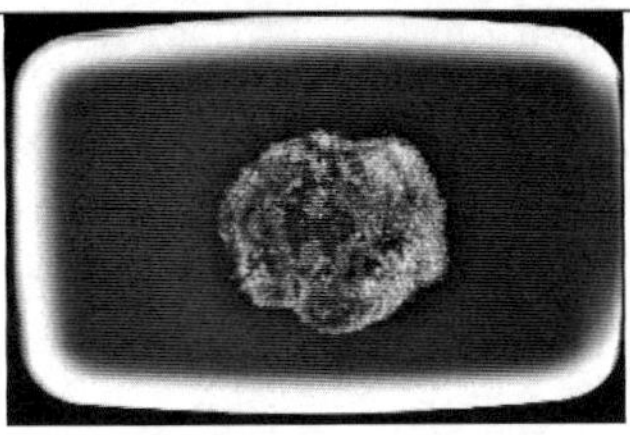	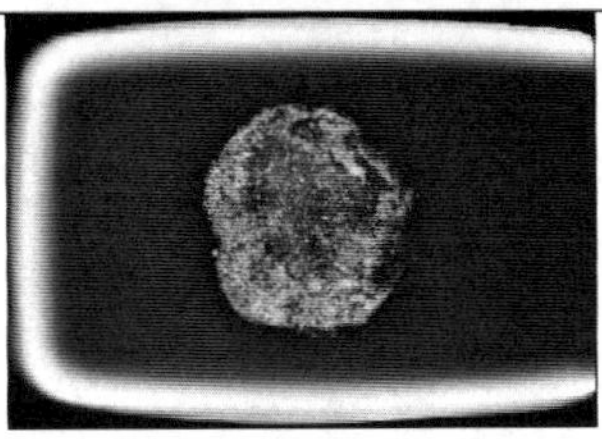
159.1	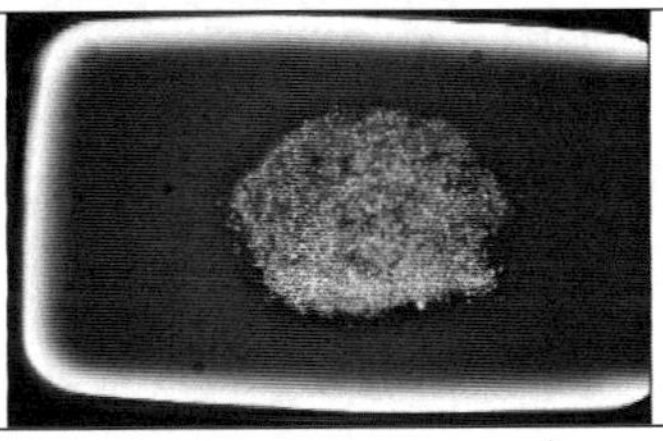	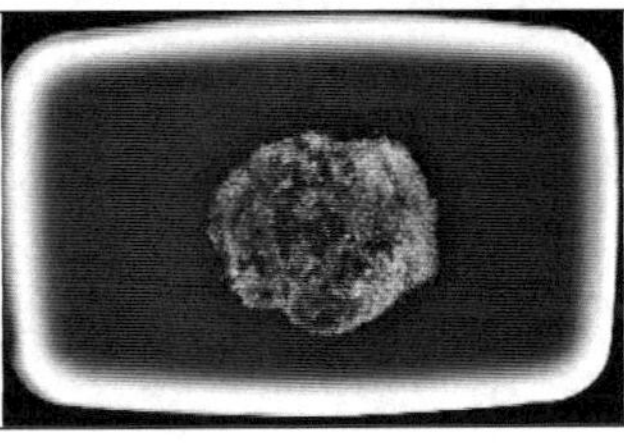	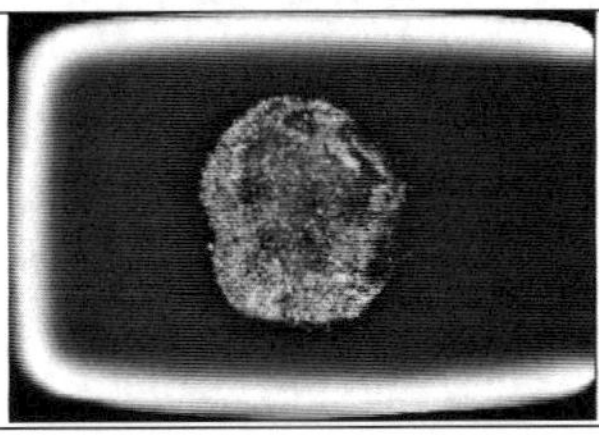

The depth of the calcifications from the surface of the fabricated gel phantoms also did not significantly affect the mammography images at exposure factors. The calcifications however did not simulate the pattern of microcalcification often observed in available mammography phantoms. Specific scoring for the calcifications also was not possible due to large calcifications of the $CaCO_3$ powders used in this study. The overall results however indicated the potential of the fabricated PVAL gel materials to be used as phantoms in the quality control of mammography.

4. Conclusion

The fabricated PVAL gel phantoms were able to visualize the calcification using the mammography imaging at all experimented compression forces and exposure factors. The exposure factors used based on the phantom thickness resulted from the compression forces did not significantly affect the visualisations of the calcification features. The depth of the calcifications also did not significantly affect the visualisations of the calcification features. The overall results indicated the potential of PVAL gel materials to be developed as phantoms for the quality control of mammography imaging.

5. References

[1] Nalawade Y V 2009 Evaluation of breat calcifications. *Indian J. Radiol. Imaging*, **19**(4) 282-286.

[2] Sickles E A 1986 Breast calcifications: mammographic evaluation. *Radiology*, **160**(2) 289-293.

[3] Xu J G 1989 Investigation on calcifications in the breast. *Chinese Journal of Radiology*. **23**(4) 198-200.

[4] Villers G, Mortier M, De Potter C, Kunnen M 1995 Breast calcifications. *J. Belge. Radiol.*, **78**(1) 11-17.

[5] Argo W P, Hintenlang K, Hintenlang D E 2004. A tissue-equivalent phantom series for mammography dosimetry. *J. Appl. Clin. Med. Phys.*, **5**(4) 112-119.

[6] Almeida C D, Coutinho C M, Dantas B M, Peixoto J E and Koch H A 2012 A new mammographic dosimetric phantom, *Radiat. Prot. Dosim.*, **15**(1) 196-198.

[7] Geise R A, Palchevsky A 1996 Composition of mammographic phantoms. *Radiology*, **198**(2) 347-350.

[8] Sivaraman A, Ganti S S, Nguyen H X, Birk G, Wieber A, Lubda D and Banga A K. 2017 Development and evaluation of polivynil alcohol based topical gel. *J Drug. Deliv. Sci. Tec.*, **39** 210-216.

[9] Mano I, Goshima H, Nambu M and Lio M. 1986 New polivynil alcohol gel for MRI phantoms. *Magn. Reson. Med.*, **3(6)** 921-6.

[10] Okoh F U, Yusof M F M, Abdullah R and Kabir N A 2020 Polyvinyl alcohol (PVA) based bolus material for high energy photons and electrons, *IOP Conf. Ser. Mater. Sci. Eng.*, **785** 012043.

[11] Koukou V, Fountous G, Martini N, Sotiropoulou P, Michail C, Kalyvas N, Valais I, Bakas A, Kounadi E, Kandarakis I and Nikiforidis G 2014 Optimization of breast cancer detection in Dual Energy X-ray Mammography using a CMOS imaging detector. *J. Phys. Conf. Ser.*, 574, 012076.

[12] Kotre C J 2010 X-ray absorptiometry of the breast using mammographic exposure factors: application to units featuring automatic beam quality selection. *Br. J. Radiol.*, **83**(990) 515-523.

[13] Alkhalifah K and Brindhaban A 2018 Investigation of exposure factors for various breast composition and thicknesses in digital screening mammography related to breast dose. *Med. Princ. Pract.*, **27** 211-216.

[14] Aslund M, Cederstrom B, Lundqvist M and Danielsson M (2005) AEC for scanning digital mammography based on variation of scan velocity, Med. Phys., **32**(11) 3367-74.

[15] Holland K, Sechopoulos I, Mann R M, den Heeten G J, van Gils C H and Karssemeijer N 2017Influence of breast compression pressure on the performance of population-based

mammography screening, *Breast Cancer Res.,* **19** 126.
[16] Groeneweld A, Herbst C P and Rae W I D 2009 The relationship between compression force, image quality and radiation dose in mammography. *SAJR,* **13**(4) 86-92.
[17] Suhaimi S A A, Mohamed A, Ahmad M and Chelliah K K 2015 Effects of Reduced Compression in Digital Breast Tomosynthesis on Pain, Anxiety, and Image Quality. *Malays. J. Med. Sci.,* **22**(6) 40–46.

Acknowledgement
The authors would like to acknowledge the financial aid of USM Bridging Grant no. 304.PPSK.6316324 for the financial support of this study. The authors also would like to acknowledge the Radiology Department, Hospital Universiti Sains Malaysia (HUSM) for the use of mammography unit in this study.

Effect of electrolyte concentration on the synthesis of graphene by electrochemical exfoliation process

Suhaila Hani Ilias[1,a)], Julie Andrianny Murshidi[1], Kok Kuan Ying[1]

[1]Material Technology Group, Industrial Technology Division, Malaysian Nuclear Agency, Bangi, 43000 Kajang Selangor, Malaysia.

[a]suhaila_hani@nuclearmalaysia.gov.my

Abstract. Graphene can be produced from graphite rod of a battery by electrochemical exfoliation process in acidic electrolytes. This process is simple, safe, fast, cheap and produces graphene in large scale. Parameters to be controlled during the process are the concentration of the electrolytes and potential for exfoliation. The synthesized graphene was characterized by Raman Spectroscopy and Field Emission Scanning Electron Microscopy (FESEM). From the results, it was found that graphene synthesized with 90:10 volume ratio had better yield, lowest interlayer distance, less oxidation and minimum defect of graphene.

1. Introduction

Graphene as a single layer of sp^2-hybridized carbon, is rightly proclaim for its superlative material properties. It has been demonstrated both as a novel building block and as an additive for a wide range of multifunctional materials [1][2]. In recent years, exfoliation of graphene has been investigated through various processes such as by mechanical, chemical, thermal, electrical or electrochemical[3][4]. These are known as top-down processes. Electrochemical exfoliation is a promising process for synthesizing graphene from graphite. It is cheaper, greener and potentially capable of mass production with few layers of graphene. In this process, an applied voltage drives ionic species to intercalate into graphite lattices where they form gaseous species that expand and exfoliate into individual graphene sheets [5]. The quality and quantity of the synthesized graphene (graphene layer numbers, O contents, lateral size, edge or sheet defects) can be obtained by controlling the process parameters (applied electrical potential, current, processing time, composition of electrolytes) [6][7].

Thus, in order to investigate the effect of electrolyte acidity concentration by using different H_2SO_4:KOH volume ratios (90:10, 80:20 and 70:30) at room temperature, this paper describes specifically the quality and microstructural properties of the synthesized graphene using Raman Spectroscopy and FESEM.

2. Experimental

The starting reagents were H_2SO_4 (SigmaAldrich, 95%), KOH (Scharlau) and IPA (HmbG Chemicals, 99.8%). Deionized water was used throughout the experimental process. The required graphite rod was obtained from a dry cell battery with a diameter of 0.8 cm, a length of 5.7 cm and weight of 4.90 g as shown in Figure 2(a).

2.1 Electrochemical exfoliation.
Electrolytes with different H_2SO_4:KOH volume ratios of 90:10, 80:20 and 70:30 were prepared by mixing 0.1 M H_2SO_4 and 0.1 M KOH solution by using a magnetic stirrer. Prior to electrochemical exfoliation process, graphite rod was pre-treated by immersing in the electrolyte for 24 hours. All processes were carried out at room temperature.

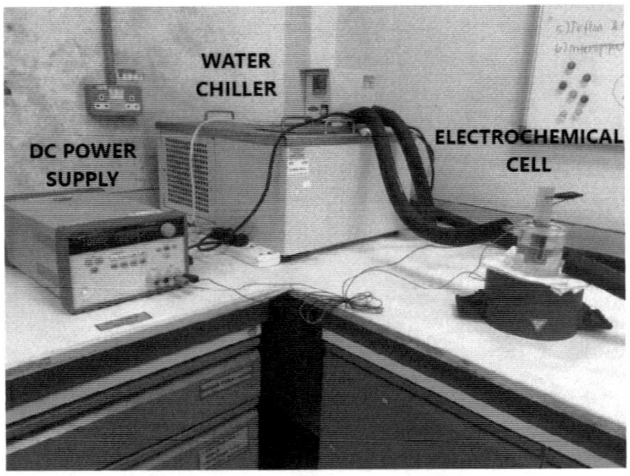

Figure 1. Electrochemical exfoliation set-up for graphene synthesis.

Electrochemical exfoliation process was carried out in an electrochemical cell. The cell was composed of a graphite rod and a platinum (Pt) plate immersed in an electrolyte. It was connected to a DC power supply as shown in Figure 1, with graphite rod acts as anode (using positive bias) and Pt plate as cathode (using negative bias). Exfoliation was conducted by applying +2.5 V (low voltage) for 5 min followed by +10 V (high voltage) for 10 min. The sample was left to settle for 5 hours before filtration process. The graphene fragments at the top of the solution were filtered using a vacuum pump and PTFE filter membrane with pore size of 0.2 μm. During the filtration process, graphene fragments were rinsed with IPA and deionized water. The filtered samples were dried overnight.

2.2 Material characterization
Raman Spectroscopy measurements were performed using Renishaw InVia Reflex spectrometer in the range of 100 to 3300 cm^{-1} with excitation laser of 532 nm to analyse the number of layers and defect of the synthesized graphene. The graphene powders were dropped cast onto a glass slide for measurement. Renishaw WiRE 5.0 software was used for the analysis, background correction and peak fitting.

The microstructural observations were conducted on Carl Zeiss GeminiSEM 500 FESEM using a secondary electron detector operating at 15 kV. The microscope was coupled with an energy dispersive X-Ray spectrometer (EDS) for elemental analysis. Samples were not coated prior to imaging in the FESEM. Samples were prepared by sprinkling the powders onto a silicon wafer fixed to a FESEM sample holder.

3. Results and discussion.
The mechanisms behind electrochemical exfoliation depend principally on the type of potential applied: anodic or cathodic. Anodic exfoliation involves the intercalation of anions as well as other co-intercalating species involved in the reaction mixture into the graphite lattices.

For anodic potential, a positive bias withdraws electrons from the graphite (anode), creating a positive charge. On the other hand, a negative bias allows electrons to the Pt (cathode), creating a negative charge. This negative charge drives the reduction of water molecules at the cathode, producing hydroxyl ions (OH^{-1}). OH^- can attack graphite (anode) at the edges and grain boundaries [8]. This oxidation at anode leads to the expansion of graphite layers at the edges, opening up the lattices for intercalation by sulfate ions (SO_4^{2-}) and also possibly more water molecules [9]. After the electrochemical intercalation and expansion of graphite, the gaseous species (e.g. SO_2 and O_2) produced (e.g. from oxidation of water molecules and from reduction of SO_4^{2-} anions) will then exfoliate and separate the graphite lattices into graphene [10].

Figure 2(b) shows a reduction in size of graphite rod after electrochemical exfoliation with a diameter of 0.5 cm, a length of 5.6 cm and weight of 4.18 g. The initial low voltage applied to the graphite rod helps to expand the graphite lattices to allow the OH^- enters between the graphite lattices. When a higher voltage is applied, the graphite is exfoliated into two types of fragments. Sediment at the bottom of the electrolyte consisting of thick graphite sheets while the sample floating on the surface of the electrolyte is predominantly graphene of thinner thickness.

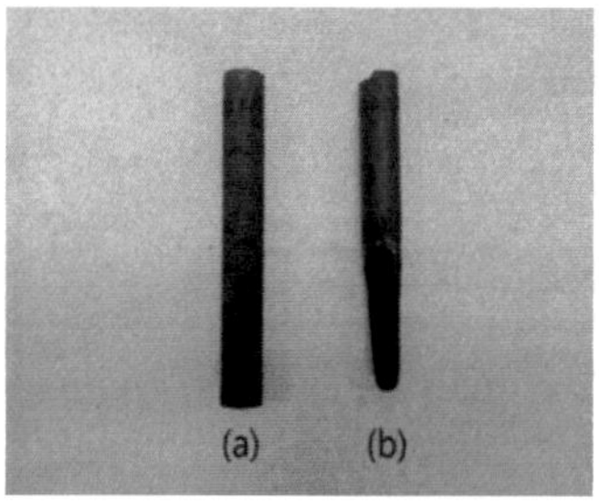

Figure 2. Graphite rod (a) before and (b) after exfoliation.

Figure 3 shows the Raman spectrum for all the samples. From the spectra, D peak is located at ~1360 cm^{-1}, G peak is located at ~1590 cm^{-1} and 2D peak is located at ~2700 cm^{-1}. The intensity of the D peak depends on the disorder of the graphene. The G peak represents the E_{2g} phonons vibrations mode in the centre of the Brillouin Zone and is a characteristics peak of carbon material [5]. The I_D/I_G ratio, which gives information about the amount of defects in graphene, was found to be 0.42, 0.75 and 0.94 for sample with H_2SO_4:KOH volume ratios of 90:10, 80:20 and 70:30 respectively. This ratio allows to distinguish graphene from multi layered graphite [11]. The intensity I_{2D}/I_G ratio, which represents the number of graphene layers, for 90:10, 80:20 and 70:30 with H_2SO_4:KOH volume ratio was 1.15, 0.19 and 0.15 respectively. Based on these results, electrolyte concentration by using 90:10 H_2SO_4:KOH volume ratio showed the highest number of graphene layers with minimum defects.

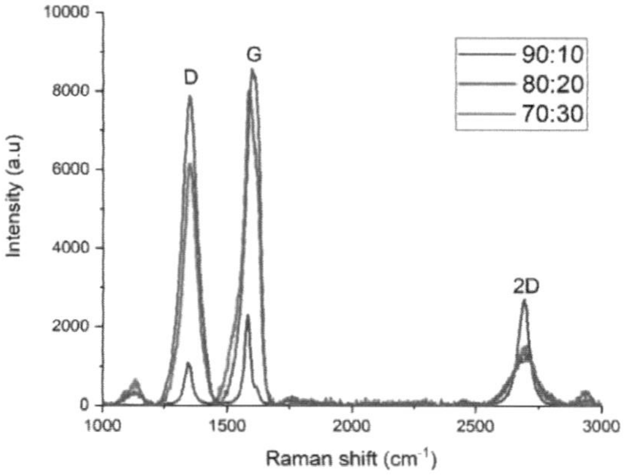

Figure 3. Raman spectra of samples with H_2SO_4:KOH volume ratios of 90:10, 80:20 and 70:30.

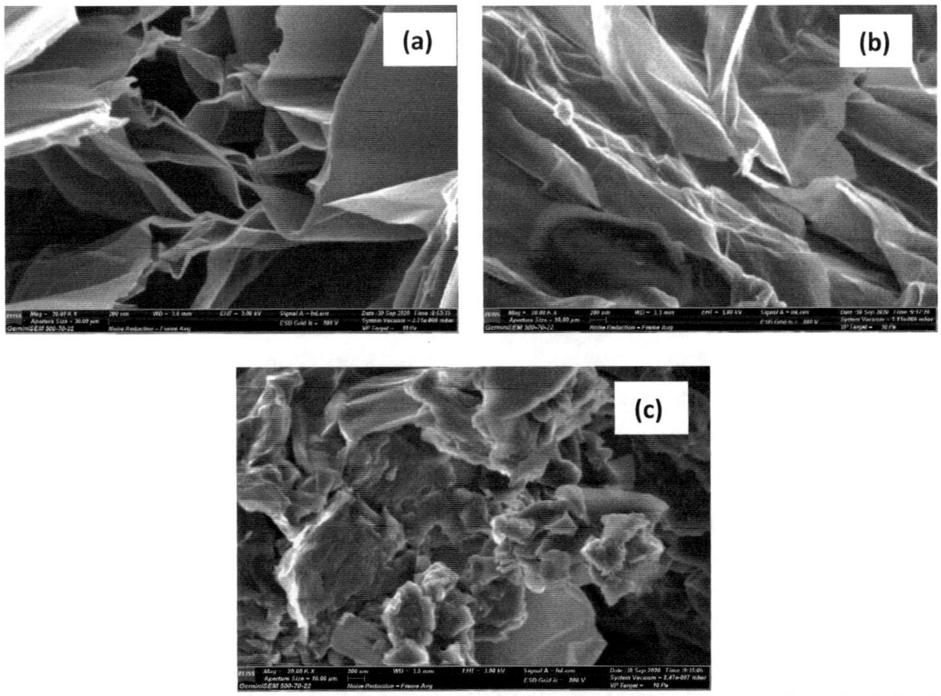

Figure 4. FESEM micrographs of samples with H_2SO_4:KOH volume ratios of (a) 90:10, (b) 80:20 and (c)70:30.

The surface morphology and cross section of the samples were characterized by using FESEM. From Figure 4, it can be seen that the lateral size of graphene sheets is quite large with almost similar average diameters of 200 nm for the samples. The layering of graphene in image of 90:10 is more prominent compared to 80:20 and 70:30. Images shown in (a) and (b) exhibit both layering and folded structures while images in (c) shows platelet-liked crystalline form of carbon, which is similar to graphite.

Table 1 shows the summarized data of EDS analysis of the synthesized graphene in terms of carbon content (atomic %), oxygen content (in atomic %), and the ratio of carbon to oxygen composition. Oxygen content increases, whereas carbon content and C/O ratio decreases when using H_2SO_4:KOH volume ratio from 90:10 to 70:30 indicating that less chemical oxidation occurs when using high concentration of acidic electrolyte during electrochemical exfoliation.

Table 1. EDS analysis of the synthesized graphene samples.

Sample	C (%)	0 (%)	C/O ratio
90:10	87.0	13.0	6.7
80:20	83.8	16.2	5.2
70:30	72.5	27.5	2.6

4. Conclusions.

In conclusion, we have successfully synthesized graphene by electrochemical exfoliation of graphite rod employing different concentration of acidic electrolyte. Electrochemical exfoliation with higher concentration of electrolyte (H_2SO_4:KOH volume ratio of 90:10) leads to better yield, less oxygen content and minimum defect of graphene. Selection of a suitable electrolytes is thus important to tune the quantity and quality of the graphene. The present study provides a proficient approach to synthesize cost effective and high-quality graphene.

5. References

[1] Edwards R S, Coleman K S 2013 Graphene synthesis: relationship to applications *J. Nanoscale* **5**(1) 38-51.
[2] Dreyer D R, Ruoff R S, Bielawski C W 2010 From conception to realization: an historial account of graphene and some perspectives for its future *J. Chem. Int. Ed.* **49** 9336-9344 (Angew).
[3] Jiao L Y, Wang X R, Diankov G, Wang H L, Dai H J 2010 Facile synthesis of high quality graphene nanoribbons. *J. Nat. Nano-technol* **5** 321-325.
[4] Cai M, Thorpe D, Adamson D H, Schniepp H C 2012 Methods of graphite exfoliation *J. Mater. Chem.* **22** 24992.
[5] Achee T C, Sun W, Joshua T Hope, Quitzau S G, Sweeney C B, Shah S A, Habib T, Green M J 2018 High-yield scalable graphene nanosheet production from compressed graphite using electrochemical exfoliation (www.nature.com/scientificreports) **8** 14525.
[6] Liu F, Wang C, Sui X, Riza M A, Xu M, Wei L, Chen Y 2019 Synthesis of graphene materials by electrochemical exfoliation: Recent progress and future potential (doi:10.1002/cey2).
[7] Kurniawan T, Anwar M A, Oktiani R, Ragadhita R, Nandiyanto A B D and Aziz M 2019 Influences of solvent types on the formation of graphene in the electrochemical exfoliation method. *J. Mater. Phys. and Mech.* **42** 151-157.
[8] Htwe Y Z N, Chow W S, Suda Y, Thant A A, Mariatti M 2019 Effect of electrolytes and sonication times on the formation of graphene using an electrochemical exfoliation process *J. App. Surface Sci.* **469** 951-961.
[9] Parvez K, Wu ZS, Li R, Liu X, Graf R, Feng X, et al. 2014 Exfoliation of graphite into graphene in aqueous solutions of inorganic salts. *J Am Chem Soc*136: 6083–91.
[10] Su C Y, Lu A Y, Xu Y, Chen F R, Khlobystov A N, Li L J 2011 High-quality thin graphene films from fast electrochemical exfoliation ACS nano **5** 2332–9.
[11] Azam M A, Abdul Aziz M F, Zulkapli N N, Omar G, Munawar R F, Mohd Suan M S, Safie N E 2020 *J. Diamond & Related Materials* **104** 107767.

Acknowledgement

The authors gratefully acknowledge the financial support from Malaysian Nuclear Agency and IAEA (CRP contract no R23130) for this work, as well as Mr Choo Thye Foo for his assistance in FESEM characterisation.

Evaluation of nuclear medicine planar images based on partial volume effects and recovery coefficients at different activity levels of radionuclides and matrix sizes

Mohd Fahmi Mohd Yusof[1,a)], Nor Amalyna Ghazali[1], Ummi Solehah Ab Ghani[1], Ahmad Thaifur Khaizul[1], Puteri Nor Khatijah Abd Hamid[2]

[1]School of Health Sciences, Universiti Sains Malaysia, 16150 Kelantan, Malaysia.
[2]School of Physics, Universiti Sains Malaysia, 11800 Penang, Malaysia.

[a]mfahmi@usm.my

Abstract. The partial volume effect evaluations had been effectively used as method for the evaluation of SPECT images in nuclear medicine. The study focused on the use of partial volume effect method and the evaluation of recovery coefficient in the 2D planar images in nuclear medicine. The planar images were obtained by using a modified Jaszczak phantom that contains radionuclide volume of differents diameter sizes and three activity levels. The planar images were also taken at three different matrix sizes of gamma camera set up at 128 x 128, 256 x 256 and 512 x 512. The results showed that the contrast values increased exponentially at larger diameter sizes of radionuclide. The contrast also significantly increased when higher matrix sizes were used. The recovery coefficient evaluation of the planar images were singnificantly decreased when larger matrix sizes were used. There were no significant differences between the recovery coefficients between planar images obtained with and without background counts. The overall results indicated the suitability of using the partical volume efffect method to evaluate the planar images of nuclear medicine.

1. Introduction

The performances parameters of gamma camera play major role in describing the radionuclide distribution [1-3]. The important performances parameters in gamma camera system are spatial resolution and sensitivity. Besides, uniformity and high counting performances are also considered. So that, the performances of gamma camera system are defined as the sharpness and detail of the image produced, efficiency of detecting the incident radiation, ability to measure the energy of incident radiation and counting rate it can handle without significant dead time losses [1] .

The performances characteristic of the gamma camera system is related to the partial volume effect (PVE). The PVE is important in qualitative and quantitative of the image [4]. It is commonly occurred at the object with small size that near to the resolution limits of the imaging device. It appears to contain smaller activity and concentration then the actual value [5]. It is also known as spill out and spill in effect. Spill out is defined as the activity inside the region of interest (ROI) distributed over the border, hence, spill in effect is referred as the activity outside the ROI integrated into the ROI due to the limited spatial resolution [4]. Thus, the spatial resolution degraded caused the activity concentration inside the ROI to be either underestimated or overestimated. The ratio between the apparent activity concentrations to the true activity concentration is known as recovery coefficient

(RC). Besides RC, image contrast also plays important roles in both qualitative and quantitative of the image quality. Image contrast is an important indicator of how well a system is performing with respect to detection of small lesion. Image contrast here is defined as the ratio of signal change of an object interest to the signal level in surrounding part of the images [6].

This study focused on the evaluation of the gamma camera performance for planar imaging in nuclear medicine based on the PVE. The contrast and of the planar images were evaluated at different matrix sizes of the gamma camera and different activity levels.

2. Methodology

2.1. Simulation of Radionuclide with Different Sizes

A modified phantom from the readily available Jasczczak was used to simulate radionuclide of different sizes as shown in Figure 1(a) [7-9]. A set of cylindrical vials with internal diameters between 1.3 and 3.5 cm was used to contain radionuclide as shown in Figure 1(b). The set of vials were placed into the Jasczczak phantom filled with water to simulate the density of human soft tissue. The vials were filled with ^{99m}Tc of three different activity levels of 3, 10 and 18 μCi while the Jaszczak phantom was filled with water and an amount of ^{99m}Tc with activity ratio of 1:12 to the activity used in the vials [7].

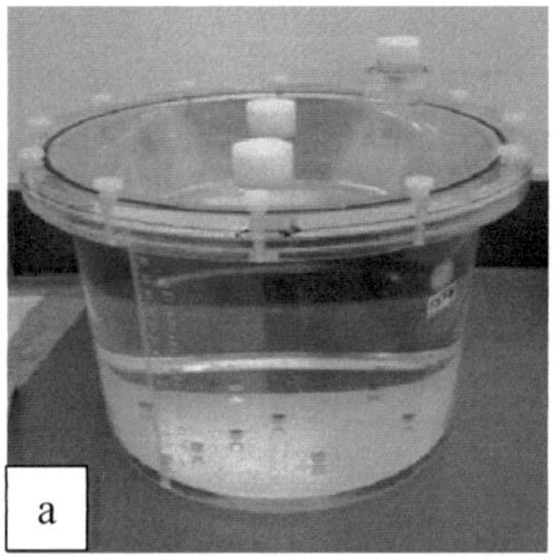

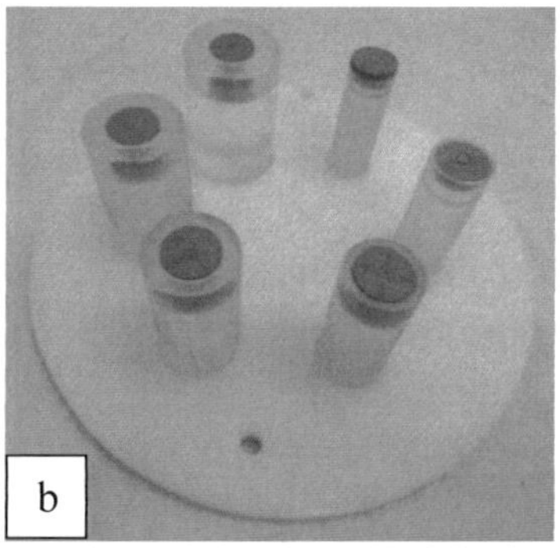

Figure 1. (a) The Jaszczak phantom used in the study.and (b) the cylindrical vials of different diameters used in this study

2.2. Measurement of contrast and recovery coefficient

The prepared phantom was scanned by using gamma camera model NM/CT 670 Pro at planar imaging mode. The phantom containing radionuclides with three activity levels were scanned each at three different matrix sizes of 128 × 128, 256 × 256 and 512 × 512. The contrast of SPECT/CT image was evaluated based on the volume of interest (VOI) on each of vials and the around the background region. The contrast of SPECT/CT image was calculated based on the equation:

$$contrast = \frac{c_{vial} - c_{background}}{c_{background}} \quad (1)$$

where c_{vial} is count per pixel of vial and $c_{background}$ is the count per pixel in background region. The contrast values at different activity level and matrix sizes were analysed. The recovery coefficient was calculated based on the equation:

$$Recovery\ coefficient = \frac{A_a}{A_t} \quad (2)$$

where A_a is apparent activity of radionuclide concentration radionuclide and A_t is actual activity of concentration radionuclide [1]. The apparent activity concentration of radionuclide was calculated by using equation:

$$Apparent\ activity, A_a = \frac{R_{voi}}{S_{vol}} \times \frac{VS}{T_{acq}} \quad (3)$$

where R_{voi} is count per pixel of volume of interest vial, S_{vol} is volume sensitivity, VS is voxel size and T_{acq} is total time of scanning [9][10]. The count per pixel for each vial was measured by manually drawn the VOI at each vial based on the CT image using the Volumetric MI Evaluation software on the Xeleris workstation. The value of volume sensitivity is 2.667 cps/µCi. While the value voxel size for matrix size 128 x 128 is 4.42 x 4.42 x 4.42 mm^3 was based from the department's recent quality assurance for the gamma camera [7]. The actual activity concentration of radionuclide was calculated by using the equation:

$$Actual\ activity, A_t = A_o e^{-\lambda t} \tag{4}$$

where A_t is actual activity concentration of radionuclide, A_O is initial activity concentration of radionuclide source (^{99m}Tc), λ is decay constant of radionuclide and the t is time elapsed from the initial activity. The value of the decay constant of radionuclide source was 2.5 x 10^{-5} s^{-1}.

3. Results and discussion

The planar images taken at different activity levels and matrix sizes are shown in Figure 2 to Figure 4. The results showed that all the images of the radionuclide at different sizes and activity levels are clearly visualised at all matrix sizes. However, the images taken by using smaller matrix size appeared to be brighter and sharper compared to higher matrix sizes.

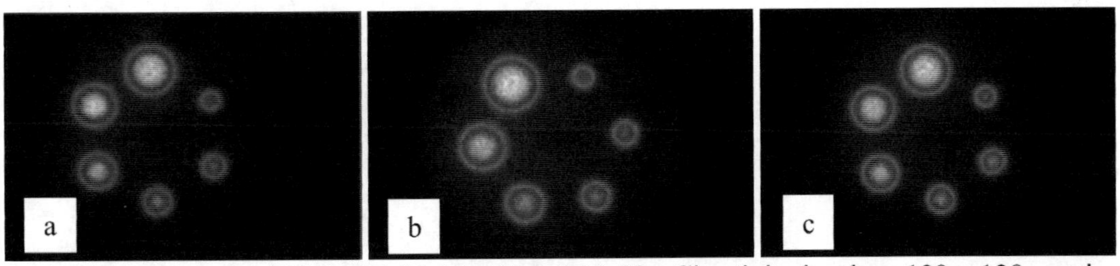

Figure 2. Planar images of (a) 3 µCi (b) 10 µCi and (c) 18 µCi activity levels at 128 × 128 matrix size.

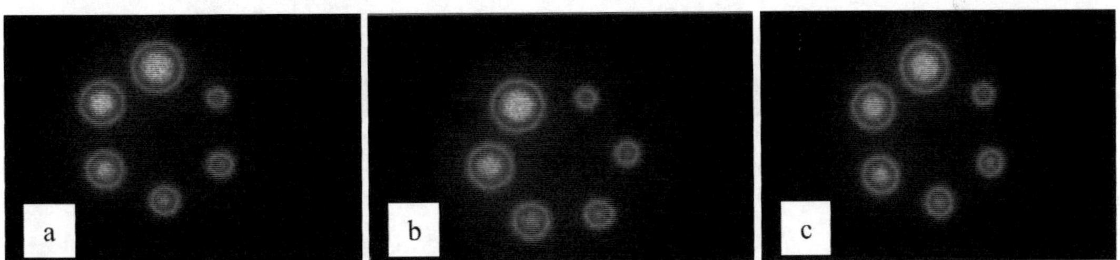

Figure 3. Planar images of (a) 3 µCi (b) 10 µCi and (c) 18 µCi activity levels at 256 × 256 matrix size.

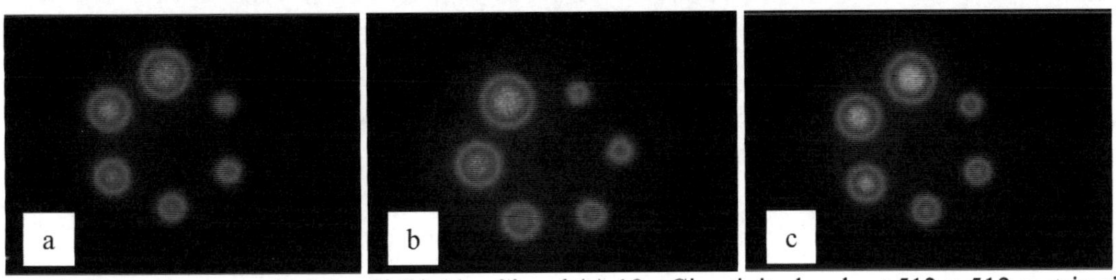

Figure 4. Planar images of (a) 3 µCi (b) 10 µCi and (c) 18 µCi activity levels at 512 × 512 matrix size.

The measured contrast values at 3 µCi, 10 µCi and 18 µCi activity levels at different matrix sizes are shown in Figure 5. The results showed that the contrast values increased at bigger sizes of radionuclide at all activity levels and matrix sizes [4,7]. The results also showed that the contrast values increased when higher activity levels were used. At low activity level (3 μCi), higher matrix sizes provided the highest contrast values whereas at high activity level (18 μCi), the lower matrix sizes provided higher contrast values [7].

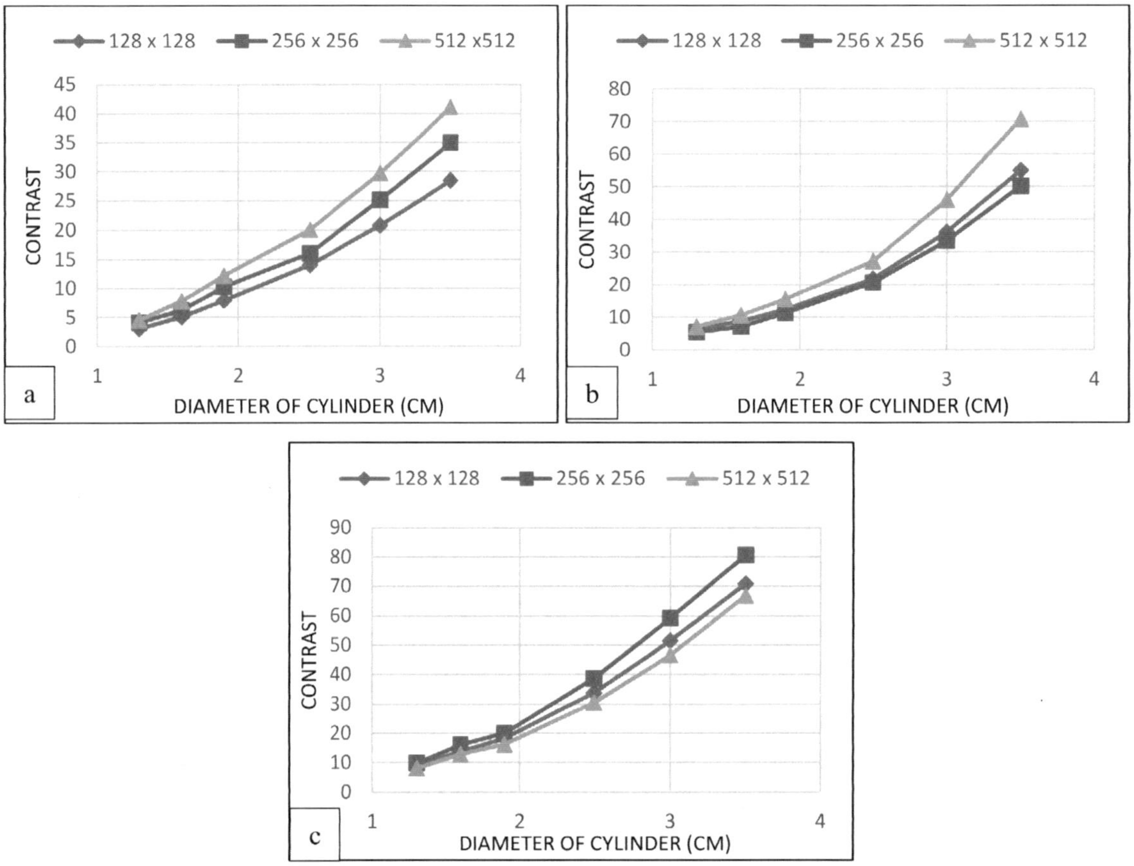

Figure 5. The contrast of planar images taken at (a) 3 (b) 10 and (c) 18 μCi by using different matrix sizes.

The correlation test between the sizes of radionuclide and matrix sizes is shown in Table 1. The correlation test showed positive relationship between diameter of cylinders and the image contrasts on different matrix sizes used indicated by the linear regression (R) values obtained from the test which were ranging from 0.971 to 0.993. The R values obtained that was close to the value of 1 indicated strong relationship between the sizes of the radionuclide and images contrast.This observation could be the result of count collected being higher at larger diameter of cylinder compared to smaller diameter of cylinder. The matrix sizes were also influenced the image [11]. The matrix size or voxel size give the significant effect on image quality and visibility of structures and object within the body. The choice of matrix size depends on several factors which are spatial resolution, smallest object of interest in image, time to perform the processing step and amount of storage available [12]. This result is in agreement with the previous work by Grotch and Erwin [13] that suggested that the ideal matrix size used should be less than 1/3 of expected full half width maximum (FWHM) resolution.

Table 1. Correlation test between contrast values at different activity levels and matrix sizes used.

Activity (μCi)	Matrix size	Mean ± SD	R value
3	128 × 128	13.21 ± 9.9	0.993
	256 × 256	16.13 ± 11.9	0.988
	512 × 512	19.23 ± 14.0	0.992
10	128 × 128	23.42 ± 18.9	0.972
	256 × 256	21.36 ± 17.5	0.971
	512 × 512	29.58 ± 24.60	0.971
18	128 × 128	32.86 ± 24.2	0.987
	256 × 256	37.37 ± 27.78	0.988
	512 × 512	30.14 ± 22.7	0.983

The recovery coefficients of the planar images at different activity levels and matrix sizes used are illustrated in Figure 6. The results showed that the activity levels and matrix sizes used in planar imaging did not significantly affect the RC pattern. In 128 × 128 matrix size, the effect of spill-over occurs when the diameter of the radionuclide exceeded 3.0 cm indicated by the RC value of more than 1. The RC in larger matrix sizes showed value of lower than 1 indicating close value between the actual and apparent activity of radionuclide [7,9].

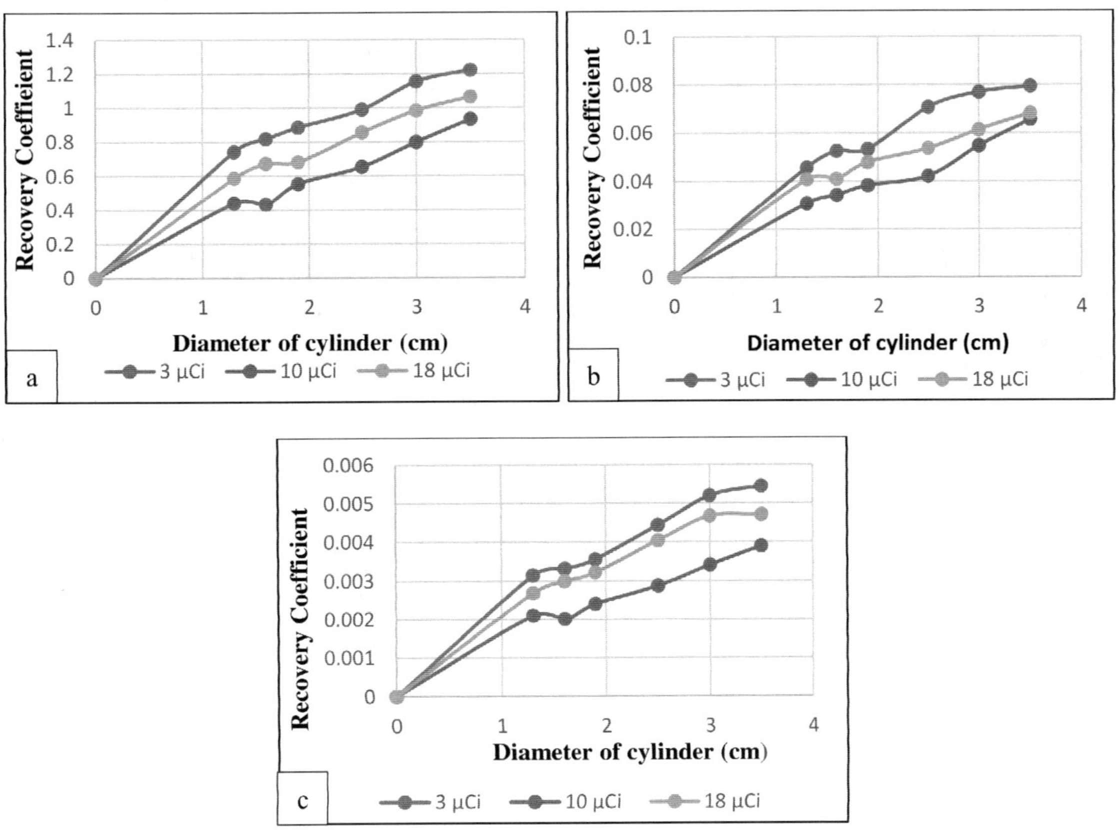

Figure 6. The recovery coefficient at different activity levels and matrix sizes of (a) 128 × 128, (b) 256 × 256 and (c) 512 × 512.

4. Conclusion

The contrast in planar images in nuclear medicine is influenced by the matrix size used rather than the activity level of the radionuclide. When lower activity levels are used, the large matrix sizes are preferred as they provided larger pixel values. The recovery coefficients in planar images did not significantly influenced by the activity levels of the radionuclide and matrix sizes.

5. References

[1] Cherry S R, Sorenson J A and Phelps M E 2012 *Phys. in Nucl. Med. 4th ed.* (Philadelphia: Saunders Elsevier).
[2] Erlandsson K, Buvat I, Pretorius P H, Thomas B A and Hutton B F 2012 A review of partial volume correction technique for emission tomography and their application in neurology, cardiology and oncology *Phys. Med. Biol.* **57** 119-159.
[3] Sjogreen K, Ljungberg M, and Strand S 2009 Parameter influencing volume and activity quantitation in SPECT. *Acta Oncol,.* **35**(3) 323-330.
[4] Ritt P, Vija H and Hornegger J 2011 Absolution quantification in SPECT. *J. Nucl. Med.* **38** 69-77.
[5] Willowson K, Bailey L, and Baldoc, C 2008 Quantitative SPECT reconstruction using CT-derived corrections. *Phys. Med. Biol.*, **53**(12) 3099–3112
[6] Saad I E, Helal N, El-din H M, and Moneam R A 2012 Evaluation of varying acquisition parameter on the image contrast in SPECT studies. *IJRRAS* **13**(2) 485-491.
[7] Yusof M F M, Ghani U S A, Ghazali N A, Khaizul A T and Idris A W 2020 Evaluation of contrast and recovery coefficients as performance parameters in planar and SPECT imaging. *IOP Conf. Series: Mater. Sci. Eng.,* **785** 012046.
[8] Hamid P N K A, Yusof M F M, Tajuddin A A, Hashim R and Zainon R 2018 Design and evaluation of corn starch-bonded *Rhizophora* spp. particleboard phantoms for SPECT/CT imaging. *IOP Conf. Series: Mater. Sci. Eng.,* **298** 012041.
[9] Hamid P N K A, Yusof M F M, Tajuddin A A, Hashim R and Zainon R 2018 Evaluation of image quality and recovery coefficient of corn starch-bonded *Rhizophora* spp. Particleboards as phantom for SPECT/CT imaging. *IOP Conf. Series: Mater. Sci. Eng.* **555** 012031.
[10] IAEA 2014 Quantitative nuclear medicine imaging concepts, requirements and methods *Human Health Reports No. 9* (Vienna: International Atomic Energy Agency).
[11] Sprawls P 2014 *Optimizing Medical Image Contrast, Detail and Noise in the Digital Era* **2**(1) 41-48.
[12] Madsen M T 1994 Computer acquisition of nuclear medicine images. *J. Med. Technol.* **22**(1) 3–12.
[13] Groch M W and Erwin W D 2000 SPECT in the year 2000: basic principles. *J. Nucl. Med. Technol.* **28**(4) 233–44.

Acknowledgement

The authors would like to thank the Nuclear Medicine, Radiotherapy and Oncology Department, Hospital University Sains Malaysia (HUSM) for the use of the gamma camera facility in this study.

Effect of electron irradiation to the photocatalytic activity of the titanium dioxide fibers

C R Che Hak[1], S Ridzuwan[2], S S Zainal Abidin[1], N U Saidin[1], S A Ahmad Fuzi[1]

[1]Material Technology Group, Industrial Technology Division, Malaysian Nuclear Agency, Bangi, 43000 Kajang, Selangor, Malaysia.
[2]Department of Physics, Faculty of Science, Universiti Putra Malaysia, 43400 Serdang Selangor, Malaysia.

rohaida@nuclearmalaysia.gov.my

Abstract. The main crystal phase of titanium dioxide (TiO_2) anatase, play the important role in the efficiency of photocatalysis. In this study, electron irradiation from 200-1,000 kGy has been used to alter the anatase-to-rutile phase ratio in TiO_2 particles. For photocatalytic reaction, the TiO_2 particles are loaded into the polymer fibers through electrospinning technique and then the ability of the fibers to photo degrade the methylene blue dye is measured. XRD analysis shows the changes in the crystal structure of the TiO_2 particles and the phase transformation from anatase to rutile after radiation. Morphology studies confirmed that the TiO_2 particles are incorporated well into the fibers and distributed homogeneously without beadings and agglomerations. The photocatalytic activity shows radiation at 1,000 kGy possess the highest degradation of methylene blue dye, resulting from the highest anatase phase content.

1. Introduction

Titanium dioxide (TiO_2) has been widely used as pigments, adsorbents, catalyst supports, filters, coatings, and dielectric materials [1]. In recent years, researchers are more focused on the use of TiO_2 semiconductors for photocatalytic activities with huge potential photovoltaic applications since TiO_2 has the properties of chemical stability, low toxicity, and low pollutant load [2]. The crystal phase of TiO_2 plays the important role in the efficiency of photocatalysis. TiO_2 exists in three different crystal forms; rutile (tetragonal), anatase (tetragonal) and brookite (tetragonal), with rutile is the most stable phase [3]. Among them, anatase phase has the most excellent photocatalytic activity compared to the others [4]. The rutile phase usually exhibits less activity than that of the anatase because of its lower surface affinity and a higher rate of recombination of photogenerated charge pairs [5]. However, the mixture of anatase and rutile TiO_2 offers a better effect on carrier transfer in photocatalytic applications, because the close contact between the anatase phase and rutile phase contributes to a smooth transfer of charge between the two phases [6]. This behavior has been shown by P25 TiO_2 powder which contains 20% rutile and 80% anatase, and it has good photocatalyst properties [7].

The photo-response and photoactivity of TiO_2 relies on the energy bandgap (E_g) of the material. TiO_2 is basically a poorly visible light absorber with E_g is around 3.2 eV, however, for an efficient photocatalytic reaction to take plase, TiO_2 requires E_g comparable to the energy of the photons of the visible light which is less than 3.0 eV. A various approach has been done to improve the photo-response and photoactivity of TiO_2. One of the well-established method is utilizing the high energy electrons to alter the E_g properties of the TiO_2. This method utilizes the accelerated electrons generated by electron beam accelerators without the use of radioactive isotope. Previous study has shown that the E_g of TiO_2 powder was altered depending on the amount of electron irradiation dose. At a low dose of irradiation (between 4- 20 kGy), the direct E_g decreases from 3.33 eV to 2.4 eV while after a high dose of irradiation exposure (between 100 – 1,000 kGy), the direct E_g increases from 3.0 eV to 4.0 eV. These results also showed that anatase transformed to rutile at high dose irradiation with the ratio of 20:80 (anatase:rutile) at dose of 700 kGy [8]. This phase transformation is occurred due to the presence of δ (O-Ti-O) bending mode of rutile in irradiated TiO_2 that led to the oxygen vacancies [9].

Besides the crystal phase and the energy band gap (E_g), the efficiency of photocatalysis is also depends on the surface properties of TiO_2. It is because the photocatalytic reaction takes place only when photoinduced electrons and holes are available on the surface [10]. Commonly, TiO_2 is used as bulk form whereby TiO_2 powder contains nanoparticles dispersed in solution to form a suspension. When used in suspension, TiO_2 aggregates rapidly due to its small size (4 nm to 30 nm) and its aggregates may cause scattering of the light beam resulting in loss of catalytic efficiency [11]. In addition, it is a must to separate the powdered TiO_2 after the photocatalytic reaction is completed to recover the catalysts. This post treatment separation is normally difficult and energy consuming, thus it is impractical for use in water-treatment plants [12]. To solve these problems, the TiO_2 particles were immobilised on solid supports to ease the post-treatment problems and to facilitate the renewable use of the catalyst [11]. To immobilize the TiO_2 particles, it can be assembled onto different substrates and fabricated into different types of thin films [12,13].

In this study, TiO_2 was bounded in a polymer fibers which is prepared via electrospinning technique, to prevent its particles aggregated. The advantage of fibrous photocatalyst is to have a high surface-to-volume ratio and 3-dimensional open structure that provides more surface actives for photocatalytic reaction to occur [14].. Recently, Riaz *et al.* has proved that the TiO_2 nanofibers calcined at 650°C, with the anatase/rutile phase ratio of 83:17 exhibited the best photocatalytic performance under UV irradiation [15]. Ismaya *et al.* found that the addition of TiO_2 nanoparticles in the polymer solution decreases the solution viscosity and increases the conductivity of the fibers. The decrease in fiber diameter will help in the increment of the active site thus increase the photocatalytic ability of the fibers [16]. Nevertheless, Ismaya *et al.* only considered the relationship between the ratio of anatase:rutile and photocatalytic activity, but not focus on the distribution of TiO_2 on the surface of the fiber. In this work, we prepared the polymer fibers using electrospinning method loaded with TiO_2 nanoparticles with different anatase:rutile ratios. The objective of the work is to study the photocatalytic reactions of the TiO_2 fibers and discussed the performance with regards to the crystal phase structure of TiO_2 adjusted by electron irradiation.

2. Materials and methods

In this study, the readily synthesized TiO_2 particles were irradiated with a 3 MeV electron beam (EPS 3000) at a dose in the range of 0 kGy to 1,000 kGy. After irradiation, the TiO_2 powder was characterized using the X-ray diffractometer (XRD, XPert Pro MPD, Ag-Kα irradiation, λ= 0.561 Å) for phase and crystallinity studies.

2.1. Preparation of titania fibers by electrospinning

The TiO_2 fibers were prepared using the electrospinning technique. A known amount of irradiated TiO_2 powder and surfactant were mixed together and stirred in 3 ml of deionized water before being ultra-sonicated for 3 hours. After that, PVA powder was added into the sonicated solution and stirred for 20 minutes at 80 °C yielding a milky viscous solution. The solution was transferred to the plastic syringe and the needle was connected to a high direct current (d.c) voltage at 15 kV and ground to a metal plate, covered with aluminum foil. The electrospinning was done at room temperature and the TiO_2 fibers were collected on the aluminum foil at the working distance of 14 cm. The rate of the spinning was 0.03 ml/h. The morphology of the fibers were then characterized using field emission scanning electron microscope (FESEM, GeminiSEM 500, Carl Zeiss) and the titanium (Ti) element was determined using energy dispersive X-ray (EDX, X-Max 80, Oxford Instrument) that attached to the same microscope.

2.2. *Experiment for photocatalytic activity of titania fibers*

Methylene blue (MB) dye was used as a reference to the organic pollutant to study the photocatalytic activity of TiO_2 fibers. Prior to the photocatalytic experiment, the calibration curve for MB dye solution with concentration of 1 ppm to 5 ppm was generated and the R2 obtained was 0.996. A 12.25 cm^2 sheet containing TiO_2 fibers was placed flat in a 100 mL beaker and each beaker was filled with 50 mL of MB solution of 2.5 ppm. The beakers were then placed 12 cm away from the UV light source (400 watts) in a black box. During the experiment, a sample solution of 5 mL was taken at interval of 30 minutes for 2 hours and every hour for 5 hours. The concentration of the MB dye solution was measured with the UV-Vis spectrometer (Lambda 35, Perkin Elmer).

3. Results

3.1. *Phase characterization*

The XRD was used to study the phases present in TiO_2 powder, PVA fibers and TiO_2 loaded PVA fibers (TiO_2 fibers). Figure 1(a) shows the diffraction patterns for TiO_2 powder before and after electron irradiation. In the TiO_2 powder, there are two phases present which are anatase (ICSD: 01-070-6826) and rutile (ICSD: 01-083-2242). The peaks at $2\theta = 9.077°$, $2\theta = 13.545°$, $2\theta = 16.970°$, $2\theta = 19.129°$, and $2\theta = 21.735°$ correspond to the anatase phase. An intense peak at $2\theta = 15.704°$ corresponds to the rutile phase for non-irradiated TiO_2 powder. As the irradiation dose increased to 200 kGy, all peak intensity reduces and become stable up to 800 kGy irradiation. The increase in the irradiation dose from 200 kGy to 800 kGy does not give in a significant change towards the number of peaks and phase present. Figure 1(b) shows the comparison of diffraction patterns for TiO_2 fibers and PVA fibers respectively. For TiO_2 fibers, the peaks are identified at $2\theta = 25.8°$, 38.15°, 48.53°, and 55.13° which can be indexed to (011), (020), (015), and (024) plane respectively. This pattern is identified as anatase phase (ICSD: 01-070-6826) that has a tetragonal structure. The broad peak at $2\theta = 25.8°$ is also observed on the pure PVA fibers, indicates that there is texturization occurred during electrospinning process [17]. Texturization is a mechanism of PVA changed its crystal structure to a preferred orientation to accommodate the presence of TiO_2 and surfactant during electrospinning process [18].

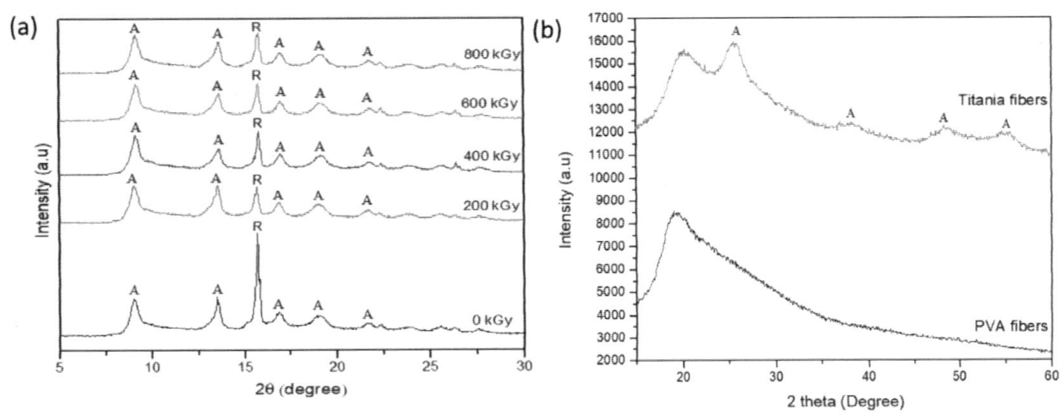

(a) (b)

Figure 1. (a) XRD spectra for TiO_2 powder irradiated with electron at doses of 0 kGy to 800 kGy; (b) XRD spectra for pure PVA fiber and TiO_2 loaded PVA fiber.

From XRD data, the crystallite size of the phase present was estimated using the Debye-Scherrer equation given by Equation 1:

$$t = 0.9\lambda/\beta \cos\theta \quad (1)$$

where λ is the X-ray wavelength, β is the full-width half maximum (FWHM) of the peak and θ is the Bragg's angle.

The crystallite size of the phase present is as tabulated in Table 1. Based on the calculated crystallite size, for the rutile phase, it can be observed that the crystallite size decreases for 0 kGy to 400 kGy samples and increases for samples of more than 400 kGy and the opposite occur for anatase phase. There is only a small difference in the ratio between anatase and rutile phase for different irradiation dose samples. The reduction in crystallite size by electron beam irradiation process plays an important role as it can allow the TiO_2 to be incorporated well in fibers and provide a larger surface area to volume ratio for the photocatalytic reaction to occur [19].

Table 1. The crystallite size of anatase and rutile phase irradiation process with different doses.

Dose (kGy)	Crystallite size (nm)		Anatase:Rutile
	Anatase	Rutile	
0	9.4	45.5	68:32
200	11.0	36.4	65:35
400	10.6	25.9	63:37
600	9.1	30.3	69:31
800	7.9	38.5	76:24
1,000	12.1	17.0	80:20*

* value obtained from Rohaida C.H C *et al.* 2019 *AIP Conf. Proc.* 2068 020010 [9].

3.2 Morphology studies and elemental analysis

The morphology of the fibers is observed using FESEM. Figure 2(a) and 2(b) shows the FESEM images of a pure PVA fiber and TiO_2 fiber respectively. Both fibers are randomly oriented with no beadings are visible. Morphology of the pure PVA fiber is clear and smooth surface and the average diameter is 197 nm. Meanwhile, morphology of TiO_2 loaded fibers is uneven surface, and the average of its diameter is 393 nm, about double the size of PVA pure fibers. The presence of TiO_2 particles in the PVA solution is believed to increase the viscosity and the conductivity of the solution, thus produce larger size fibers [21]. The good thing is surface of the fiber does not show any agglomeration of TiO_2 particles indicating the nanoparticles were well dispersed on the fibers. Figure 2(c) shows at 100,000 times magnification, the surface looks porous and homogeneous, indicating the well dispersion of TiO_2 nanoparticles on the fibers' surface.

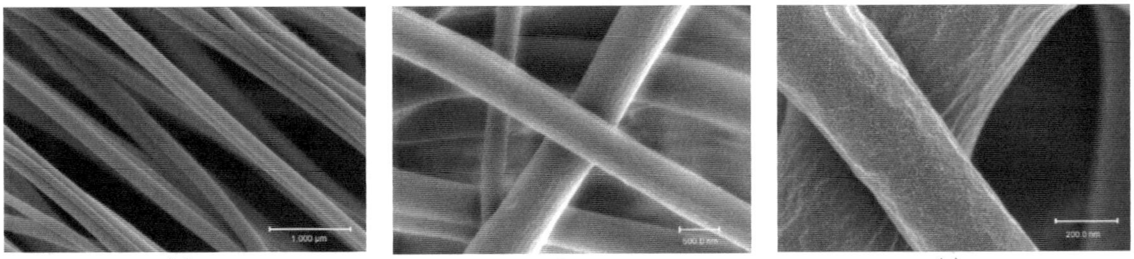

(a) (b) (c)

Figure 2. FESEM images (a) Pure PVA fibers, Scale: 1.0 µm, (b) TiO_2 loaded fibers, Scale: 500 nm, (c) TiO_2 loaded fibers, Scale: 200 nm.

Spot elemental identification using EDX was performed on the fibers to identify the presence of titanium (Ti) and other elements and to determine their concentration (weight percent, wt%) throughout the fibers. Figure 3(a) illustrates the 2 different spots were chosen to identify the concentration (wt%) of Ti; spot 30 and 32 which represents location on the fiber and outside of the fiber respectively. From the EDX spectra shown in Figure 3(b) and (c), the concentration of Ti is determined at 16.23 wt% at the outside of the fiber and 6.22 wt% on the fiber itself. This indicates that not all TiO_2 nanoparticles is incorporated into fiber during electrospinning. The free particles are restored on the surrounding of the fibers. To check the distribution of TiO_2 particles on the surface of fibers, spot elemental identification was carried out at 4 different spots (spot 1,2,3 and 4) on one of the fiber as shown in Figure 3(d). Graph in Figure 3(e) shows there are 4 elements presence on the fibers' surface; carbon (C), oxygen (O), titanium (Ti) and sulfur (S). Sulfur is basically from the surfactant. The graph indicates that the concentration of each element at the 4 different spots. The wt% of Ti in each spot is in the range of 15.26 to 18.08 wt%, almost identical to each point. These results proved that the TiO_2 particles are distributed homogeneously throughout the fibers.

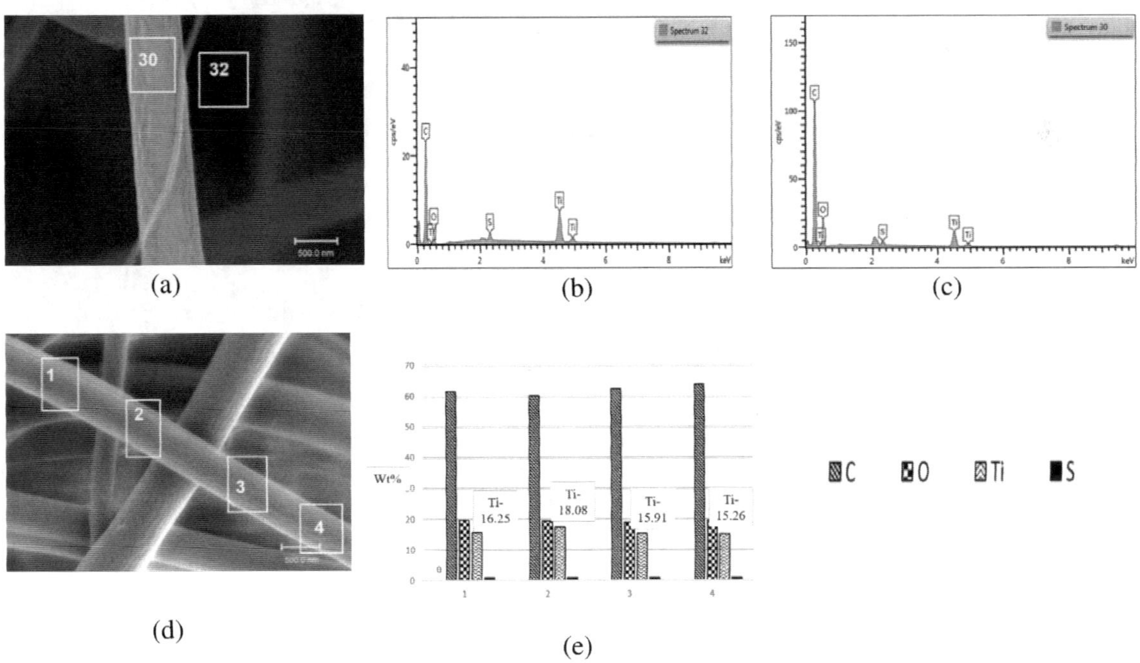

(a) (b) (c)

(d) (e)

Figure 3. EDX results (a) elemental analysis at spot 30 & 32, Scale: 500 nm, (b) Spectrum spot 32, Ti = 16.23 wt% (c) Spectrum spot 30, Ti = 6.22 wt%, (d) elemental analysis at spot 1 - 4, Scale: 500 nm.

3.3 *Photocatalytic activity of titania fibers*

The concentration of the methylene blue (MB) dye after the photocatalytic reaction was determined using absorbance measurement with UV-Vis spectrometer. Figure 4(a) and 4(b) show the concentration of MB dye decrease significantly over time in all fibers for 2 and 5 hours respectively. Figure 4(b) shows that the decrement is very rapid for the first one hour and then gradually decrease afterwards. This result is similar to the one that reported previously whereby a rapid increase in photocatalytic degradation of methylene blue in the first 30 minutes of the experiment and becomes slower after the next 30 minutes [22]. Figure 4(c) shows that the fibers loaded with 1,000 kGy irradiated TiO_2 exhibited the highest percentage of dye degradation which is 78.4% from the initial concentration after 2 hours. This is followed with fibers loaded with 800 kGy irradiated TiO_2 which shows degradation of 77.4 %. For fibers loaded with TiO_2 irradiated at 200 - 600 kGy, the percentage of degradation is about the same to the non-irradiated TiO_2 loaded fibers. This means that, TiO_2 irradiated at above 800 kGy provide improved photocatalytic reaction than the others, which corresponds to the crystal phase of the TiO_2 after irradiation as shown in Table 1. At 800 and 1,000 kGy of irradiation dose, the ratio of anatase:rutile in TiO_2 is 76:24 and 80:20 respectively, indicating that TiO_2 with the highest anatase content possessed the highest degradation of MB dye. These results proved that the anatase:rutile phase ratio is important for photocatalytic performance as reported by previous researches that the mixed-phase TiO_2 with a higher amount of anatase exhibit a higher photocatalytic reaction than the individual phase either rutile or anatase [23].

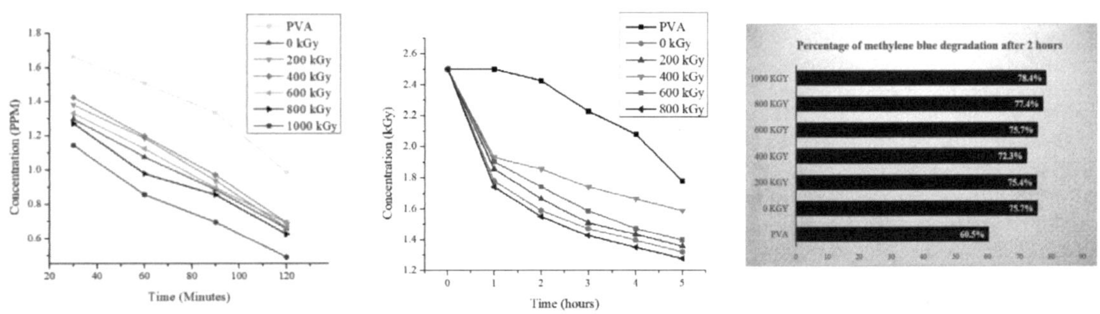

(a) (b) (c)

Figure 4. The photocatalytic reaction of fibers (a) the degradation of MB dye solution in 2 hours, (b) the degradation of MB dye solution in 5 hours, (c) the percentage of MB dye degradation after 2 hours.

4. Conclusion

The electron irradiation on TiO_2 powder from 200 to 1,000 kGy dose leads to the changes in the crystallite size as well as the anatase-to-rutile phase ratio. TiO_2 particles are successfully incorporated into PVA fibers. The irradiation dose does affect the photocatalytic activity of TiO_2. 1,000 kGy dose exhibit the highest degradation of MB dye which is nearly 80% in 5 hours of immersion time, attributed by the highest anatase phase content in this sample. Therefore, the amount of electron irradiation dose modifies the crystallite size of TiO_2, hence change the anatase-to-rutile phase and as a result affect the photocatalytic reactions when exposed to UV light.

5. References

[1] Thamaphat K, Limsuwan P and Ngotawornchai B 2008 *J. Nat. Sci.* **42** 357.
[2] Zhang Q, Gao L and Guo J 2000 *J. of the Eur. Ceramic Soc.* **20**(2) 2153.
[3] Fox M A and Dulay M T 1993 *Chem Rev.* **93** 341.
[4] Luttrell T, Halpegamage S, Tao J, Kramer A, Sutter E and Batzill M 2014 *Sci. Rep.* **4** 404.
[5] Zhang J, Zhou P, Liu J and Yu J 2014 *Phys. Chem.* **16** 20382.

[6] Mutuma B K, Shao G N, Kim W D and Kim H T 2015 *J. Colloid Interface Sci.* **442** 1.
[7] Rui Z, Wu S, Peng C and Ji H 2014 *Chem. Eng. J.* **243** 254.
[8] Rohaida C H C, Salwa Z A S, Ubaidah S N, Aishah A F S, Yusof A and Yusof M S M 2017 Materials Science Forum **888** 297.
[9] Rohaida C H C, Aishah A F S and Yusof A 2019 *AIP Conf. Proc.* **2068** 020010.
[10] Bai S, Jiang W, Li Z and Xiong Y 2015 *Chem Nano Mat* **1** 223.
[11] Bhattacharyya S K and Ray M B 2004 *Catalysis Today* **98**(3) 431.
[12] Nawi M A, Jawad A H, Sabar S and Ngah W S W 2011 Desalination **280**(1-3) 288.
[13] Zhou X S, Li LJ, Lin Y H and Nan C W 2008 *J. of Electroceramics* **21**(1–4) 795
[14] Su C, Shao C and Liu Y 2011 *J. Colloid Interf. Sci.* **359** 220.
[15] Riaz A, Qi H, Fang Y, Xu J and Zhou C *et al.* 2006 *J. Mater. Res.* **31** 3036
[16] Nasikhudin I, Ismaya E P, Diantoro M, Kusumaatmaja A and Triyana K 2017 *IOP Conf. Ser.: Mater. Sci. Eng.* **202** 012011.
[17] Costa R G, Ribeiro C and Mattoso L H 2010 *J. of Nanosci. and Nanotechnol.* **10**(8) 5144.
[18] Ojha S 2017 *Electrospun Nanofibers* **1** 239.
[19] Li W, Ni C 2004 *J of Appl. Phys.* **96** 6663.
[20] Reddy K M, Manorama S V and Reddy A R 2002 *Mater. Chem. and Phys.* **78** 239
[21] Tyas Ari W W, Kusumaatmaja, A, Roto C and Triyana K 2016 AIP Conf. Proc. 150010 1755
[22] C Ramli, Asim Z A, Isahak N, Emdadi W N, Z Ahmad Ludin, N Yarmo M A 2014 *The Scientific World J.* **1.**
[23] Zhang H, Yu M and Qin X 2019 *Nanomaterials* **9**(4) 535.

Acknowledgments
Authors would like to thank Alurtron unit for providing the beam time for electron irradiation, to Universiti Putra Malaysia for allowing their student to perform internship at our laboratory and to Agensi Nuklear Malaysia for supporting this research work.

Analysis of radioactivity concentration in natural rubber latex and radiation prevulcanized natural rubber latex (RVNRL)

Sofian Ibrahim[1,a)], Nurrul Assyikeen Md. Jaffary[1], Norfaizal Mohamed[1] and Hasni M. Ali[1]

[a]*Malaysian Nuclear Agency, 43000 Kajang, Selangor, Malaysia.*

[a]*sofian_ibrahim@nm.gov.my*

Abstract. Radiation vulcanization of natural rubber latex (RVNRL) is an alternative technology to conventional sulphur vulcanization. To eliminate some public members sceptical view toward the radioactivity of the prevulcanized latex produced using gamma radiation, RVNRL was prepared at several radiation doses and then their radioactivity concentration were measured using gamma counting technique. The activity concentration level of Ra-226, Th-232, and K-40 in RVNRL found to be not significantly different from non-vulcanized natural rubber latex (NRL). The reported radioactivity proved that using gamma radiation in the RVNRL preparation could not ionized and produce radioactive materials from NRL minerals. Hence, it is concluded that RVNRL is safe to be used as the main raw material in latex dipped products industry.

1. Introduction

Natural rubber latex (NRL) is a non-elastic sticky milky fluid that was harvested from *Hevea brasiliensis.* It has a spherical shape with a diameter between 0.02 - 3 μm, and protected by a layer of the complex film consisting of proteins and lipids as shown in Figure 1. The rubber particles contain almost 99 % of the rubber hydrocarbon chain with the *Cis*-1,4-Polyisoprene isomer [1].

To improve its elasticity and thermal stability, the NRL will undergo the vulcanization process. At present, three popular vulcanization processes are being used in NRL industries, namely sulphur, radiation and peroxide vulcanizations.

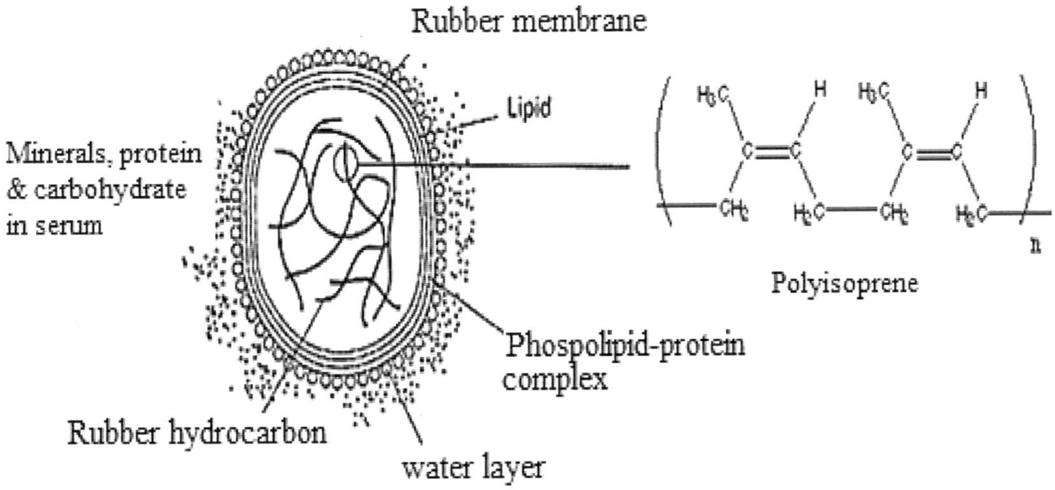

Figure 1. Structure of natural rubber latex [2].

The production of prevulcanized natural rubber latex by radiation technology is not new in Malaysia. It was first introduced in the mid-1990s by the Malaysian Nuclear Agency. Radiation vulcanization is an alternative technology to the conventional crosslinking process by using sulphur and its other chemicals ingredient. The use of radiation do not only produced radiation prevulcanized natural rubber latex (RVNRL) that can be used for the production of nitrosamines free products, it also able to excludes type I and IV allergies that caused by chemicals and high protein content in the products i.e. glove, finger coat, balloon, condom etc. [1].

When it comes to the use of nuclear technology in the industries, some public members are very sceptical about the products produced. Some of them though the products will become radioactive and emit radiation that might give health risk. The though is certainly not true because the products exposed to radiation for sterilization, decontaminations and crosslinking purpose, will not turn into radioactive material. In addition, the public are unaware that natural radiation is also present around them at background level i.e. soil and food (Table 1 and Table 2). Study on the acceptance of irradiated products like food has been studied by [4]. Most of the respondent has a neophobia and negative though on the irradiated food and had low intent to purchase irradiated food. In the field of medical, the used of neutron by the source of boron, ^{10}B (n,α) ^{7}Li to treat cancer has showed that the patients became radioactive and their ^{24}Na in blood can be detected via a Na-I scintillation counter [5]. Society's perception should not be underestimated, and therefore, the radioactivity level and concentrations for Radium-226 (Ra-226), Thorium-232 (Th-232) and Potassium-40 (K-40) will be investigated and measured to fend off such misconceptions.

Table 1: Comparison of radioactivity concentration for of Ra-226, Th-232, and K-40 in soil

Location	Level	Activity concentration (Bq/kg)			References
		Ra-226	Th-232	40K	
Malaysia	Minimum	38	63	170	UNSCEAR 2000
	Maximum	94	110	430	
	Average	67	82	310	
Thailand	Minimum	11	7	7	UNSCEAR 2000
	Maximum	78	120	712	
	Average	48	51	230	
China	Minimum	2	1	9	UNSCEAR 2000
	Maximum	440	360	1800	
	Average	32	41	440	
Japan	Minimum	6	2	15	UNSCEAR 2000
	Maximum	98	88	990	
	Average	33	28	310	
India	Minimum	7	14	38	UNSCEAR 2000
	Maximum	81	160	760	
	Average	29	64	400	
Worldwide	Average	35	30	400	UNSCEAR 2000

Recently, Toyen *et al.* [1] from Thailand reported that the radioactivity concentrations for Ra-226, Th-232 and K-40 in the Thailand's NRL were 29.9 ± 1.1, 95.2 ± 1.2 and 96.2 ± 1.0 Bacquerel/kilogram (Bq/kg) respectively (Table 2). The reported value proved that natural radionuclide present in NRL, even in small concentration. As for Malaysia, based on the UNSCEAR report (2000), the average of natural radioactivity concentrations for Ra-226, Th-232 and K-40 in soil were 67, 82 and 310 Bq / kg respectively as showed in Table 1. We are confident that the rubber trees and NRL produced in Malaysia also consist of radioisotopes with specific radioactivity concentrations. Hence, for this study, radioactivity level in raw NRL produced in Malaysia (Jelebu, Negeri Sembilan) were measured and RVNRL prepared at various radiation doses will be analyzed using the gamma counting technique

Table 2. Radioactivity concentration of Ra-226, Th-232, and K-40 in agriculture/crops/commercial products.

Type of agriculture/crops/ commercial products	Level	Activity concentration (Bq/kg)			Location	References
		Ra-226	Th-232	40K		
Rice (originated from cultivated fields tilled with phosphogypsum)	Minimum Maximum Average	0.36 1.98 1.53			Thessaloniki, Northern Greece	(Papastefanou et al., 2006)
Pasta	Average	0.068	0.078	0.049		(Pearson et al., 2016)
Vegetables	Minimum Maximum Average	46.05 68.83,	9.29 50.57	325.89 530.52	Turkey	(Bolca et al., 2007)
Cosmetic products (natural clay based)	Minimum Maximum Average	57.8 168.0 115.6	34.8 75.5 49.7	996.3 210.5 658.4	Greek	(Papadopoulos et al., 2014)
Tobacco	Average	4.56	3.94	1289.53	Egypt	(Shyti et al., 2019)
Natural rubber latex	Average	29.9	95.2	96.2	Thailand	Toyen & Saenboonruang (2020)
RVNRL (prepared at 12 kGy)	Average	30.8	89.1	95.4	Thailand	Toyen & Saenboonruang (2020)
RVNRL (prepared at 24 kGy)	Average	31.2	89.7	93.2	Thailand	Toyen & Saenboonruang (2020)

2. Materials and methods

2.1. Materials

The latex utilized in this work is a high ammonia type of latex (HA latex) supplied by Titi Latex Sdn Bhd, Malaysia. The sensitizer used were hexanediol diacrylate (HDDA) supplied by Allnex, China. The stabilizer used was potassium laurate supplied by Tiarco Chemical (M) Pt. Ltd., Malaysia and the antioxidant used was Aquanox Lp supplied by Aquaspersion (M) Pt. Ltd., Malaysia. These materials were used as received.

2.2. Preparation of RVNRL compounding formulations

A typical latex compounding formulation for RVNRL preparation is given in Table 3. The sensitizer, stabilizer, antioxidant, and water homogenized were used to form an emulsion before it was added into the latex and was stirred gently. Once the addition of the emulsified materials completed, the latex mixture was stirred for three hours [7]. It was then transferred into a one-liter screw-capped plastic container and irradiated with gamma rays produced by Cobalt-60 source at a dose of 10, 20, and 30 kGy. After 6 hours of irradiation, the RVNRL was formed into a film by the coagulant dipping method.

Table 3. Compounding formulation of standard RVNRL.

Materials	Part per hundred (phr)
NR Latex (62% Total Solid Content)	100
Stabilizer	0.06
HDDA	2.50
Antioxidant	2.50
Water	Add to 52% Total Solid Content

2.3. Measurement of radioactivities

The radioactivities measurement for Non-vulcanized natural rubber latex (control) and RVNRL were measured by using gamma counting technique. The rubber films stacks in 350ml screw-cap geometry container were sealed, and incubated for 21 days. This is to allow Ra-226 decay to Rn-222 and achieve secular equilibrium. The entire sample spectrum individually measured with the well-calibrated gamma ray spectrometry consisting of a high-purity germanium (HPGe) setup and multichannel analyzer of 16,384 channels. The detector used is a coaxial 3 inches diameter closed end, closed facing window geometry with vertical dipstick (500 – 800 um) operated at 3,500 HV bias supply [8].

The activities of Ra-226 were calculated through its progeny energy peaks i.e. Pb-214 and Bi-214, for Th-232 (Ra-228) were calculated through Ac-228 peak energy, while K-40 was calculated *via* directly its energy peak [9]. All radioactivity calculations applied for density corrections [10].

3. Results and discussion

The radioactivities for Ra-226, Th-232, and K-40 in both non-vulcanized NRL and RVNRL were presented in Table 4. The activity concentration of Ra-226, Th-232 and K-40 from NRL produced at Jelebu, Negeri Sembilan found to be far lower compared to NRL planted in Thailand (Table 2) [6]. This varies happened due to geographical differences (minerals in soil) and tree clones. Aside from that, it is noticeable the presence of relatively high K-40 in NRL and RVNRL. As expected, potassium is the essential element and the most abundant inorganic mineral nutrient in plant cells due to its functions as a stabilizer in metabolism and an osmoticum contributing to cellular hydrostatic pressure and growth [11].

According to Makuuci 2003, the radiation energy more than 10 Mega electron volt (MeV) are needed to activate materials to become radioactive, while in industrial applications typically used energies from 100 Kilo electron volt (KeV) up to 1 MeV . In the case of gamma energy used to vulcanized NRL, 1.17 MeV are used [12]. Data obtained in Table 4 showed the radioactivity concentration in all samples relatively unchanged, proved that radiation vulcanization technique does not increase the radioactive nuclide of NRL minerals. This finding has proven that in terms of radiological safety and radioactivities concentration, RVNRL is safe to be use as main raw material in latex dipped products industry.

Table 4: Radioactivities of Ra-226, Ra-228 and K-40 in NRL and RVNRL.

Sample	Radioactivity level (Bq/kg)		
	Ra-226	Th-232	K-40
NRL (control)	< 0.50	1.10±0.08	34±11
RVNRL 10 kGy	< 0.50	1.34±0.10	36±12
RVNRL 20 kGy	< 0.50	2.98±0.22	29±9.8
RVNRL 30 kGy	< 0.50	1.03±0.08	33±11

4. Conclusion

The radioactivity concentration level of Ra-226, Th-232, and K-40 in RVNRL prepared at a radiation dose of 10, 20 and 30 kGy is not significantly different from non-vulcanized NRL. This finding proved that using gamma radiation to prepare RVNRL will not ionize and produce radioactive materials from NRL minerals. Hence, we conclude that RVNRL is safe for the user as the main raw material in latex dipped products industry.

5. References

[1] Ma'zam M S 2005 Prevulcanized latices: course on latex concentrate production, prevul canized and other modified lattices (Malaysia: Lembaga Getah Malaysia).

[2] Sabharwal S, Das T N, Chaudhari Y K, Bhardwaj Y K and Majali A B 1998 Mechanism of n-butyl acrylate sensitization action in radiation vulcanization of natural rubber lat ex. *Radiat. Phys. and Chem.* **51**(3) 309-315.

[3] Akademi Hevea Malaysia 2012 Kursus ujian lateks *Lecture note of Short Course on Lat ex Testing 25-28 Jun 2012* (Lembaga Getah Malaysia).

[4] Filho L, Lucia T Della, Lima R M 2017 Thoughts, attitudes and profile of Brazilian con sumers regarding food irradiation *Int. J. of Consumer Studies* **41**(5) 518–525.

[5] Fujiwara K, ;Kinashi Y; Takahashi T; Yashima H; Kurihara K; Sakurai Y; Tanaka H, Takahashi S 2013 Induced radioactivity in the blood of cancer patients following bor on neutron capture therapy *J. of Radiat. Res.* **54**(4) 769–774.

[6] Toyen D and Saenboonruang K 2020 Determination of radioactivities in gamma vulcani zed natural rubber latex (GVNRL) for the assessment of radiological safety. IOP Con f. Series: Materials Science and Engineering **773** 012011.

[7] Sofian I, Khairiah B, Chantara T R and Ali N H M 2018 Enhancing mechanical properti es of prevulcanized natural rubber latex via hybrid radiation and peroxidation vulcani zations at various irradiation. *Radiat. Effects and Defects in Solids* **174** 223-233.

[8] Yii M W; Assyikeen M. J N, Zaharudin A 2011 Radiation hazard from natural radioacti vity in the sediment of the east coast peninsular Malaysia exclusive economic zone (E EZ). *The Malaysian J. of Analyt. Sci.* **15**(2) 202-212.

[9] Yang Y X ; Wu X M; Jiang Z Y; Wang W X; Lu J G; Lin J; Wang L.M. and Hsia Y F 2005 Radioactivity concentrations in soils of the Xiazhuang granite area, China *Appl. Radiat. Isot.* **63** 255-259.

[10] Yii M W and U'yun W M Z 2016 Ambient radioactivity and radiological studies in the vicinity of Lynas Rare-Earth Processing Plant, Gebeng Industrial Estate, Kuantan, Pa hang *Technical Report ScienceFund Project Nuclearmalaysia/L/2016/96.*
[11] Dreyer I, Uozumi N2011 Potassium channels in plant cells *FEBS J.* **278** 4293-4303.
[12] IAEA 2005 <http://www-naweb.iaea.org/napc/iachem/Brochgammairradd.pdf>. [16 Jul y 2020].
[13] Bolca M; Saç M M; Çokuysal B; Karali T; Ekdal E 2007 Radioactivity in soils and vari ous foodstuffs from the Gediz River Basin of Turkey *Radiat. Measur.* **42**(2) 263–270.
[14] El-Reefy HI, Sharshar T, Zaghloul R and Badran H M 2006 Distribution of gamma-ray emitting radionuclides in the environment of Burullus Lake: I. Soils and vegetation, *J. Environ. Radioact.* **87** 148-169.
[15] Makuuci K 2003 *An Introduction to Radiation Vulcanization of Natural Rubber Latex* (Bangkok: T RI Global Co. Ltd.).
[16] Pairu I; Manshol W Z W; Chai C K, Sofian I, Saadiah S, Sharifah I 2017 Safety evaluation test on human for radiation prevulcanised natural rubber latex (RVNRL). . *Sains Nuklear Malaysia.* **29**(1) 28-36.
[17] Papadopoulos A; Giouri K; Tzamos E; Filippidis A, Stoulos S 2014 Natural radioactivit y and trace element composition of natural clays used as cosmetic products in the Gre ek market *Clay Minerals* **49**(1) 53–62.
[18] Papastefanou C; Stoulos S; Ioannidou A, Manolopoulou M 2006 The application of phosphogypsum in agriculture and the radiological impact *J. of Environ. Radioact. 89* (2) 188–198.
[19] Pearson A J; Gaw S; Hermanspahn N; Glover C N 2016 Natural and anthropogenic radi onuclide activity concentrations in the New Zealand diet *J. of Environ. Radioact.* **151** 601–608.
[20] Shyti M; Cfarku F; Bërdufi I; Bylyku E, Hoxha R 2019 Natural radioactivity in cigarette tobacco and radiation dose induced from smoking. *AIP Conf. Proc.* **2075**(1) 91–95.
[21] United Nations Scientific Committee on the Effects of Atomic Radiation Report 2000 <https://www.unscear.org/docs/publications/2000/UNSCEAR_2000_Report_Vol.I.p df.> [1 July 2020].
[22] Yii M W, Lorensu H J K and Asri R Mohd 2019 Radionuclide concentration in surface soil of Nuclear Malaysia's Bangi complex and its potential radiation risk *Nuclear Technical Convention 2019* (Bangi, Selangor, Malaysi: Agensi Nuklear Malaysia).

Acknowledgments
Sincere words of thanks to Malaysian Nuclear Agency and Titi Latex Sdn Bhd, Malaysia, for their cooperative effort in making this project successful.

IOP Conf. Series: Materials Science and Engineering 1106 (2021) 012018 doi:10.1088/1757-899X/1106/1/012018

Evaluation of radium and caesium containment performance of palm oil fuel ash-supplemented cementitious backfill for borehole disposal: Kinetic investigations

E Phillip[1,2*], K S Khoo[1], M A W Yusof[2] and R O Abdel Rahman[3]

[1]Department of Applied Physics, Faculty of Science and Technology, National University of Malaysia (UKM), Bangi, Selangor, Malaysia
[2]Nuclear Malaysia, Bangi, Selangor, Malaysia
[3]Hot Lab. Center, Atomic Energy Authority of Egypt, P.O. No. 13759, Cairo, Egypt

*esther@nuclearmalaysia.gov.my

Abstract. Radium and caesium have been identified among the major radionuclides found in disused sealed radioactive sources (DSRSs) and these DSRSs are potentially planned to be disposed in borehole disposal. In borehole disposal, cementitious backfill is commonly used as engineered barrier for containment of radionuclides. Present study evaluates the potential use of palm oil fuel ash (POFA)-supplemented cementitious backfill to contain radium and caesium radionuclides. In this respect, kinetic investigations were carried out to understand the nature of the sorption process onto both Ordinary Portland Cement (OPC) and POFA-supplemented cementitious material. The results showed that the sorption process for both cementitious backfill materials reached equilibrium within 48 hours and 24 hours for radium and caesium, respectively. For both backfill materials, results from kinetic analysis of radium and caesium indicated that the sorption process followed pseudo-second order model thus suggesting chemisorption reaction. Meanwhile, pore diffusion and film diffusion were identified as the main mechanisms controlling the sorption process. It was also found that the sorption capacity was higher for POFA-supplemented cementitious material (6 – 60 Bq g-1) compared to OPC (1 – 44 Bq g^{-1}) indicating valuable containment of radium and caesium by POFA-supplemented cementitious backfill.

1. Introduction

Sealed radioactive sources (SRSs) are widely used in numerous applications particularly in medicine, industry and research. In many cases, once the sources decay to a level at which they are no longer applicable, or are damaged to a point where they no longer able to serve their purposes these sources are deemed disused and referred to as disused sealed radioactive sources (DSRSs). Nevertheless, these DSRSs may still potentially caused hazards to humans and the environment thus requires safe and proper management. DSRSs contain a wide array of radionuclides with a broad spectrum of activity and are divided into different categories. For example, categories 3-5 DSRSs include but not limited to industrial gauges, smoke detectors, lightning conductors and Ra needles containing Am-241, Ra-226, Cs-137, Sr-90, Co-60 [13]. Short-lived radionuclides are easily managed by utilizing decay storage meanwhile longer-lived radionuclides in particular Ra-226 require long-term management such as by disposal. Realizing the need for safe, secure and permanent management of DSRSs in its member states, the IAEA is actively promoting the implementation of narrow diameter borehole disposal for DSRSs especially for countries without any geological repositories.

The borehole disposal concept adopts a multi-barrier system that encompasses geological and engineered barrier system. The engineered barrier system is basically made up of waste package, waste containers, and backfill. Similar to other disposal concepts, the main purpose of the backfill, that surrounds the waste container, is to limit the release of radionuclide contaminants to the geosphere upon the failure of the waste container[12][15]. In many disposal systems, commonly used backfill materials include soil, rock, clay, bentonite and cement. In borehole disposal, cement has been proposed as the backfill material. Cement and cementitious materials are favored materials to be used. Apart from the physical and mechanical benefits provided these materials also contain different minerals with high sorptivity for many radionuclides to provide for effective radionuclides containment in a longer period[10]. Additionally, the alkaline cement environment makes the radionuclides less soluble and mobile[7]. Thus, retention and retardation of radionuclides migration can be achieved.

Many radionuclides bind strongly to cementitious matrices[3][4][36] and therefore sorption properties can be exploited for containment of radionuclide contaminants. In hardened cement paste, calcium silicate hydrate (C-S-H) (chemical composition: CaO-SiO_2-H_2O) is the predominant phase responsible for most metals and radionuclides binding[5][21][36]. Pozzolanic reaction that occurs when a supplementary cementitious material (SCM) is added to a cement blend resulted in higher amounts of C-S-H phases as some of the $Ca(OH)_2$ formed during early hydration reacts with Si from the SCM to produce more C-S-H[10].

Various types of cement-based materials of varying SCM are used and proposed as backfill materials to fulfil the physical, mechanical and chemical requirements of the said application, for example, cement containing fly ash and silica fume[17]. Cement containing palm oil fuel ash (POFA) is slowly gaining attention as a green product to support environmental sustainability in the construction field. POFA is a by-product generated from the palm oil processing industry that contains a large amount of silica[34] and has a high potential to be used as cement replacement[35]. Some publications reported the feasibility of using POFA as SCM to enhance the physical and mechanical properties of cement[24][25]. On the other hand, the sorptivity of contaminant by POFA was attributed to its silica-based structure e.g. silanols (Si-OH) and siloxanes (Si-O-Si)[44]. This study is dedicated to demonstrate the potential of POFA-supplemented cementitious material as backfill for radionuclides containment based on their sorption performance compared to the Portland cement backfill. The sorption performance was analyzed from the kinetic point of view. The radionuclides of interest are Ra and Cs; both are major radionuclides found in DSRSs. Many studies have been dedicated to investigating the containment of Cs in cement systems due to its solubility in the high pH environment of the cement-based materials[27][19][2][6]. Despite Ra was reported as the second most important divalent elements (M^{2+}) in radioactive waste disposal due to its long half-life that is critical from the environmental perspective[22], yet investigations on Ra retention into cement-based materials are comparatively scarce[36].

2. Experimental

2.1. Materials

The cement used in this study is Ordinary Portland cement (OPC) type CEM I 42.5N manufactured by Tasek Corporation Berhad. The OPC sample was sieved to obtain particles of less than 250 μm prior to its use. POFA sample was collected from Jugra Palm Oil Mill in Banting, Selangor, Malaysia. The chemical compositions of the OPC and POFA were determined using EDXRF (Shimadzu), and the results are listed in Table 1. The POFA sample was subjected to a series of drying, heat-treatment, grinding and sieving processes to obtain the final POFA sample with particles sizes less than 250 μm.

Table 1. Chemical composition of raw materials OPC and POFA.

Chemical composition	%	
	OPC	**POFA**
SiO_2	21	54
CaO	60	5
Al_2O_3	11	0
K_2O	2	32
Fe_2O_3	5	4
P_2O_5	0	5
SO_3	1	0

Cubes of hardened cement paste were prepared from OPC alone as well as the mixture of OPC and POFA. An optimum mixture design of 30% OPC replacement by POFA and water to cement ratio of 0.4 was used. In this study, one-year-old backfill materials were used, where the backfill cubes were crushed, ground and sieved to obtain powdered samples of less than 250 μm and these samples were used. Meanwhile, cement-equilibrated water was prepared by agitating cementitious materials in distilled water for 28 days followed by decantation.

The solution containing Ra and Cs radionuclides were prepared from Ra-226 and Cs-134 obtained from Eckert & Ziegler. Working solutions of Ra and Cs were 6 Bq mL^{-1} and 10 Bq mL^{-1}, respectively. All the working solutions were prepared by dilution in cement-equilibrated water; the chemical composition of the pore water solutions for both samples is listed in Table 2. Radioactivity of Ra and Cs was measured using a gamma spectrometer (Canberra). The radioactivity of Ra was detected at energy 186 keV and Cs at 605 keV.

Table 2. Chemical composition of the pore water samples used in the sorption experiments (a) OPC pore water solution (b) POFA-supplemented cementitious material pore water solution.

Chemical composition	%	
	Cement-equilibrated water (OPC)	**Cement-equilibrated water (POFA-supplemented cementitious material)**
Ca	96	61
K	3	13
Mg	0	13
Na	1	13

2.2. Sorption experiments

Batch sorption experiments were performed to investigate the kinetic sorptive behavior of Ra and Cs onto the cementitious materials (OPC and POFA-supplemented cementitious material). Samples containing radionuclides-pore water solution and the cementitious materials in powdered form at a fixed dosage of 0.1 g mL^{-1} were continuously agitated, to prevent agglomeration, on an orbital shaker at 150 rpm. Experiments were conducted for periods ranging from 5 mins to 4 days at room temperature (25±2°C). The aliquots were centrifuged at 4000 rpm for 20 minutes to separate the supernatant and precipitate. The radioactivity of the supernatant samples containing unabsorbed Ra and Cs were analyzed using a gamma spectrometer. All experiments were done in triplicates. The amount of sorbed radionuclide per unit mass of sorbent, q (Bq g^{-1}), at any time, t, is given as Equation (1):

$$q_t = \frac{(C_o - C_t)V}{m} \quad (1)$$

where,

C_o is the initial activity concentration of sorbate in the bulk solution (Bq mL^{-1});
C_t is the activity concentration of sorbate in the bulk solution at time t (Bq mL^{-1});
V is the volume of the bulk solution (mL);
m is the mass of the sorbent (g).

2.3. *Kinetic analysis models*

2.3.1. Pseudo-first-order model. Linearized form of pseudo-first-order model of Lagergren [18] is given as Equation (2):

$$\log(q_e - q_t) = \log q_e - \frac{k_1}{2.303}t \tag{2}$$

where,

q_e is the amount of radionuclide sorbed at equilibrium (Bq g^{-1});
q_t is the amount of radionuclide sorbed at time t (Bq g^{-1});
k_1 is the equilibrium rate constant of pseudo-first-order sorption (min^{-1});
t is the time (min).

2.3.2. Pseudo-second-order model. Linear form of the pseudo-second-order model is given as Equation (3):

$$\frac{t}{q_t} = \frac{1}{k_2 q_e^2} + \frac{1}{q_e}t \tag{3}$$

where,

q_e is the amount of radionuclide sorbed at equilibrium (Bq g^{-1});
q_t is the amount of radionuclide sorbed at time t (Bq g^{-1});
k_2 is the equilibrium rate constant of pseudo-second-order sorption (Bq $g^{-1}min^{-1}$).

The values of q_e and k_2 were determined from the slope and intercept of the linear plot t/q_t against t, respectively. Initial sorption rate, h (Bq g^{-1} min^{-1}) is defined as Equation (4):

$$h = k_2 q_e^2 \tag{4}$$

2.3.3. Elovich's model. Elovich's model is used to describe reaction rates that increase exponentially with the adsorption, which implies multilayer sorption[1][3][43]. The model is used to describe different mechanisms, i.e. bulk and surface diffusion, as well as chemisorption. Linearized form [32] of Elovich's model is given as Equation (5):

$$q_t = \frac{\ln a_e b_e}{b_e} + \frac{1}{b_e}\ln t \tag{5}$$

where,

q_t is the amount of radionuclide sorbed at time t (Bq g^{-1});
a_e is the initial sorption rate (Bq g^{-1} min^{-1});
b_e is the desorption constant related to the activation energy for sorption reaction (Bq g^{-1}).

2.3.4. Intraparticle diffusion model (IPD). IPD model introduced by Weber and Morris (1963) is widely used in analysing sorption mechanisms[23] and is given as Equation (6):

$$q_t = k_{ipd}t^{1/2} + \theta \tag{6}$$

where,

q_t is the amount of radionuclide sorbed at time t (Bq g^{-1});
k_{ipd} is the IPD rate constant (Bq g^{-1} min$^{1/2}$);
θ is a constant related to the boundary layer thickness (Bq g^{-1}).

2.3.5. Bangham's model. Bangham's model assumes that intraparticle (pore) diffusion is the only rate-controlling step in a sorption process and its equation is expressed as Equation (7):

$$log\left[log\left(\frac{C_o}{C_o-C_s q_t}\right)\right] = log\left[\frac{k_b C_s}{2.303V}\right] + \alpha \log t \tag{7}$$

where,

C_o is the initial activity concentration of the radionuclide (sorbate) (Bq L^{-1});
C_s is the weight of cementitious backfill material (sorbent) used per liter of solution (g L^{-1});
q_t is the amount of radionuclide sorbed at time t (Bq g^{-1});
k_b is the Bangham's constant;
α is also a constant with value <1.

k_b and α were obtained from the intercept and slope of the linear plots of $log\left[log\left(\frac{C_o}{C_o-C_s q_t}\right)\right]$ against log *t*.

2.3.6. Boyd's model. The outcome from the IPD model was further analysed using the film diffusion model of Boyd's to determine the actual rate limiting step involved in the sorption process. This model assumes that the main resistance to diffusion is the boundary layer surrounding the adsorbent particle[23]. The Boyd's model is expressed as Equation (8):

$$F(t) = 1 - \frac{6}{\pi^2} \exp\left(1 - B_t\right) \tag{8}$$

where,

F(t) is the fraction of radionuclide sorbed at time *t;*
B_t is the mathematical function of *F(t).*

F(t) is given as Equation (9):

$$F(t) = \frac{q_t}{q_e} \tag{9}$$

where,

q_t is the amount of radionuclide sorbed at time *t* (Bq g^{-1});
q_e is the amount of radionuclide sorbed at equilibrium (Bq g^{-1}).

The following approximations are used to compute the value of B_t:

For *F(t)* values > 0.85, equation (10) is used:

$$B_t = 0.4977 - \ln\left(1 - F(t)\right) \tag{10}$$

For *F(t)* values < 0.85, Equation (11) is used:

$$B_t = \left(\sqrt{\pi} - \sqrt{\pi - \left(\frac{\pi^2 F(t)}{3}\right)}\right)^2 \tag{11}$$

3. Results and discussions

3.1. Effect of contact time

Experimental results showing the effect of contact time on the batch sorption of Ra and Cs onto the cementitious materials are depicted in Figure 1 (a) and (b) respectively. Typical sorption curves depicting the rapid sorption process in the beginning and the slow sorption process that followed were observed for both Ra and Cs. The rapid sorption process could be attributed to the vast availability of active sorption sites at the beginning of the reaction and as sorption progressed more sites become occupied leading to slow sorption that occurred in the longer period until equilibrium was reached [9]
.

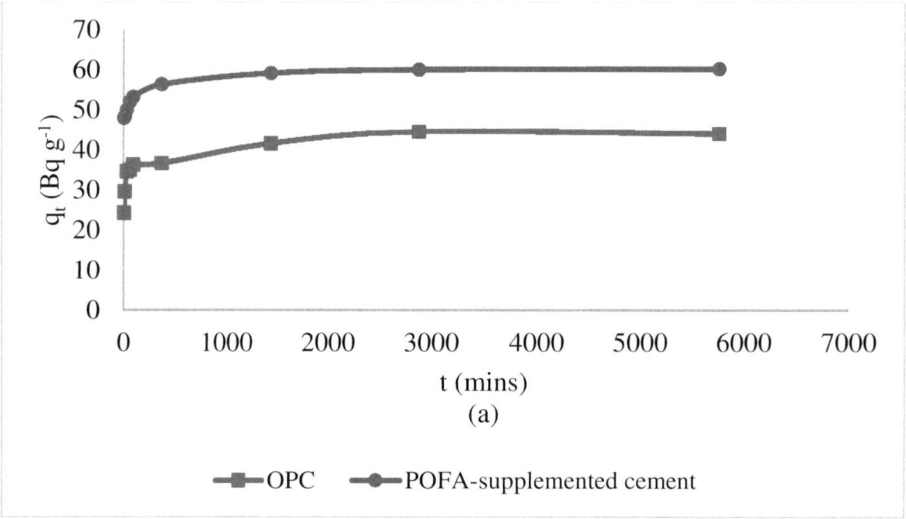

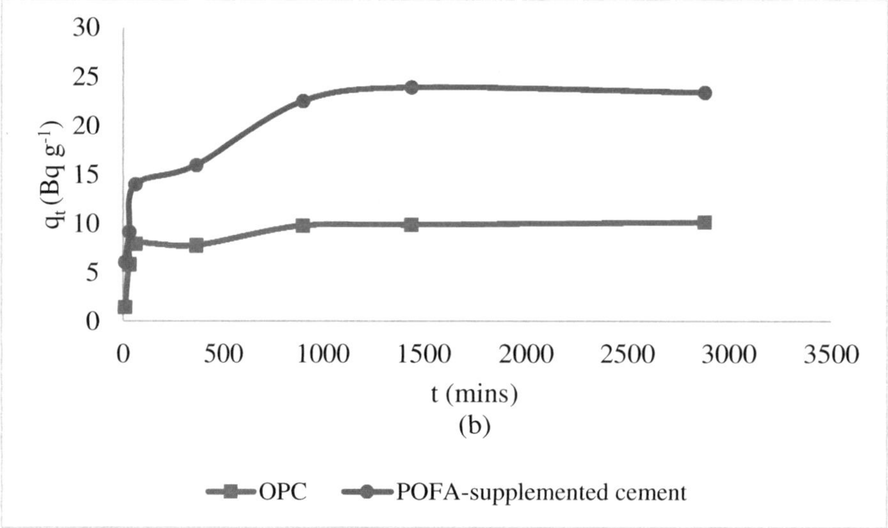

Figure 1. Effect of contact time on sorption of radionuclides by cementitious backfill materials: at fixed sorbent dosage, 0.1 g mL^{-1} at room temperature (25±2°C) (a) Ra with initial activity concentration, 6 Bq mL^{-1}; (b) Cs with initial activity concentration 10 Bq mL^{-1}.

It is also noted that the sorption capacity of Ra and Cs by POFA-supplemented cementitious material is higher than OPC for the entire period investigated thus indicating the contribution of POFA to the sorption of Ra and Cs. The sorption capacity, given by the values of q_t, for Ra sorption by POFA-supplemented cementitious material was in the range of 48 – 60 Bq g^{-1} compared to 24 – 44 Bq g^{-1} for OPC. Meanwhile, the values of q_t for Cs sorption by POFA-supplemented cementitious material ranged from 6 – 24 Bq g^{-1} and for OPC the values of q_t were lower ranging from 1–10 Bq g^{-1}. According to [5][21][29][36], radionuclides retention on cement was primarily controlled by C-S-H phases. Ochs *et al.* [27] specifically pointed out to the role of C-S-H as the main sorbing phase for Cs in hardened cement paste. In this study, it was very likely that the sorption mechanisms involved surface sorption and incorporation of Ra and Cs in the interlayer of C-S-H that might include a substitution of Ca^{2+} [36][37] or complexation on deprotonated silanol groups [15]. The presence of more C-S-H phases due to pozzolanic reaction in the POFA-supplemented cementitious material could have led to the higher sorption capacity of Ra and Cs. Another reason could be due to the presence of free SiO^- from POFA. At high pH, negative surface charge on SiO^- increased and led to a stronger attraction for positive such as Ra and Cs ions. This could also be explained the rapid sorption in the beginning and as the reaction progressed more SiO_2 was consumed and hence the gradual and finally steady sorption as interaction with C-S-H dominated. Yusof *et al.* [44] explained the mechanism of As sorption on POFA via the complexation of SiO^-.

It was mentioned that the sorption of alkaline earth metals by cementitious materials are greatly affected by the presence of Ca [24] as well as other alkalis (such as Na and K), where these cations decreased the sorption of other divalent ions on C-S-H [21]. The sorption process for both samples reached equilibrium within 48 hours and 24 hours for Ra and Cs, respectively. In comparison to findings by Volchek *et al.* [38] that investigated the sorption of Cs by Portland cement mortar coupons, the equilibrium in our study was achieved much earlier.

In order to elucidate the sorption mechanism, several kinetic models were tested using the experimental data. Kinetic models, namely pseudo-first order, pseudo-second order, Elovich's, IPD, Bangham's and Boyd's models, were used to investigate the reaction rates and controlling sorption mechanisms between the radio-contaminants (Ra and Cs) and the proposed cementitious backfill materials.

3.2. Determination of reaction rates

3.2.1. Pseudo-first order model. Pseudo-first order model illustrated by linear plots of log (q_e-q_t) against t gave the values of q_e from the intercept and k_1 from the slope as well as the values for correlation coefficient, R_2. The parameter values are shown in Table 3. It was observed that the values of q_e calculated were inconsistent with those obtained experimentally. Moreover, it was also observed that the model was valid only during the early phase of the sorption process for Cs sorption similar to findings reported by[3][9][11][38]. Subsequently, the use of this model to represent the reaction kinetics throughout the reaction time would lead to underestimating the value of sorbed radionuclides. Therefore, the model was deemed unsuitable to be used for prediction of Ra and Cs sorption for the entire sorption period.

3.2.2. Pseudo-second-order model. As the sorption data was not successfully applied using the pseudo-first order model, it was further treated with pseudo-second order model. This model asssumes that the uptake rate is second order with respect to the available sites[11]. Pseudo-second order model is depicted in Figure 2 (a) and (b) as plot of t/q_t against t and the fitting parameter values are shown in Table 3. The high correlation coefficient, R^2, values (R^2=1) and comparable experimental and calculated q_e values suggested that the sorption process could be well explained by this model. The findings also proposed that the rate limiting step for the sorption process was chemisorption that involved valency forces through sharing of electrons between sorbent and sorbate or covalent forces from exchange of electrons between sorbent and sorbate [11]. Similar findings were obtained by Volchek *et al.*[38].

The equilibrium rate constant of pseudo-second-order sorption, k_2, values obtained for OPC (Ra: 1.52E-03 Bq g^{-1} min^{-1}; Cs: 1.83E-03 Bq g^{-1} min^{-1}) were higher than that of POFA-supplemented cementitious material (Ra: 1.41E-03 Bq g^{-1} min^{-1}; Cs: 6.38E-04 Bq g^{-1} min^{-1}). However, the value for Ra sorption by OPC was just slightly higher compared to POFA-supplemented cementitious material. The initial sorption rate, h, values were higher for sorption by POFA-supplemented cementitious materials (Ra: 5.13E+00 Bq g^{-1} min^{-1}; Cs: 1.96E-01 Bq g^{-1} min^{-1}) than OPC (Ra: 2.65E+00 Bq g^{-1} min^{-1}; Cs: 1.96E-01 Bq g^{-1} min^{-1}) indicating rapid initial sorption by POFA-supplemented cementitious materials.

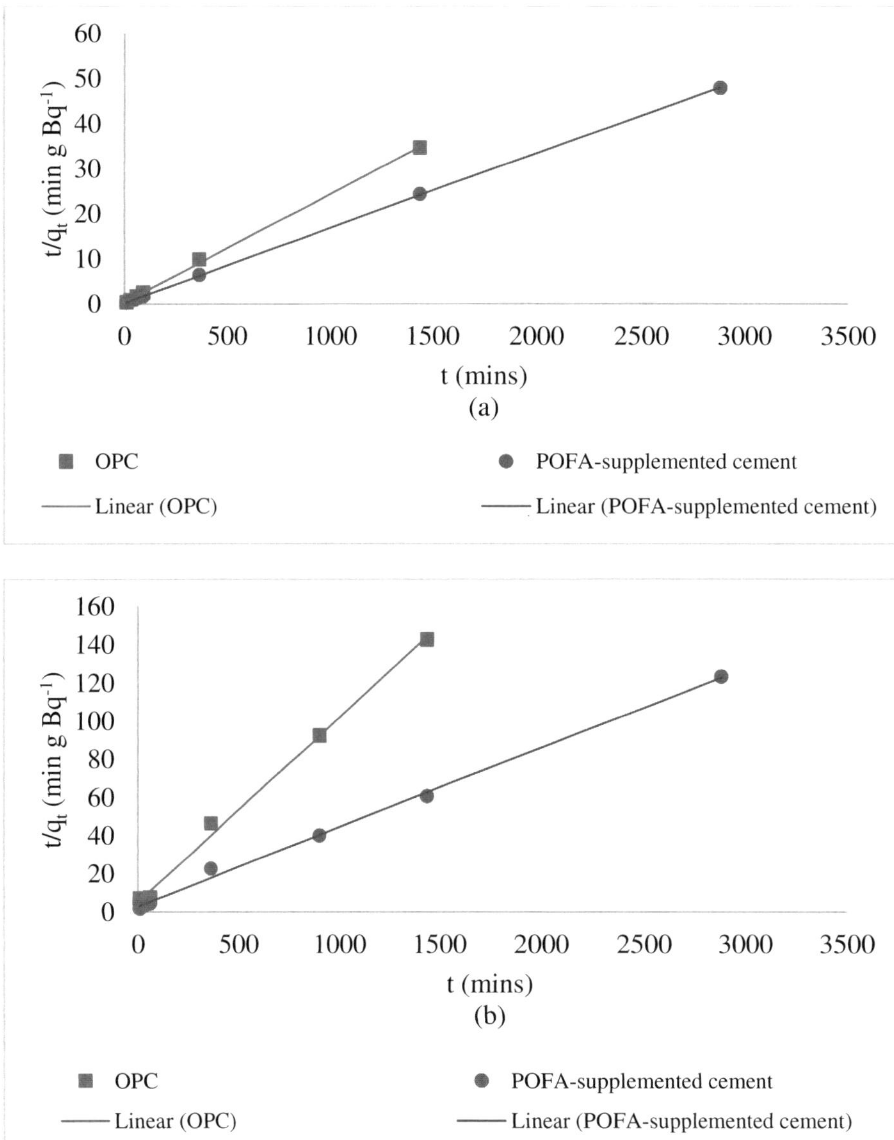

Figure 2. Pseudo-second-order model for the sorption of radio-contaminants by OPC and POFA-supplemented cementitious materials (a) Ra (b) Cs.

3.2.3. Elovich's model. The initial sorption rate, a_e, and the desorption constant, b_e, were calculated from the intercept and slope of the linear Elovich's plot q_t against ln t. The results obtained were presented in Figure 3 (a) and (b) and Table 3. A linear trend was noted for the fast sorption then the data deviated, which implied that the sorption in the fast period proceeded via exponential coverage on the surface[1][3]. The good correlation coefficient, R^2, values obtained indicated the applicability of this model to the sorption of Ra and Cs. Nevertheless, it was noticed that the model suited sorption by POFA-supplemented cementitious material (R^2= 0.95 – 0.98) better than the OPC (R2=0.81 – 0.92). It was also observed that the sorption was more rapid for POFA-supplemented cementitious material (Ra:1.24E+10 Bq g^{-1} min^{-1}; Cs:2.27E+00 Bq g^{-1} min^{-1}) compared to the OPC (Ra: 2.67E+04 Bq g^{-1} min^{-1}; Cs: 1.95E+00 Bq g^{-1} min^{-1}). It was also observed that the b_e values for Ra was higher for POFA-supplemented cementitious material (5.11E-01 Bq g $^{-1}$) compared to the OPC (3.95E-01 Bq g $^{-1}$). However, the b_e values for Cs was higher for OPC (7.67E-01 Bq g $^{-1}$) in comparison to the POFA-supplemented cementitious material (3.06E-01 Bq g $^{-1}$).

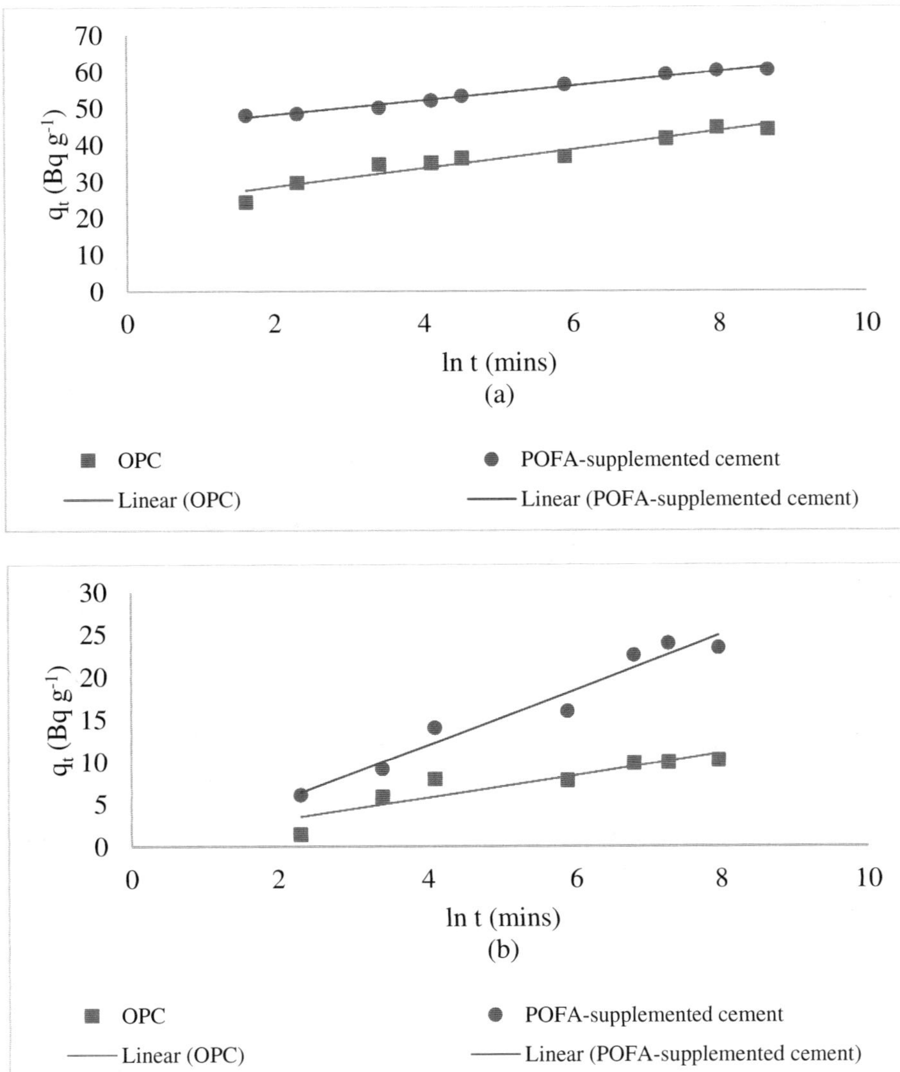

Figure 3. Elovich's model for the sorption of radio-contaminants by OPC and POFA-supplemented cementitious materials (a) Ra (b) Cs.

Table 3. Kinetic parameters obtained from various models.

Rate model	Parameter	OPC		POFA-supplemented cementitious material	
		Ra	Cs	Ra	Cs
Pseudo-first order	k_1 (min^{-1})	9.21E-04	4.61E-03	1.61E-03	2.53E-03
	q_e (Bq g^{-1})	1.23E+01	5.45E+00	9.69E+00	1.59E+01
	R^2	0.79	0.90	0.95	0.95
Pseudo-second order	k_2 (Bq g^{-1} min^{-1})	1.52E-03	1.83E-03	1.41E-03	6.38E-04
	q_e (Bq g^{-1})	4.18E+01	1.03E+01	6.02E+01	2.40E+01
	q_e (Bq g^{-1}) (Experimental)	4.46E+01	9.87E+00	6.01E+01	2.39E+01
	H (Bq g^{-1} min^{-1})	2.65E+00	1.96E-01	5.13E+00	3.67E-01
	R^2	1.00	1.00	1.00	1.00
Elovich's	a_e (Bq g^{-1} min^{-1})	2.67E+04	1.95E+00	1.24E+10	2.27E+00
	b_e (Bq g^{-1})	3.95E-01	7.67E-01	5.11E-01	3.06E-01
	R^2	0.92	0.81	0.98	0.95

3.3. Controlling sorption mechanism

3.3.1. Intraparticle diffusion model (IPD). IPD model is illustrated by plot of q_t against $t^{1/2}$ (Figure 4 (a) and (b)) that gives the value of k_{ipd} from the slope and θ from the intercept. The k_{ipd} and θ values listed in Table 4 referred to the results obtained for the entire contact time investigated. However, it was noted that the results illustrated in (Figure 4 (a) and (b)) showed multi-linear plots instead of a single linear plot that passed through the origin indicating that the sorption processes were controlled by multiple mechanisms[40]. Therefore, the mechanism involved was not limited to pore diffusion alone but could also by other mechanism such as film diffusion. The plots obtained could be divided into three distinct linear regions; each representing a controlling mechanism or several simultaneous controlling mechanisms[33]. The first distinct region that occurred during the initial sorption process demonstrated external diffusion through the boundaries that involved rapid transportation of radionuclides from the liquid phase to the solid boundary of the cementitious materials. This model assumes that external resistance to mass transfer (film diffusion) is only significant during the early stage of a sorption process[23]. The second region was attributed to IPD and the final region illustrates the equilibrium stage in which sorption took place slowly as the surface coverage approached saturation.

It was noticed that the boundary layer thickness increased as sorption progressed indicating large film diffusion resistance.

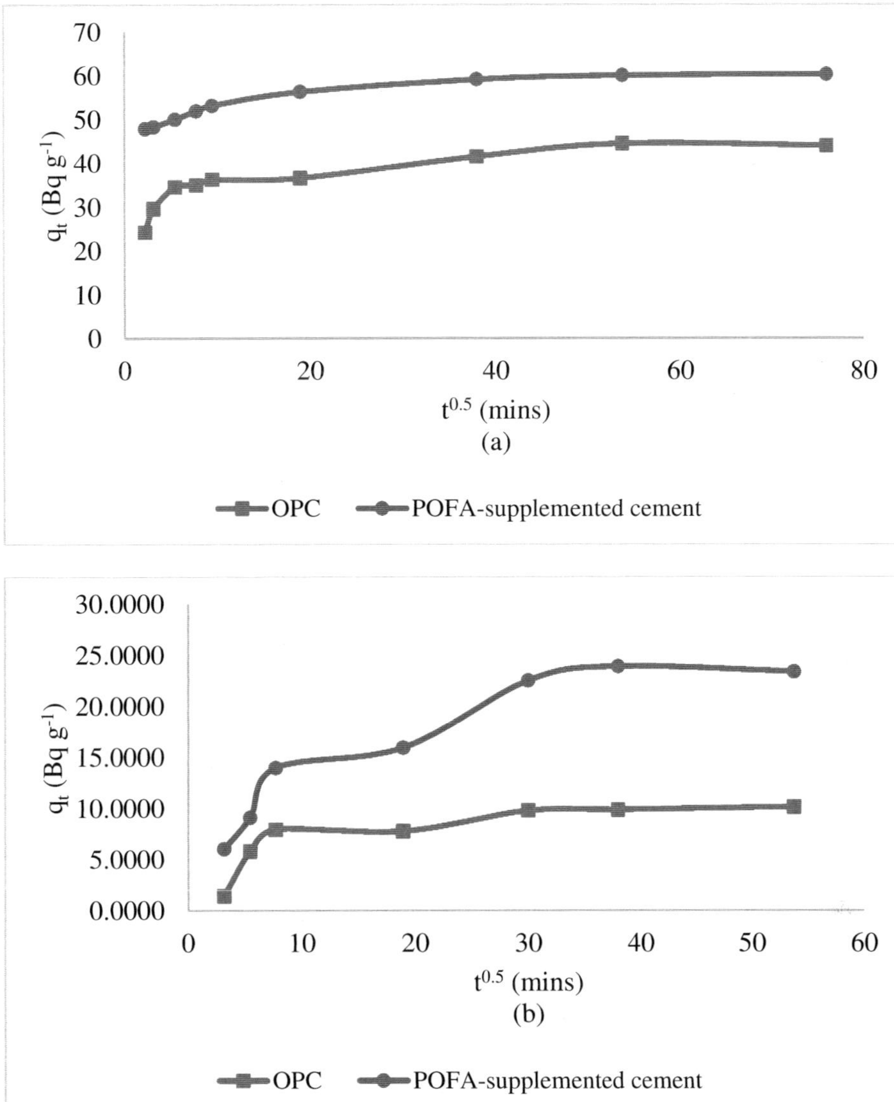

Figure 4. IPD model for the sorption of radio-contaminants by OPC and POFA-supplemented cementitious materials (a) Ra, (b) Cs.

3.3.2. Bangham's model. Bangham's model is described by plotting the values of $log\left[log\left(\frac{C_o}{C_o-C_s q_t}\right)\right]$ against log t. Results obtained are tabulated in Table 4. A desired linear fit was however not obtained for the double logarithmic plots as shown by the correlation coefficients, R^2, (R^2= 0.82 – 0.96) thus indicating that the sorption kinetics were not limited to only pore diffusion.

3.3.3. Boyd's model. Boyd's model is demonstrated by plot of B_t against t (Figure 5 (a, b)). The value of q_e used is obtained from the data fitting using pseudo-second order model similar to the approach adopted by Malash *et al.*[23]. .

The linear plots obtained did not pass through the origin thus indicated that the rate limiting step for the sorption process was governed by film diffusion (external mass transfer) or chemical reaction [9][16][30].

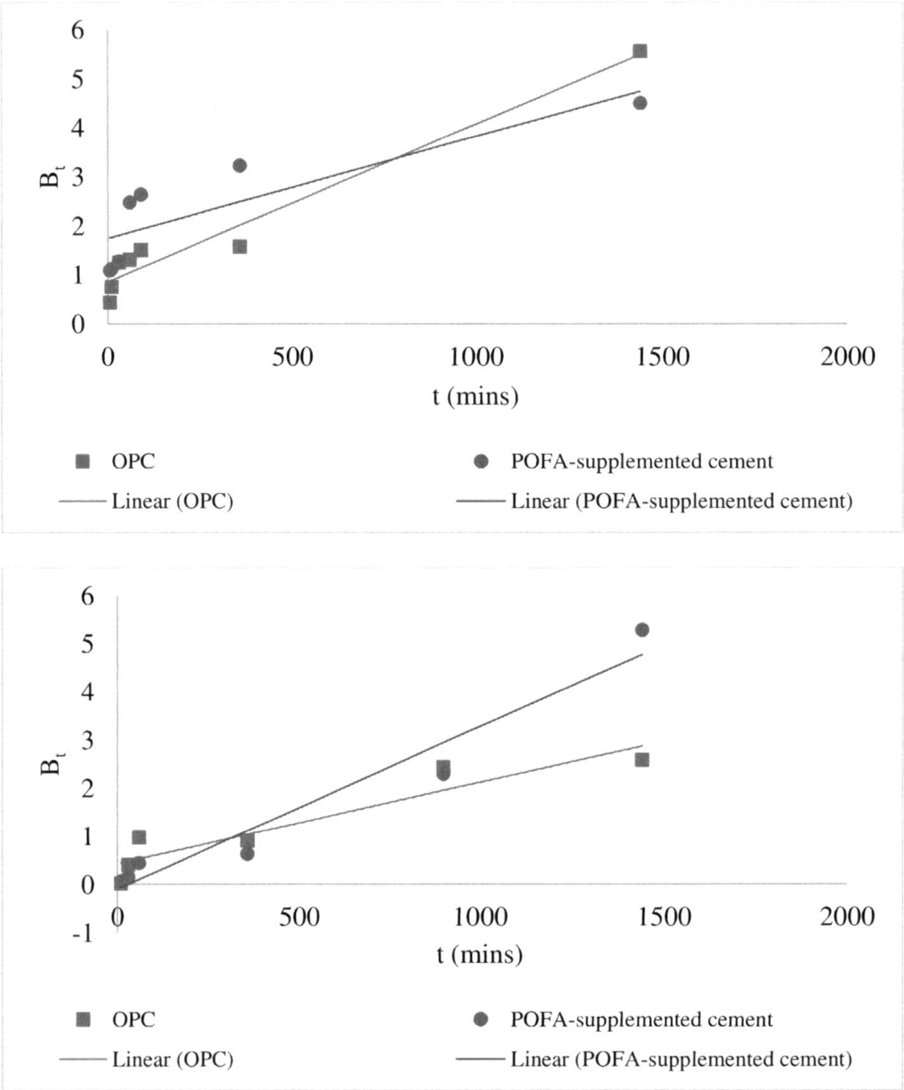

Figure 5. Boyd's model for the sorption of radio-contaminants by OPC and POFA-supplemented cementitious materials (a) Ra, (b) Cs.

Table 4. Mechanistic parameters obtained from various models.

Mechanistic model	Parameter	OPC		POFA-supplemented cementitious material	
		Ra	Cs	Ra	Cs
Intra particle diffusion	k_{ipd} (Bq g^{-1} min$^{1/2}$)	2.12E-01	1.27E-01	1.69E-01	3.43E-01
	θ (Bq g^{-1})	3.12E+01	4.68E+00	5.01E+01	8.72E+00
	R^2	0.71	0.60	0.80	0.82
Bangham's	k_b	1.00E-04	1.00E-04	3.00E-04	4.00E-04
	α	1.17E-01	1.14E-01	1.11E-01	2.11E-01
	R^2	0.93	0.82	0.96	0.93

4. Conclusion

The kinetic of Ra and Cs sorption by POFA-supplemented cementitious material was demonstrated and compared with OPC. For Ra, sorption equilibrium was achieved after 48 hours. Meanwhile, for Cs, equilibrium was reached after 24 hours. Ra showed higher sorption by both cementitious materials in comparison to Cs. The sorption capacity of Ra ranged from 24 – 60 Bq g^{-1} for both cementitious materials. Meanwhile, the sorption capacity of Cs ranging from 1 – 24 Bq g^{-1} showed much lower values compared to Ra. The sorption capacity by POFA-supplemented cementitious material also exhibited higher values with approximately 5 – 25 Bq g^{-1} more for POFA-supplemented cementitious material compared to OPC for both Ra and Cs thus indicating additional sorption by C-S-H phases resulted from pozzolanic reaction when POFA was added as SCM. Of the kinetic models applied, pseudo-second order model best described the kinetic data suggesting chemisorption reaction. IPD model and Bangham's model suggested that pore diffusion mechanism was not the only rate limiting step that controlled the sorption. Furthermore, Boyd's model suggested that other mechanism such as film diffusion might also occur.

5. References

[1] AbdelMoamen O A, Ismail I M, Abdelmonem N and AbdelRahman R O 2015 Factorial design analysis for optimizing the removal of cesium and strontium ions on synthetic nano-sized zeolite *J. of the Taiwan Inst. of Chem. Engineers* **000** 1–12.

[2] Abdel Rahman R O and Ojovan M I 2016 Recent trends in the evaluation of cementitious material in radioactive waste disposal *Handbook of Environ.Eng.* **17** (Switzerland: Springer) Chapter 9.

[3] Abdel Rahman R O, Abdel Moamen O A, Abdelmonem N and Ismail I M 2019 Optimizing the removal of strontium and cesium ions from binary solutionson magnetic nano-zeolite using response surface methodology (RSM) and artificial neural network (ANN) *Enviro.Res.* **173** 397–410.

[4] Abdel Rahman R O, Zin El Abidin D H A and Abou-Shady H 2013 Assessment of strontium immobilization in cement–bentonite matrices *Chemical Engineering Journal* **228** 772–780

[5] Aggarwal S, Angus M J and Ketchen J 2000 Sorption of radionuclides onto specific mineral phases present in repository cements *NSS/R312, AEA-D and R-0395 Technical Report.*

[6] Arayro J, Dufresne A, Zhou T, Ioannidou K, Ulm J, Béland L K P R 2018 Thermodynamics, kinetics, and mechanics of cesium sorption in cement paste: A multiscale assessment *Phys. Rev. Mater. 2.*

[7] Batchelor B 2006 Overview of waste stabilisation with cement *Waste Manage.* **26** 689–698.

[8] Boyd G E, Adamson A W and Myers Jr L S 1947 The exchange adsorption of ions from aqueous solutions by organic zeolites, II: Kinetics *J. Am. Chem. Soc.* **69** 2836-2848.

[9] Gasser M S, Mekhamer H S and Abdel Rahman R O 2016 Optimization of the utilization of Mg/Fe hydrotalcite like compounds in the removal of Sr(II) from aqueous solution *J. of Environ.Chem. Eng.*

[10] Glasser F P 2013 Cements in radioactive waste disposal *IAEA-TECDOC-CD—1701* (Companion CD)) (International Atomic Energy Agency (IAEA)).

[11] Ho Y S and McKay G 1999 The sorption of lead(II) ions on peat *Water Res.* **33** 578-584

[12] IAEA 2001 (a) Performance of engineered barrier materials in near surface disposal facilities for radioactive waste *IAEA-TECDOC-1255.*

[13] IAEA 2011 BOSS: Borehole disposal of disused sealed sources: A technical manual *IAEA-TECDOC-1644*

[14] IAEA 2001 (b) Technical considerations in the design of near surface disposal facilities for radioactive waste *IAEA-TECDOC-1256.*

[15] Iwaida T, Nagasaki S and Tanaka S 2000 Sorption study of strontium onto hydrated cement phases using a sequential desorption method *Radiochim. Acta* **88** 483–486.

[16] Jeong S H and Park K 2008 Drug loading and release properties of ion-exchange resin complexes as a drug delivery matrix *Int. J. Pharm.* **361** 26-32.
[17] Johnston H M and Wilmot D J 1992 Sorption and diffusion studies in cementitious grouts *Waste Manage.* **12** 289-297.
[18] Lagergren S 1898 About the theory of so-called adsorption of soluble substances *Kungliga Svenska Vetenskapsakademiens* **24**(4) 1-39.
[19] Li K and Pang X 2014 Sorption of radionuclides by cement-based barrier materials *Cement and Concrete Research* **65** 52–57.
[20] Lothenbach B and Nonat A 2015 Calcium silicate hydrates: Solid and liquid phase composition *Cement and Concrete Research* **78(A)** 57-70.
[21] Lovgren L 2005 Effects of complexing compounds on sorption of metal ions to cement *SKB* R-05-72
[22] Ochs M, Giffaut E P I 2006 Caesium sorption by hydrated cement as a function of degradation state: Experiments and modelling *Waste Manage.* **26** 725-732.
[23] Malash G F and El-Khaiary M I 2010 Piecewise linear regression: A statistical method for the analysis of experimental adsorption data by the intraparticle-diffusion models *Chem. Eng. J.* **163** 256-263.
[24] Missana T, García-Gutiérrez M, Mingarro M and Alonso U 2017 Analysis of barium retention mechanisms on calcium silicatehydrate phases *Cement and Concrete Res.* **93** 8–16.
[25] Nagaratnam B H, Rahman M E, Mirasa A K, Mannan M A and Lame S O 2015 Workability and heat of hydration of self-compacting concrete incorporating agro-industrial waste *J. of Cleaner Produc.*
[26] Nagaratnam B H,Mannan M A, Rahman M E,Mirasa A K, Richardson A and Nabinejad O 2019 Strength and microstructural characteristics of palm oil fuel ash and flyash as binary and ternary blends in self-compacting concrete *Constr. and Building Mater.* **202** 103–120.
[27] Ochs M, Mallants D and Wang L 2015 Radionuclide and metal sorption on cement and concrete, topics in safety, risk, reliability and quality (Switzerland: Springer).
[28] Phillip E, Khoo K S, Yusof M A W and Abdel Rahman R O 2021 Assessment of POFA-cementitious material as backfill barrier in DSRS borehole disposal: Ra confinement *J.of Environ. Manage.***280**
[29] Pointeau I 2000 Etude mécanistique et modélisation de la retention de radionucléides par les phases de silicate de calcium des ciments hydrates *Thèse de l'Université de Reims-Champagne Ardennes* (France).
[30] Reichenberg D 1953 Properties of ion exchange resins in relation to their structure. III. Kinetics of exchange *J. Am. Chem. Soc.* **75** 589-598.
[31] Sand R, Zeldovich Y B1934The catalytic oxidation of carbon monoxide on manganese dioxide *Acta Phys. Chem. USSR* **1** 554.
[32] Sparks D L 1989 Kinetics of soil chemical processes (New York: Academic Press).
[33] Tan K L and Hameed B H 2017 Insight into the adsorption kinetics models for the removal of contaminants from aqueous solutions *J. of the Taiwan Inst. of Chem. Engineers* **74** 25-48.
[34] Tangchirapat W, Saeting T, Jaturapitakkul C, Kiattikomol K, Siripanichgorn A 2007 Use of waste ash from palm oil industry in concrete *Waste Manage.* **27** 81-88.
[35] Tangchirapat W, Tangpagasit J, Wae-kum S, Jaturapitakkul C, Kiattikomol K and Siripanichgorn A 2003 A new pozzolanic material from palm oil fuel ash *KMUTT Res. and Dev. J.***26** 459-473.
[36] Tits J, Iijima K, Wieland E and Kamei G 2006 The uptake of radium by calcium silicate hydrates and hardened cement paste *Radiochim. Acta* **94** 637–643.
[37] Tits J, Wieland E, Dobler J P and Kunz D 2004 The uptake of strontium by calcium silicate hydrates under high pH conditions: An experimental approach to distinguish adsorption from co-precipitation processes *Proc. of the Mater. Res. Soc. Symp.* **807** 689–694.
[38] Volchek K, Miah M Y, Kuang W, DeMaleki Z and Tezel F H 2011 Adsorption of cesium on

cement mortar from aqueous solutions *J. of Hazard. Mater.* **194** 331-337.
[39] Weber W J and Morris J C 1963 Kinetics of adsorption on carbon from solution *J. Sanit. Eng. Div.* **89** 31-60.
[40] Wu F C, Tseng R L and Juang R S 2009 Initial behavior of intraparticle diffusion model used in the description of adsorption kinetics *Chem. Eng. J.* **53** 1-8.
[41] Yang G, Chen H, Qin H and Feng Y 2014 Amination of activated carbon for enhancing phenol adsorption: effect of nitrogen-containing functional groups *Appl. Surf. Sci.* **293** 299-305.
[42] Yang W, Ding P, Zhou L, Yu J, Chen X and Jiao F 2013 Preparation of diamine modified mesoporous silica on multi-walled carbon nanotubes for the adsorption of heavy metals in aqueous solution *Appl. Surf. Sci.* **282** 38-45.
[43] Yurtsever M and Sengil A 2009 Biosorption of Pb(II) ions by modified quebracho tannin resin *J. Hazard. Mater* **163**(1) 58-64.
[44] Yusof M S M, Othman M H D, Wahab R A, Jumbri K, Razak F I A, Kurniawan T A, Samah R A, Mustafa A, Rahman M A, Jaafar J and Ismail A F 2020 Arsenic adsorption mechanism on palm oil fuel ash (POFA) powder suspension *J. Hazard. Mater* **383.**

IOP Conf. Series: Materials Science and Engineering 1106 (2021) 012019 doi:10.1088/1757-899X/1106/1/012019

The effect of various electrospinning parameter on preparation of alumina nanofibers

Umar Abdillah[1], Hafizal Yazid[2,a)], Sahrim Ahmad[1], Nurulizzati Makhtar[1,3], Siti Zaubidah[1,3], Ruey Shan Chen[1], Nur Syafiqaz[1]

[1]Department of Applied Physics, Faculty of Science and Technology, UKM, 43600 Bangi, Selangor, Malaysia.
[2]Material Technology Group, Industrial Technology Division, Malaysia Nuclear Agency, 43000, Kajang, Selangor, Malaysia.
[3]School of Applied Physics and Material Sciences, Faculty of Applied Physics, Universiti Teknologi MARA, Shah Alam, Selangor, Malaysia.

[a]hafizal@nm.gov.my

Abstract. Alumina nanobers were successfully prepared via an electrospinning technique with combination of aluminum oxyhydroxide (AlOOH) as alumina precursor and Polyvinylpyrrolidone (PVP) as the polymer precursor. To produced alumina precursor, Aluminum Isopropoxide (AIP) was dissolved in distilled water under the constant stirring process at 80°C. Electrospinning solution was prepared by mixing Aluminum Oxyhydroxide (AlOOH) and Polyvinylpyrrolidone (PVP) aqueous solution for 2 hours under room temperature. Different concentrations of polymer precursor and various electrospinning parameters such as applied voltage, distance tip to collector, and flowrate of sol-gel were studied. The as-spun fibers were calcined at 1200°C and characterized by SEM. Results showed that the average fiber diameter decreases with decreasing of Polyvinylpyrrolidone (PVP) solution concentration, increasing of applied voltages and decreasing of electrospinning flow rate. At 10%wt Polyvinlpyrrolidone (PVP) concentration with 17.5v of applied voltages and 0.03 ml/min of flow rates, calcined fibers showed an average diameter of 100-200nm.

1. Introduction

Over the past decade, the fabrication of ceramic metal oxide composites has gained much attention due to their unique properties and wide range applications such as high-temperature insulation, supercapacitor, optical, and as brake-disc in automotive brake system[1]. Alumina which also calls as Aluminum Oxide (Al_2O_3) is ceramic metal oxide where its superiority has opened up opportunities for the discovery of outstanding materials and applications[2]. Due to its high temperature capacity, alumina is widely used in high-temperature applications such as blast furnace slag component[3]. Moreover, good abrasion resistance of alumina making it useful for abrasive material[4]. Solid alumina has been used in manufacturing of orthopedic implant and electrical insulator while alumina fiber is widely used in manufacturing of fire retardation components[5]. For future applications of alumina, Bacciarini[6] reported that metal matrix composite reinforced alumina powder shows promising features for airframe applications in the aerospace industry[6].

Electrospinning is a process of producing nanofibers and microfibers by which involves an electrohydrodynamic process. It is a unique method used to produces long and short ultrafine fibers with a simple setup[7]. The basic apparatus of electrospinning includes a high voltage power supply, pump, syringe, and wire grounded collector. Around the beginning of the electrospinning process, high voltage power is connected to the syringe before the sol-gel is injected. Due to the voltage applied, an electrical field is generated from syringe tip to the collector. As the sol-gel is injected from the syringe, an electrical repulsive force directs the fiber in form of jet onto the collector. The fibers then are calcined at high temperatures to remove the polymer precursor and the desired ceramic phase is obtained. The ultrafine fibers prepared by this process demonstrate a brilliant performance and tremendous mechanical properties[8].

Enhancement of mechanical properties of fibers has resulted in a major interest in nanomaterials. In nanofiber composite, the mechanical properties of fiber-reinforced composite depend extensively on the quality of bonding between fiber and matrix[9]. Sethuramalingam[10] in his study revealed that increasing the area of the contacted surface between fibers and matrix enhanced mechanical properties of the composites[10]. This can be achieved by various approaches includes reduction of diameter sizes of the fibers. Through an electrospinning process, fiber diameter can be minimized by different parameters setting such as increasing the voltage applied and increasing the distance of an electric field traveled by the polymer[11]. Reducing polymer concentration also resulted in smaller fiber diameters. The effect of different parameter settings on the preparation of alumina nanofibers through an electrospinning process has been reported. Despite that, most of the previous researches on the fabrication of alumina nanofibers are focused on the preparation of fiber in form of continuous strand [12].

In some applications, short fibers demonstrate better performance above the continuous fibers by offering high level of stiffness to weight ratio[13]. Moreover, short fibers much more versatile in term of processibility which it can be used in most extrusion techniques with thermoplastics[14]. However, the preparation of short fibers through the electrospinning process usually required a post-spinning process. Most of the previous researches utilized the conventional method to shortening the fibers such as grinding and milling[15]. One of the major limitations of these post-processing methods is the fiber must be reasonably brittle to fracture. Therefore, the nonconventional method to fabricate short fibers by which may eradicate the need for post-processing process is studied.

In this study, attempts have been made to prepare short alumina nanofibers by electrospinning technique by studying the effect of different precursors concentration and other processing parameters on morphology of nanofibers. Initially, aluminum isopropoxide (AIP) which could be easily dissolved in ethanol was selected as the alumina precursor. Meantime, high molecular weight Polyvinylpyrrolidone (average Mw 1,300,000) was used as the polymer precursor.

2. Experiment

2.1. Materials

Aluminum isopropoxide (AIP) with a reagent grade of 99 % from Sigma Aldrich was used as the source of alumina precursor. For polymer precursor, high molecular weight Polyvinylpyrrolidone (Mw 1300000) was purchased from Sigma Aldrich alongside nitric acid (HNO3) and Ethyl Acetoacetate which also obtained from Sigma Aldrich.

2.2. Preparation of sol-gel

To simplify the condition for analysis, a typical procedure was selected, different concentration of polymer solutions was prepared by adding Polyvinylpyrrolidone (PVP) into ethanol with stirring for 2hours at 50°C. As preparation of alumina precursor, initially a number of grams of Aluminum isopropoxide (AIP) was added into 10ml of deionize water under stirring process. Then, a few drops of Ethyl Acetoacetate were added and the stirring process was continued for 16hours. Both solutions then were transferred onto glass beaker for mixing process of 2hours at 50°C. A numbers of different concentration were studied includes 6%, 8%, 10% and 12%.

2.3. Electrospinning process

The solution was transferred into a plastic syringe equipped with a 23-gauge needle. The needle was connected to a voltage supply which is capable of generating DC voltage as high as 25kV. In a typical procedure the distance between the needle and collection screen was fixed at 16cm. The relative humidity was controlled between 50% and 60% at room temperature. Throughout the electrospinning process, the voltage was kept constant at 17.5kV. 0.01ml/min of flowrate is used during electrospinning of different polymer concentrations. After optimum concentration of polymer was selected, A set of different flowrates was studied includes 0.03ml/min, 0.04ml/min and 0.05ml/min.

2.4. Characterization

The morphology of fibers was examined by scanning electron microscopy (SEM).

3. Result and discussion

3.1. Effect of solution concentration

From the SEM micrographs, solution concentration had considerable influence on the morphology of the nanofibers. From the morphology images illustrated in Figure 1, it can be observed that there is formation of bead on electrospun fibers. The formation of beads is related to the instability of the jet of polymer solution which is due to the insufficient amount of viscoelastic force to overcome the repulsive forces of charge. As can be observe from the figures, beads and beaded fibers are less likely to be formed for the more viscous solutions. The diameter of the beads become bigger and the average distance between beads on the fibers longer as the viscosity increases. Meanwhile, the shape of the beads gradually changes from spherical to spindle-like. At higher concentration of polymer (12%wt), the stack fibers are formed. Due to its large diameter size, the concentration is not suitable for fabrication of nanofibers. Based on the morphological result, 10%wt polymer concentration was selected.

3.2. Effect of different flowrates setting

From the SEM micrographs illustrated in Figure 2 until Figure 4, the influence of flowrates on the morphology of the nanofibers can be observed. At flow rate of 0.03 ml/min (Figure 2), 0.04ml/min (Figure 3) and 0.05ml/min (Figure 4), the bead formation on fibers were reduced due to the feeding rate was proportional to the electrospinning speed. In this case, electrospinning process was accomplished under stable condition, lesser probability of instability and lesser falling droplets of the unspun polymer jet. Even though no bead found at the flowrates of 0.04ml/min and 0.05ml/min of fibers, the fibers have bigger diameters. The distribution of fiber diameters was affected by flow rate variations. Obviously, when the flow rate is increased, fiber diameter distribution became wider which also resulted in the selection of 0.03ml/min of flowrate as the most optimum for fabrication of short alumina nanofibers.

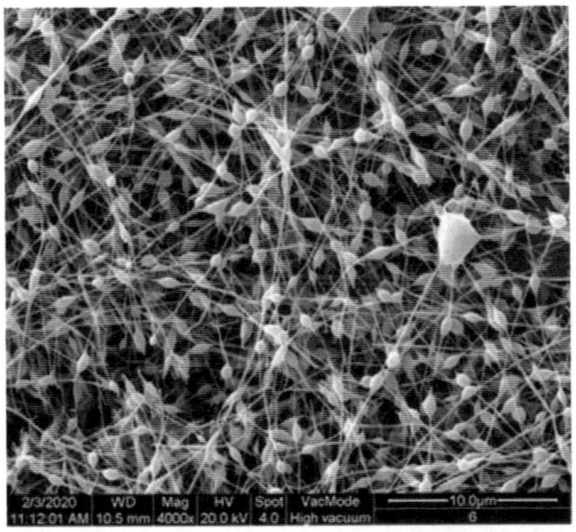

(a) Fibers from 6%wt polymer concentration

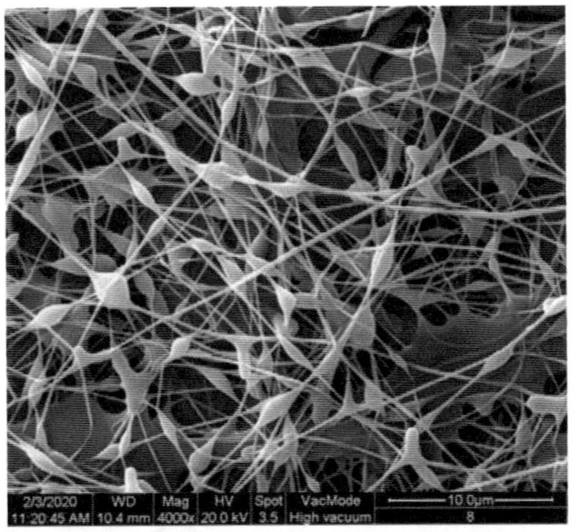

(b) Fibers from 8%wt polymer concentration

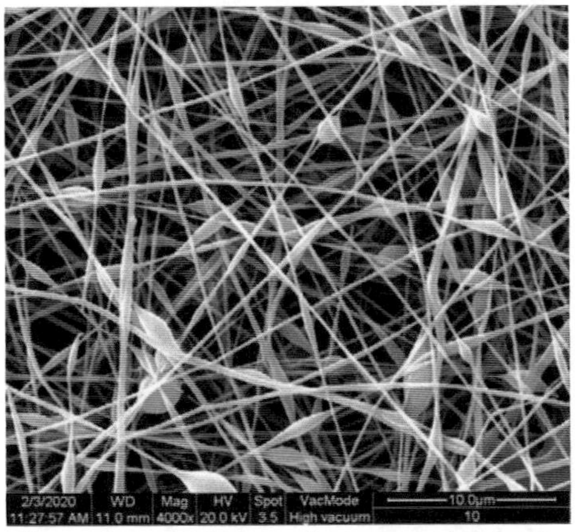

(c) Fibers from 10%wt polymer concentration

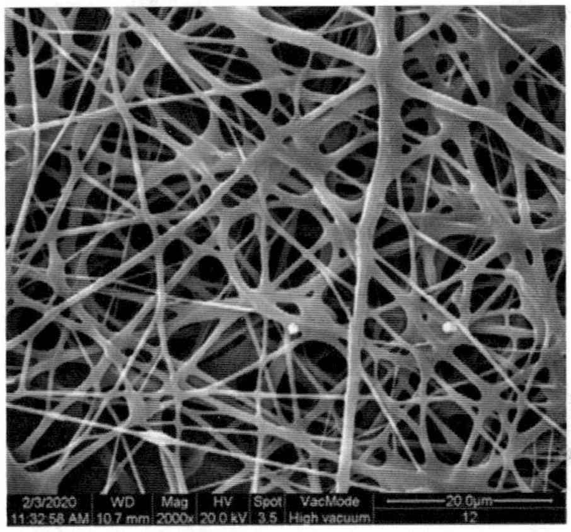

(d) Fibers from 12%wt polymer concentration

Figure1. SEM micrographs of electrospun fibers.

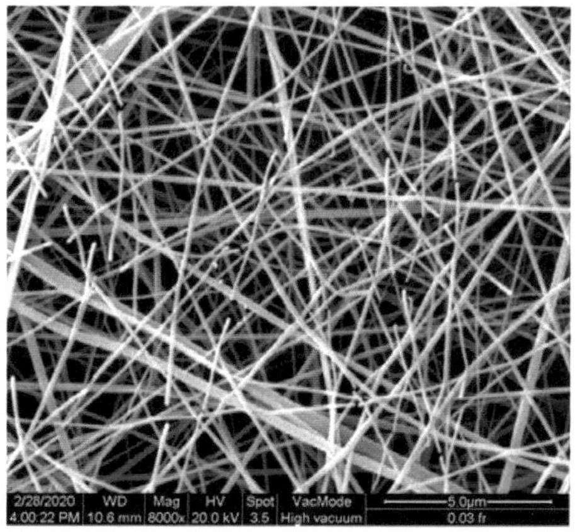

Figure 2. Fibers from 0.03ml/min of flowrate.

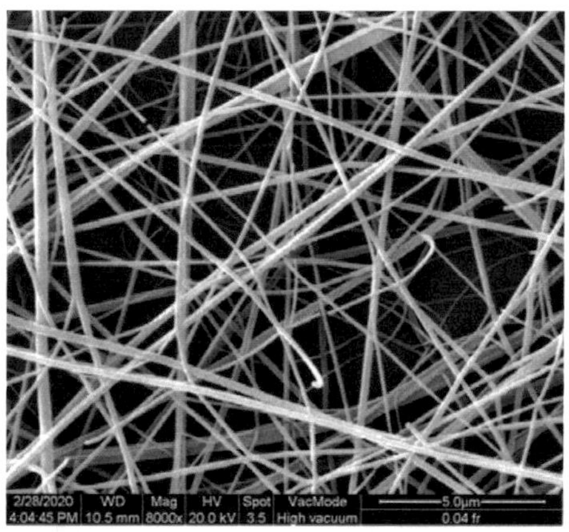

Figure 3. Fibers from 0.04ml/min of flowrate.

Figure 4. Fibers from 0.05ml/min of flowrate.

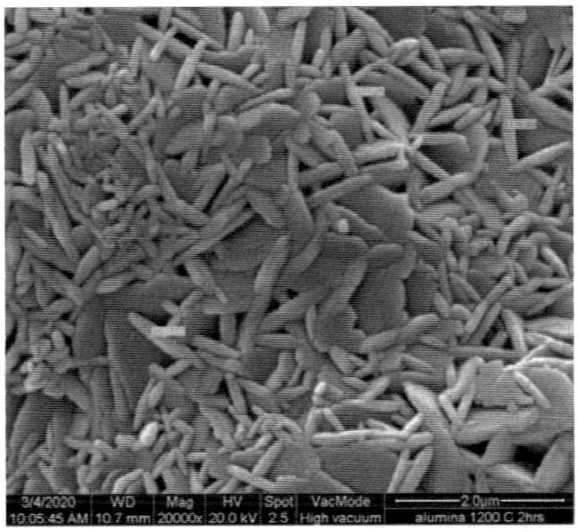

Figure 5. Short alumina nanofibers with an average diameter of 100-200 nm.

4. Conclusion

In this study, the effect of flow rate and polymer concentration on the morphology of electrospun nanofibers was investigated. It was found that the fiber diameter distribution and the morphology of the produced nanofibers were all influenced by the flow rate and concentration of the polymer solution. Moreover, beaded fibers are less likely to be formed for the more viscous solutions. However, at higher concentration of polymer (12%wt), the stack fibers are formed. Due to its large diameter size, the concentration is not suitable for fabrication of nanofibers. Reduction of beaded fibers has been achieved from all set of flowrates studied (0.03 ml/min, 0.04ml/min, and 0.05ml/min). However, the fibers of 0.04ml/min, and 0.05ml/min of flowrates have bigger diameters. Obviously, when the flow rate was increased, fiber diameter distribution became wider which also resulted in the selection of 0.03ml/min of flowrate as the most optimum for fabrication of short alumina nanofibers. In a nut shell, alumina short nanofibers have been successfully fabricated as shown in Figure 4. At 10%wt Polyvinylpyrrolidone (PVP) concentration with 17.5V of applied voltages and 0.03 ml/min of flow rates, calcined fibers showed an average diameter of 100-200 nm.

5. References

[1] Emarati M M S M 2020 Efficient one-step fabrication of superhydrophobic nano-TiO2/TMPSi ceramic composite coating with enhanced corrosion resistance on 316L *Ceramics Int.* 1652-1661.

[2] Kageyama A T T S R K S T O N 2018 Synthesis and characterization of a silica-alumina composite membrane and its application in a membrane reactor *Separation and Purification Technol.*, **195**, 437-445,.

[3] Zuo C W Y L H 2017 Dissolution behavior of a novel Al2O3-SiC-SiO2-C composite refractory in blast furnace slag *Ceramics Int.* **43**(9) 7080-7087.

[4] Li W. D. B. L. J. X. M 2019 Morphological evolution and grinding performance of vitrified bonded microcrystal alumina abrasive wheel dressed with a single-grit diamond *Ceramics Int.* **45**(16) 19669-19678.

[5] Gopal G M V 2019 9 - Zirconia-alumina composite for orthopedic implant application, in *Woodhead Pub. Ser. in Biomat.*

[6] Bacciarini C M V 2014 Aluminium AA6061 matrix composite reinforced with spherical alumina particles produced by infiltration: perspective on aerospace applications *J. of Metallurgy* **2014** 1-10.

[7] Li J T M Y X M M Dan 2006 Electrospinning: a simple and versatile technique for producing ceramic nanofibers and nanotubes *J. of the American Ceramic Soc.* **89**(6) 1861-1869.

[8] George T Z L S G 2021 22 - Sol-gel electrospinning of diverse ceramic nanofibers and their potential applications *Electrospun Polym. and Composites* 689-764.

[9] Cho J H. S M H P F B G P K 2020 Experimental cum computational investigation on interfacial and mechanical behavior of short glass fiber reinforced dental composites *Composites Part B: Eng.* **200**.

[10] Sethuramalingam S K V C 2020 Studies on influence of silica filler and rice husk ash on the mechanical properties of vulcanized hybrid rubber composite *Materialstoday: Proc.*

[11] Akhmal M N A I D K N H N 2017 A review of evolution of electrospun tissue engineering scaffold: From two dimensions to three dimensions *J.of Eng.in Med.* (Proc. of the Inst. of Mech. Engineers Part H).

[12] Song K Z Y S Z D Q L Y L X 2020 Morphology, microstructure and mechanical properties of electrospun alumina nanofibers prepared using different polymer templates: A comparative study*J. of Alloys and Comp.* **829**.

[13] Nayak S M S K. S K 2009 Influence of short bamboo/glass fiber on the thermal, dynamic mechanical and rheological properties of polypropylene hybrid composites *Mater. Sci.and Eng.: A* **23**(1-2) 32-38..

[14] Sanomura M K Y 2003 Fiber orientation control of short-fiber reinforced thermoplastics by ram extrusion," *Polymer Composites* **24**(5) 587-596.

[15] Knotek M P M B B K M V A M M P J N E. A P B P 2012 Cryogenic grinding of electrospun poly-ε-caprolactone mesh submerged in liquid media," *Mater. Sci. and Eng C* **32**(6) 1366-1374.

Acknowledgement
The authors would like to thank the Nuclear Malaysia and Ministry of Higher Education (MOHE) for the research financial support under the research scheme of Fundamental Research Grant (FRGS) with the grant number FRGS/1/2018/STG07/MOSTI/02/01.

IOP Conf. Series: Materials Science and Engineering 1106 (2021) 012020 doi:10.1088/1757-899X/1106/1/012020

The effect of object to film distance (OFD) to industrial radiography image

R Ramli[1,a)], M S M Amin[2], M H Zasikin

[1]Institute of Sustainable Energy, Universiti Tenaga Nasional, Kajang, Selangor, Malaysia.
[2]Industrial Technology Division, Malaysian Nuclear Agency, Bangi, Selangor, Malaysia.

[a]Rogemah@uniten.edu.my

Abstract. The object to film distance (OFD) to image when using radiography testing were discussed and whether the image can be improved by IIP Console Software in Digital Radiography method were studied. The artificial defect of wooden block with different size was used as the test sample in this study. The object to film distance (OFD) to the image were capture by using XRS-4 model X-ray source and flat panel detector. The image was analyzing by using IIP Console Software. The obtained results show that irregular surface has better and sharper image compare to flat surface to 9 mm hole wooden block. Meanwhile for 15 mm hole, flat wooden block show better and sharper image compare to irregular surface

1. Introduction

Radiography testing is one of the non-destructive testing (NDT) applications. The internal structure of the object is basically equivalent to the imparted energy of radiation source. Thus, it is essential to analyze the object's details of non-destructive testing method in medical and industrial application using radiography imaging [1]. The internal structures of industrial object are guessed from a set of projection of the radiography method [1]. Radiography generates the test object's visual response on the film. The quality of the radiographic image is determined by the film form, the source to film distance, the time and angle of exposure. A radiographic image consists of varying shades of grey, ranging from white grey on one end to black grey on the other. A film showing quite bright but also very dark areas has a high contrast or a short grey scale, because there are few shades of grey from one side to the other [2].

The voltage has an effect on the radiographic image, as the energy used to make the X-rays is related to the X-rays' penetrating power. Source to film distance (SFD) is a measurement of the variation to the film to be viewed from the X-ray tube [3]. The effect of source to film distance would also cause the film density and exposure time to change accordingly. Digital imaging creates a permanent virtual copy of a sample that can be re-evaluated, sent to experts remotely but also protects the original sample against damage. It can be used to take long fracture measurements for estimating the stature particularly [4]. Image displays are required to integrate visual information and can be embedded with various technology forms. Other significant factors which can interfere with this phase have not been evaluated jointly [4]. The rapid development of digital imaging techniques needs guidance on how to produce an effective display of the electronic image. Radiographic caries detector

IOP Conf. Series: Materials Science and Engineering 1106 (2021) 012020 doi:10.1088/1757-899X/1106/1/012020

is a restricted activity local-area-contrast, it remains unknown if this high-contrast detection model is relevant to low-grey level differences [5][6].

The X-ray image from the radiographic test is computed by using image processing to measure the region of defect in terms of pixel count. An attempt to fill this void by applying image processing technique directly to the X-ray image for process parameter selection. Furthermore, the relation between the size of the defect and the tensile strength of the X-ray image is produced [7].

2. Methodology

2.1. Sample preparation

Two types of wooden block which were used in this study were flat surface and irregular surface. An artificial defect has been made by using drill press machine with 9 mm and 15 mm diameter of drill size through the center of the sample (Figure 1).

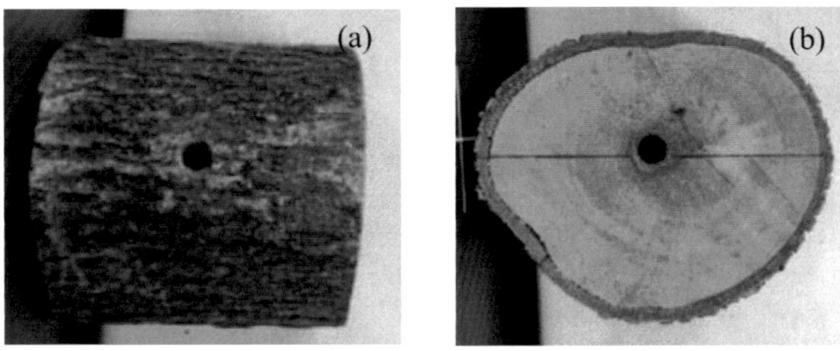

Figure 1. (a) Flat surface 9 mm hole wooden block and (b) Irregular surface 9 mm hole wooden block.

2.2. Testing method

The X-Ray source Model XRS-4 and flat model detector Model SENTINELDR were used for this project (Figure 1). IIP Console Software was used for image processing and digitalization of X-Ray images (Figure 2). This method uses X-Ray sensitive plate to collect data during object processing, which is transferred to the device directly without any need of intermediate cassette. The object to film distance were tested at a distance of 0 mm, 5mm, 10 mm, 20 mm, 25 mm, 30 mm and 35 mm.

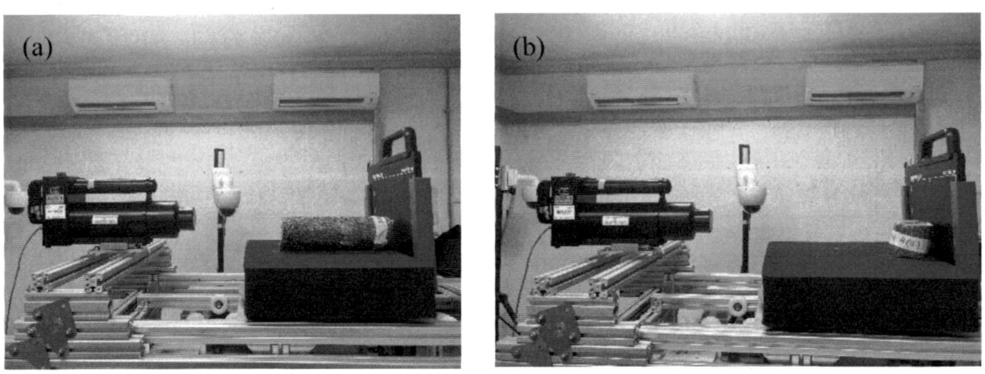

Figure 2. (a) Radiography test for flat sample and (b) radiography test for irregular sample.

IOP Conf. Series: Materials Science and Engineering 1106 (2021) 012020 doi:10.1088/1757-899X/1106/1/012020

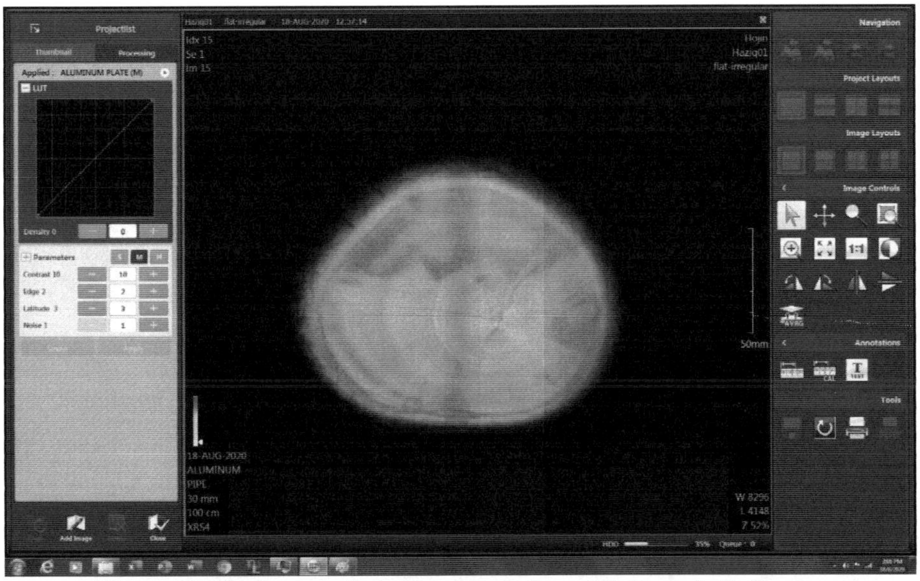

Figure 3. The image of wooden block from IIP Console Software.

3. Results and discussion

3.1 Wooden block 9 mm hole

Table 1 shows the value of defect size for flat and irregular wooden block when object to film distance (OFD) increases. From the data obtained, both defect size for flat and irregular wooden block increase significantly with increasing of OFD. This may due to the surface that come into contact with the panel detector is different. For examples at the OFD 5mm, 10mm and 35mm both defect size show in high gap values where the defect size for flat is 9.27mm irregular is 9.02mm at OFD 5mm. Defect size of flat is higher than the defect size of irregular. The same goes to defect size at OFD 35mm which defect size in flat block is 12.58mm and the defect size in irregular block is 12.06mm.

Table 1. Data of defect size and OFD for 9 mm hole.

Defect size flat (mm)	Defect size irregular (mm)	OFD (mm)
9.02	9	0
9.27	9.02	5
9.65	9.28	10
9.91	9.91	15
10.41	10.29	20
10.8	11.05	25
11.56	11.56	30
12.58	12.06	35

The defect size for flat wooden block is higher compare to irregular wooden block when the value of OFD increases and the value for both flat and irregular shape wooden block increase simultaneously with the increasing of OFD value as shown in Figure 4.

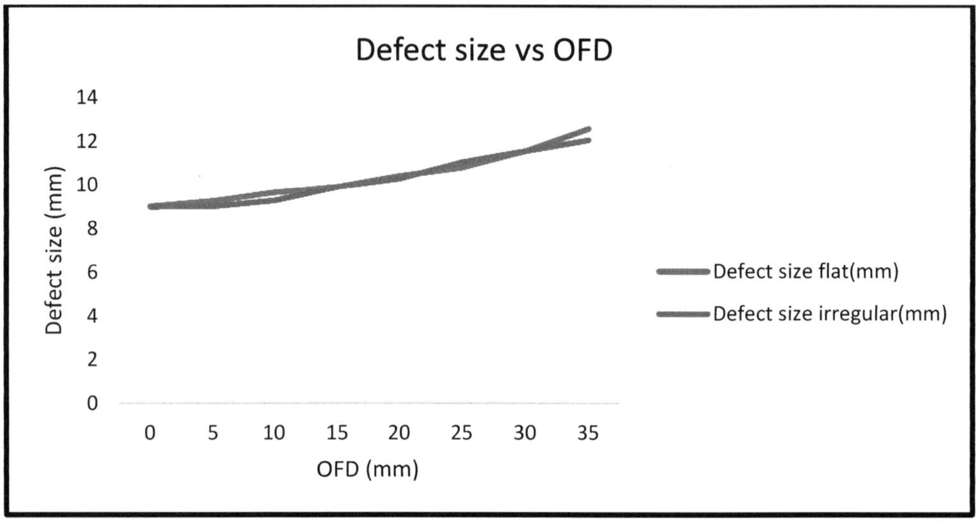

Figure 4. Graph of defect size versus OFD for flat and irregular wooden block for 9 mm hole.

The irregular surface has sharper and better image compare to flat surface wooden block as the value of gap is higher for flat surface.

3.2 Wooden block 15 mm hole

Table 2 shows the value of defect size for flat and irregular wooden block when object to film distance (OFD) increases. From the data obtained, both defect size for flat and irregular wooden block increase with increasing of OFD. The gap size of irregular surface is higher compare to flat surface as shown in Figure 5.

Table 2. Data of defect size and OFD for 15 mm hole.

Defect size flat (mm)	Defect size irregular (mm)	OFD (mm)
14.99	15	0
15.25	16.64	5
16.14	17.92	10
16.77	18.44	15
17.66	19.31	20
18.18	21.36	25
18.81	22.74	30
20.45	24.9	35

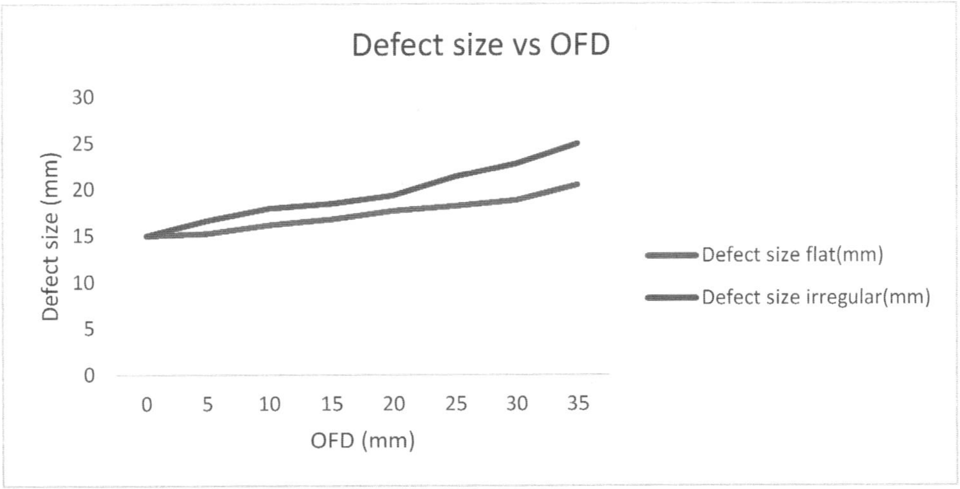

Figure 5. Graph of defect size versus OFD for flat and irregular wooden block for 15 mm hole

The flat surface wooden block give sharper and better image compare to irregular wooden block surface as the gap size for irregular is higher compare to flat surface wooden block.

4. Conclusion

This study focuses on the image and gap size value when the object to film distance increases. The value of gap size increases simultaneously with the increasing of OFD for both flat and irregular wooden surface for 9 mm and 15 mm holes. The obtained results show that irregular surface has better and sharper image compare to flat surface to 9 mm hole wooden block. Meanwhile for 15 mm hole, flat wooden block show better and sharper image compare to irregular surface.

5. References

[1] Arka D, Ayman I 2020 H Investigation of a model based reconstruction technique for neutron tomography at the PULSTAR reactor *Nucl. Inst. Meth. Phys. Res. A* **954.**

[2] Chitra P, Rani B S and Venkatraman B 2012 Study and analysis on effect of source to film distance and exposure voltage on the radiographic image 107–110.

[3] Carew R M, Viner M D, Conlogue G, Márquez-Grant N and Beckett S 2019 Accuracy of computed radiography in osteometry: A comparison of digital imaging techniques and the effect of magnification," *J. Forensic Radiol. Imaging* **19**.

[4] Lima C A S *et al.* 2020 Is the digital radiographic detection of approximal caries lesions influenced by viewing conditions? *Oral Surg. Oral Med. Oral Pathol. Oral Radiol.* **129(**2) 165–170.

[5] Haak Rand Wicht M J 2005 Grey-scale reversed radiographic display in the detection of approximal caries *J. Dent.* **33**(1) 65–71.

[6] Kim K S *et al.* 2018 Improvement of radiographic visibility using an image restoration method based on a simple radiographic scattering model for x-ray nondestructive testing *NDT E Int.* **98** 117–122.

[7] Sudhagar S, Sakthivel M, and Daniel S A A 2019 Application of image processing to radiographic image for quantitative assessment of friction stir welding quality of Aluminium 2024 alloy *Measurement* 107294.

Acknowledgement

This study was done by a final year project student from Universiti Tenaga Nasional.

IOP Conf. Series: Materials Science and Engineering 1106 (2021) 012021 doi:10.1088/1757-899X/1106/1/012021

Application of Neutron Flux from PUSPATI TRIGA Mark II Research Reactor in Enhancing Superconducting Properties of BSCCO Superconductor

Nasri A Hamid [1, a)] and Zaahidah Atiqah Mohiju[1]

[1]Nuclear Engineering and Energy Group, College of Engineering, Universiti Tenaga Nasional, Jalan IKRAM-UNITEN, 43000 Kajang, Selangor, Malaysia.

[a]Nasri@uniten.edu.my

Abstract. The PUSPATI TRIGA Mark II Research Reactor is currently been used in various applications such as Neutron Activation Analysis (NAA), Delayed Neutron Activation Analysis (DNAA), radioisotope production for medical, industrial and agricultural purposes, Neutron Radiography and Small Angle Neutron Scattering (SANS). However, there are not much research activities on utilizing neutron flux from the reactor to study properties of superconducting materials. For superconductors to be fabricated into components of electrical and magnetic devices, the sustainability of the transition temperature, TC and the transport critical current density, JC of the materials is vital to ensure the durability and effectiveness of the devices. A strong superconducting flux pinning capability of superconductors is necessary to maintain high TC and larger JC. Meanwhile, exposure of superconductors to neutron flux is able to create the self-organization of defects that are responsible in producing higher TC and JC in superconductors where the defects act as the flux pinning centres. In this paper, we report the utilization of neutron flux from the PUSPATI TRIGA Mark II research reactor in enhancing the TC and JC of bismuth-strontium-calcium-copper oxide (BSCCO) superconductor. Appropriate sample holder and shielding was fabricated as protection from radiation effect. In this work, 2212-phase BSCCO superconductor (Bi-2212 phase) samples were synthesized using the conventional solid-state reaction method. The samples underwent fast neutron irradiation with neutron flux of 2.0 × 1011 neutrons/(cm2.s) for 6 hours. Results showed that the TC of the samples improved slightly, and the JC was found to increase by more than 3 folds.

1. Introduction

The PUSPATI TRIGA Mark II research reactor known as RTP is the only nuclear research reactor in Malaysia. The reactor starts its operation in 1982 and reached its first criticality on 28 June 1982. Its compact core allows for a low critical mass and higher neutron fluxes while ensuring a high degree of safety. RTP is a pool type reactor, where the reactor core sits at the bottom of a 7-metre high Aluminium tank surrounded by a biological shield made of high-density concrete [1]. The reactor uses solid fuel elements in which the zirconium-hydride moderator is homogeneously mixed with enriched uranium with 1.0 MW thermal power [2], [3]. Demineralized water acts both as coolant and neutron moderator, while graphite acts as a reflector. The reactor was designed to effectively implement various fields of basic nuclear science studies and widely used for training and educational purposes. It incorporates facilities for advanced neutron and gamma radiation studies as well as for applications.

On the other hand, high-temperature superconductors (HTS) are widely used in various science and engineering fields such as power transmission lines, motors and generators, transformers, and Computerized Tomography (CT scan) devices. The bismuth-strontium-calcium-copper oxide (BSCCO) superconductor such as the Bi-2212 phase is a good candidate for consideration in many applications because its phase formation is highly stable with relatively high T_C and J_C [4]. In order for the superconducting materials such as the BSCCO superconductors to maintain high T_C and J_C, they must have effective flux pinning centers. Innovative technique has to be introduced to enhance the flux pinning capability of BSCCO superconductor to sustain high T_C and J_C. One of the methods to improve the flux pinning capability of superconducting materials is through neutron irradiation [5]. The effect of radiation on superconducting materials is depending on the nature and amount of pre-irradiative defects [6]. Neutron irradiation is able to produce point defects and cascades, resulted in enhancement of T_C and J_C in superconductors [7].

Zehetmayer *et al.* in their study on Hg-based high temperature superconductor discovered that the effect of neutron irradiation is more obvious at elevated temperature due to large defects created by the irradiation [8]. Aoki et al. in their work on the effect of neutron irradiation on 2223-phase BSCCO (Bi-2223 phase) and yttrium-barium-copper oxide (YBCO) superconducting tapes found that the superconducting characteristics of the tapes did not deteriorate under irradiation flux of 6.92×10^{10} neutrons/(cm^2.s) [9]. Shitamichi *et al.* stated that neutron radiation not only improve the flux pinning, but also causes the activation of the material that can be prevented by controlling the amount of impurities, using low activation elements and reduced thermal neutron fluence [10]. In addition, elimination of impurities happened when neutron irradiation introduces homogenous distribution of defects in the superconductor microstructure, and this contributed to higher Bi-2212 phase volume fraction [11]. In the YBCO superconductor that has similar characteristics and properties as the BSCCO superconductor, there were evidences of self-organization of defects due to neutron irradiation that creates a sharp distribution of defects concentration within the sink areas [6]. This directly caused defect less superconducting region with size much larger than the coherence length of the materials and promotes high T_C and J_C.

The aim of this work is to apply the electron flux from the PUSPATI TRIGA Mark II research reactor at the Malaysian Nuclear Agency in conducting research on superconducting materials. Bi-2212 phase superconductor was used as the samples with small addition of nanosized magnesium oxide (MgO) particles. The nanosized MgO particles were added to improve the texture and reducing the formation of secondary phases in the Bi-2212 phase superconductor [12]. In this work, neutron flux of 2.0×10^{11} neutrons/(cm^2.s) was utilized and the samples were irradiated for 6 hours. The transition temperature, T_C and the transport critical current density, J_C were determined using the cryogenic four-point probe equipment.

2. 2. Experimental procedure

2.1. Neutron Irradiation

Neutron irradiation of samples was conducted at the Nuclear Malaysia Agency using the PUSPATI TRIGA MARK II research reactor. Water acted as coolant of the reactor since it has good thermal and nuclear properties. Furthermore, it does not capture much neutrons throughout the irradiation process. Before irradiating the samples, there were initial preparations that need to be done such as fabrication of the aluminium sample holder, and the shielding using aluminium sheet and boron carbide powder. Estimation in details of elements' activation after termination of irradiation were also estimated. In the irradiation, neutron flux of 2.0×10^{11} neutrons/(cm^2.s) was applied and the samples were irradiated for 6 hours. Thus, total fluence of 4.32×10^{15} neutrons/cm^2 were exposed to the samples.

IOP Conf. Series: Materials Science and Engineering 1106 (2021) 012021 doi:10.1088/1757-899X/1106/1/012021

2.2. *Sample preparation and characterization*

The conventional solid-state method was used to synthesize the samples. Chemical powders were acquired from Sigma-Aldrich, USA. Molar ratio of bismuth (II) oxide (Bi_2O_3), strontium (II) carbonate (Sr_2CO_3), calcium carbonate ($CaCO_3$) and copper oxide (CuO) were mixed according to its ratio into composition of Bi: Sr: Ca: Cu = 2:2:1:2. The mixture was ground and heated at temperature of 800°C for 24 hours to remove impurities, and eventually resulted in the formation of Bi-2212 phase superconductor. The resultant powder was added with 5% weight percentage of nanosized MgO before reground and pressed with pressure of 7.0 tons. Each sample with mass of 2.0 g was palletized into a pellet with 13 mm in diameter and 3.0 mm in thickness. The samples were sintered at 840°C for 48 hours, then furnace-cooled to room temperature. Microstructure investigation was carried out using the Hitachi S3400N Scanning Electron Microscope (SEM). Table 1 shows the composition of each element in preparing each sample.

Table 1. Mass of each element in the nanosized MgO added Bi-2212 superconductor sample.

Element	Mass (g)
Bi	0.9388
Sr	0.3936
Ca	0.0900
Cu	0.2855
O	0.2875
MgO	0.0045

2.3. *Estimation of elements' neutron activation*

Stability of nucleus in the elements may change their nucleus composition with or without any exterior influence and this feature is known as radioactivity. During nuclear decay, emission of radiation may occurs. Rate of radioactive decay of each type of sample was quantified using characterization of radionuclides half-life, $T_{½}$. Since radioactive decay represents the transformation of an unstable radioactive nuclide into a more stable nuclide, which may also be radioactive, it is an irreversible event for each nuclide. Mass of elements in the samples is very crucial in estimating their $T_{½}$. For estimation of elements' neutron activation, the online WISE Uranium Project Calculator (https://www.wise.org) was used. Examples of input data and output parameters from the calculator are as shown in Figure 1.

2.4 *Sample holder and shielding*

A cylindrical aluminium sample holder of 0.05 m in diameter and 0.15 m in height was fabricated for the irradiation. Boron carbide (B_4C) powder was used as the filler in the sample holder. The filler is able to control the neutron irradiation from the reactor since it contains high boron's density, non-reactive towards chemicals and excellent refractory material. During irradiation, B_4C absorbed neutron and decayed from boron (^{10}B) to helium (^{4}He) and lithium (^{7}Li). Maximum weight of 15 grams B_4C is needed to occupy the empty spaces inside the sample holder. Radioactivity for the 30 g of Aluminium sample holder and 15 g of B_4C powder was 8.359 mCi ($^{27}Al \rightarrow {}^{28}Al$) and 1.103 pCi ($^{13}C \rightarrow {}^{14}C$), respectively.

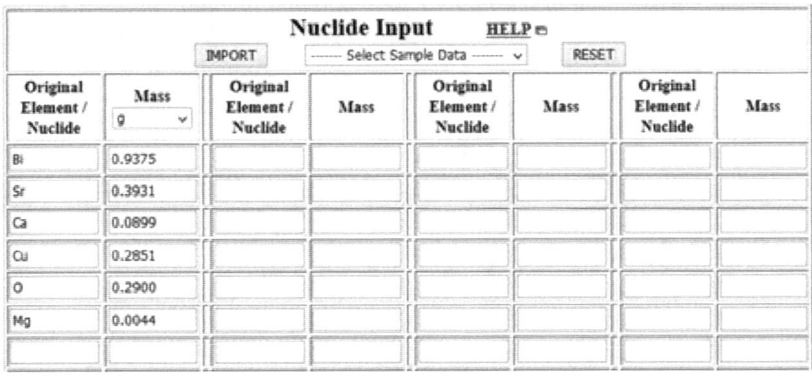

(a)

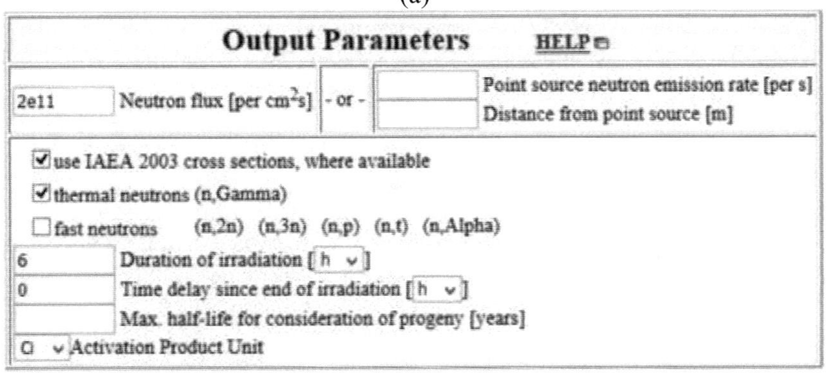

(b)

Figure 1. (a) Data input for original elements existed in the sample, and (b) output parameters according to the reactor setting.

2.5 Measurement of T_C and J_C

Determination of superconducting properties for superconductors are extremely crucial for industrial applications. Transition temperature (T_C) and transport critical current density (J_C) for each sample were determined using the cyrogenic four-point probe equipment. Silver conductive paint was applied at the joints in between the probes and surface of the sample to ensure the connections are firm. The onset transition temperature, $T_{C,onset}$ and the zero transition temperature, $T_{C,zero}$ were determined from the plotted resistivity against temperature graph. The transport critical current, I_C was determined from the I-V curve using the 1μV/cm criterion. The transport critical current density, J_C was calculated by dividing I_C with the sample cross-sectional area, A.

3. Results and Discussion

Table 2 shows the radioactivity decay of the Bi-2212 phase superconductor sample. According to the estimated calculation via WISE Uranium Project Calculator (https://www.wise.org), the highest radioactivity detected is copper-63 (Cu-63) that decayed to copper 64 (Cu-64) with radioactivity of 12.76 mCi. Particular protection such as shielding is required in handling radionuclide materials and it has to take into consideration on all the safety measures to avoid any health issue and uncontrolled circulation of radioactive materials.

Table 2. Radioactive decay of each element in the Bi-2212 superconductor sample.

Element		Mass of element	Activity	Half-life, $T_{½}$
From	To			
Bi-209	Bi-210	938.8 mg	16.8 μCi	5.012 d
Sr-84	Sr-85	2.111 mg	135.4 nCi	64.84 d
Ca-40	Ca-41	86.99 mg	9.841 pCi	140 x10^3 a
Cu-63	Cu-64	195.5 mg	12.76 mCi	12.70 h
O-18	O-19	650.7 μg	18.03 pCi	26.91 s

Nuclear radiation such as neutron flux is able to create defects in the microstructure of superconducting materials, and formed flux pinning centers that will improve their superconducting properties [7]. Nevertheless, it may also lead to reduction of T_C due to disorientation of grain boundaries that is strongly related to oxygen vacancy in the copper dioxide (CuO_2) planes and multiple lattice distortion around them [13]. Figure 2 shows the SEM micrographs of Bi-2212 phase superconductor for both the non-irradiated and neutron irradiated samples. The micrograph of sample that was exposed to neutron flux shows more random grain orientation and higher degree of texturing. However, several types of material defects such as dislocation networks, impurities and precipitates may exists within the superconducting matrix of the irradiated sample [7,11].

(a) (b)

Figure 2. SEM micrographs of (a) non-irradiated, and (b) neutron irradiated Bi-2212 superconductor samples.

Table 3 shows the T_C and J_C of non-irradiated and neutron irradiated Bi-2212 phase superconductos. The $T_{C,zero}$ for both samples are almost similar but the $T_{C,onset}$ of neutron irradiated sample increased significantly. Variation of T_C relies on the number of holes provided on the CuO_2 layers [13]. In such situation, the number of holes are artificially modified by the interaction between neutron flux and the CuO_2 layers of the Bi-2212 phase superconductor, and contributed to the enhancement of T_C.

Table 3. T_C and J_C of non-irradiated and neutron irradiated Bi-2212 superconductor samples.

Sample	$T_{C,\ zero}$ (K)	$T_{C,\ onset}$ (K)	J_C (A/cm^2)
Non-irradiated	84	100	2.86
Irradiated	85	120	10.44

In addition, Table 3 shows the results of J_C for both of the Bi-2212 phase superconductor samples. There is a significant increases of I_C in the neutron irradiated sample. Critical current, I_C and J_C can be enhanced by incorporating a high density of extended defects that act as effective pinning centres and resulted in significant super-currents flow along the CuO_2 planes [14]. The existence of a strongly linked network of percolative paths, combined with good interplanar coupling also contributed to higher J_C [15].

4. Conclusion

The potential to utilize neutron flux from the PUSPATI TRIGA Mark II research reactor to study superconducting properties of superconductor materials such as BSCCO superconductor is very encouraging. In this study, neutron flux has shown its ability to improve superconducting properties of Bi-2212 phase superconductor with neutron fluence of 4.32×10^{15} neutrons/cm^2. From the SEM analysis and results of T_C and J_C, neutron flux is able to create homogenous distribution of defects in the superconductor microstructure. There is also the possiblity that the neutron flux initiates production of holes in the CuO_2 planes of the Bi-2212 superconductor. This indicates that neutron irradiation has the ability to enhance the superconducting properties of superconductor. As such, there are potentials for superconductors to be applied in various electrical and magnetic devices and equipment.

5. References

[1] Alnour I A, Wagiran H, Ibrahim N, Hamzah S, Wee B S, Elias M S and. Karim J A 2013 Determination of neutron flux parameters in PUSPATI TRIGA Mark II research reactor, Malaysia, *J. Radioanal. Nucl. Chem.* **296**(3) (doi: 10.1007/s10967-012-2375-9) 1231-1237.

[2] Chen W K, Rabir M H, Zin M R M, Usang M D, Bayar A M J and Hamzah N S 2016 Neutron flux and power in RTP core-15 *AIP Conf. Proc.* **1704** (https://doi.org/10.1063/1.4940114) 050018.

[3] Rabir M H, Bayar A M J, Hamzah N S, Mustafa M K A, Karim J A, Zin M R M, Ismail Y, Hussain M H, Husin M Z M, Dan R M, Ismail A R, Husain N, Khan Z K A J, Yakin S R M, Saad M F and Masood Z 2018 Measurement and simulation of thermal neutron flux distribution in the RTP core", *IOP Conf. Ser.: Mat. Sci. and Eng.* **298** (doi: 10.1088/1757-899X/298/1/012029) 012029.

[4] Elschner S, Breuer F, Wolf A, Noe M, Cowey L, and Bock J Characterization of BSCCO 2212 bulk material for resistive current limiters", *IEEE Trans. Appl. Supercond.*, vol. 11, no. 1, 2001, pp.2507-2510. doi:10.1109/77.920375 7

[5] Mihalache V, Totovana A, Sandu V, Popa S, Aldica G, and Iyo A 2005The influence of neutron irradiation on $(B_{0.65}C_{0.35})Ba_{1.4}Sr_{0.4}Ca_2Cu_3O_z$ superconducting phase: the role of the grain edge *J. Superconductivity* **18**(4) (doi: 10/1007/s10948-005-0029-7) 461-467.

[6] Sandu V, Popa S, Sandu E. Di Gioacchino D, and Tripodi P 2006 Neutron-irradiated superconducting $YBa_2Cu_3(Li)O_{7-\delta}$ with improved irreversibility *J. Optoelectronics and Adv. Mat.* **8**(1) 391-395.

[7] Pande C S Microstructure aspects of high and low T_C superconductors 2000*Mater. Phys. Mech.*, **2** 1-9.

[8] Zehetmeyer M, Eisterer M, Kazakov S M, Karpinski J, Wisniewski A, Puzniak R, Daignere A, and Weber H W 2004 Effects of neutron and electron erradiation on superconducting HgBa2CuO4+δ single crystals *Physica C: Superconductivity and its Appl.*, **408-410** 30-31.

[9] Aoki T, Ueda H, Ishiyama A, Miyahara N, Kashima N, and Nagaya 2011 S Effect of neutron irradiation on high-temperature superconductors *IEEE Trans. Appl. Supercond.* **21** 3200-3202.

[10] T. Nakano S M, Terai T, Yamawaki M, and Hoshiya T 2003 Activation of high-T_C superconductors due to neutron irradiation *Physica C: Superconductivity and its Appl.* **392-396**(1) (https://doi.org/10.1016/S0921-4534(03)00990-0) 254-258.

[11] Mohiju Z A, Hamid N A, and Abdullah Y 2017 Enhancement of flux pinning properties in nanosized MgO added Bi-2212 superconductor through neutron irradiation *AIP Conf. Proc* **1799** (http://dx.doi.org/10.1063/1.4972928 040004.

[12] Ilyushechkin A, Agranvoski I, Altman I and Choi M 2010 Effect of MgO nanoparticles embedded into Bi-2212/Ag tapes on the microstructure and superconducting properties", *Materials Sci. and Eng. B* **167** (doi:10.1016/j.mseb.2010.01.030) 60–64.

[13] Kleiner R. and Buckel W., *Superconductivity – Fundamentals and Applications*. Weinheim, Geramny: Wiley-VCH, 2004, pp. 94-100.

[14] Silver T, Pan A V, Ionescu M, Qin M J and Dou S X 2002 Developments in high temperature superconductivity *Annu. Rep. Prog. Chem., Sect. C: Phys. Chem.* **98**. (https://doi.org/10.1039/B111186H) 323-373.

[15] Hamid N A and Abd-Shukor R 2001 Transport critical current density of Ag-sheathed (Tl,Cr)$Sr_2CaCu_2O_7$ high temperature superconductor tapes *Supercond. Sci. and Technol.* **14**(2) 113-116.

Acknowledgement

This research project is jointly funded through Universiti Tenaga Nasional BOLD Project Code 10436494/B/2019012 and IAEA Research Contract No. 18272. The authors thank the Malaysia Nuclear Agency for the collaboration extended in this project.

.

IOP Conf. Series: Materials Science and Engineering 1106 (2021) 012022 doi:10.1088/1757-899X/1106/1/012022

Radiation measurement of boron neutron capture therapy research facility at Triga Mark-II research reactor Malaysia

Safwan Shalbi[1], N Sazali[1,a)], Muhammad Aliff Ashraff Rosdi[1], Faridah Mohamad Idris[2]

[1]Faculty of Mechanical Engineering, Universiti Malaysia Pahang, 26600 Pekan, Pahang, Malaysia.
[2]Malaysian Nuclear Agency, 43000 Kajang, Selangor, Malaysia.

[a]azlianie@ump.edu.my

Abstract. Proper shielding ionizing radiations is a mandatory requirement in all nuclear facilities owing to the harmful effects of radiation on workers and general public. The present work is aimed to measure neutron fluxes and gamma fluxes around the Boron Neutron Capture Therapy (BNCT) research facility at the PUSPATI TRIGA Reactor. This measurement is needed in the subsequent safety assessment so as to ensure the research facility is safely operated

1. Introduction

Boron Neutron Capture Therapy (BNCT) is one of the promising methods in order to cure cancer by using neutron beam in which can be obtained commonly from research reactor or nuclear generator such as an accelerator. BNCT was the method used the combination of low energy of neutron (slow neutron) irradiation and the targeting of a tumour site with a proper boron containing compound. In general, most of the BNCT studies use neutron source from the research reactors [1]. It is upon this general observation that a BNCT facility was recently proposed to be installed at the PUSPATI TRIGA Reactor. The development of the BNCT facility in Malaysia required neutron source which only available at the thermal column of the Malaysia research reactor. The specification of neutron flux and the gamma dose rate must be considered for the development of the BNCT facility as a safety precaution for this research. Based on previous research, the thermal column is identified as a suitable place for BNCT facility. TRIGA MARK II was believed to supply a sufficient a quality neutron beam and as well to develop BNCT facility [2]. This research is thus aimed to ensure the newly developed BNCT research facility is safely operated with proper shielding design in light of this objective, a set of radiation measurements was taken during commissioning of the facility [3]. In addition, it is expected that this work provides a basic guideline on the neutron and gamma dose measurements for new nuclear research facility in Malaysia.

2. Methodology

Neutron and gamma dose measurement around the new BNCT research facility is conducted to ensure that the radiation dose around the research facility is in a safe range in both situations when the research facility is in operation as well as not in operation. This measurement used the Ludlum model 2363 survey meter and the model 42-42L PRESCILA neutron detector. This survey meter features lightweight neutron detection system, independent gamma, and neutron measurements, neutron detection range: 1 μSv/hr to 10000 μSv/hr, gamma detection range: 9 μSv/hr to 8770 μSv/hr, a scaler, and data logging capabilities. Figures 1 shows the neutron survey meter that was used in this research and Figure 2 show the portable gamma survey meter.

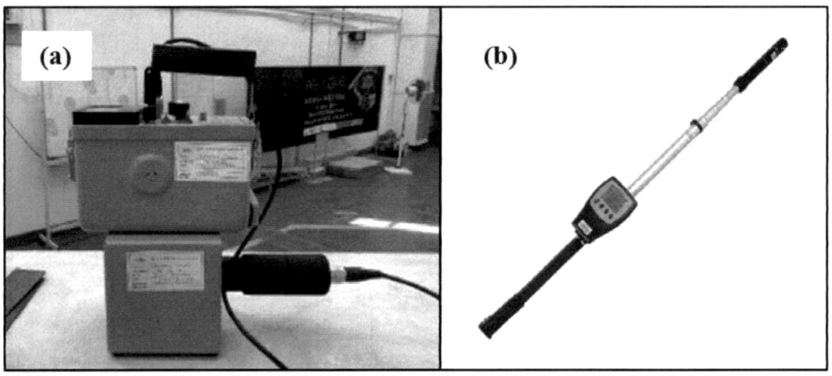

Figure 1.(a) Ludlum model 2363 survey meter with the model 42-42L PRESCILA neutron detector and (b) a portable gamma survey meter.

2.1. Dose measurement before, during and after installation of BNCT research facility
Before the measurements started, the selected locations will be marked using a tape. The survey meter battery and the calibration date are checked in order to ensure the data taken are reliable. Next, the surrounding of the measurement area is checked to make sure that it is safe from any danger. For these objectives, the dose measurement is carried out by measure the dose before, during and after the installation of collimator of BNCT research facility with 750kW reactor power without opening the beam shutter. There are two type of height for this dose measurement, first on the beamline height and another is 2-meter height from the beamline. Three measurements will be taken for each of the locations. The highest reading of the radiation dose will be selected.

IOP Conf. Series: Materials Science and Engineering 1106 (2021) 012022 doi:10.1088/1757-899X/1106/1/012022

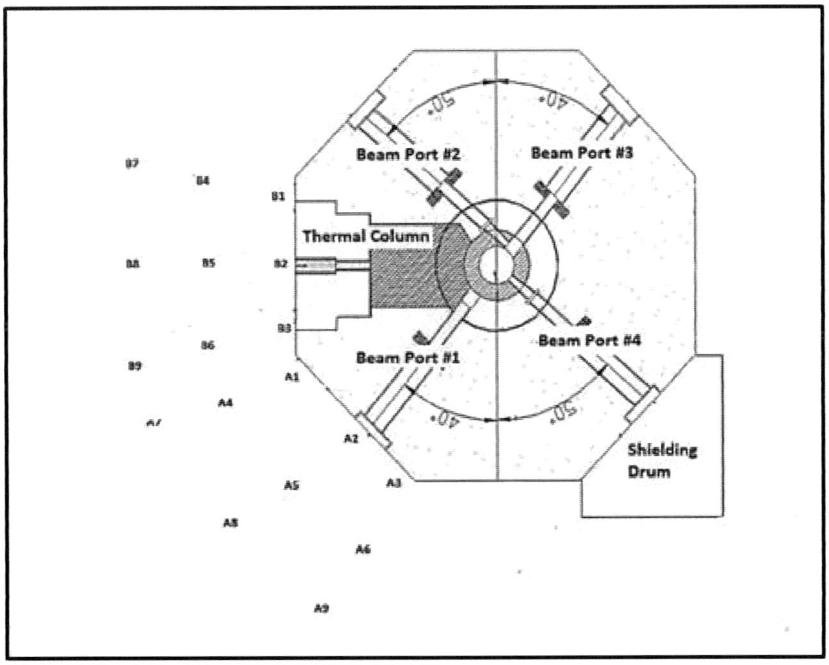

Figure 2. The position for dose measurement at BNCT research facility.

2.2. Neutron and gamma measurement at BNCT research facility during standard operation power

Once the BNCT installation in the RTP thermal column is carried out, testing and commissioning should be done to ensure that the collimator and shielding of BNCT used meets the standards. Therefore, a test was performed on the RTP thermal column to see the capability of the BNCT shielding and collimator. In addition, this test also aims to ensure that the radiation produce from the collimator in accordance with the prescribed rules and does not endanger the safety of workers in the reactor itself. This test is done by increasing the reactor power from 0 kW to a 250 kW. When the power of the reactor reaches 10 kW,50kW, 100kW and 250kW the measurements of the neutron dose, gamma dose starts to be taken at each of the marked locations the measurements will be taken for each of the locations. The highest reading of the radiation dose will be selected. After the measurements of all the doses are completed at the selected locations, the measurements are repeated for three times for average reading. Survey meters are the main instruments in this research to ensure no excess neutron radiation or gamma beam can endanger the safety of workers. Figure 3 display the position for the neutron and dose mapping for BNCT research facility.

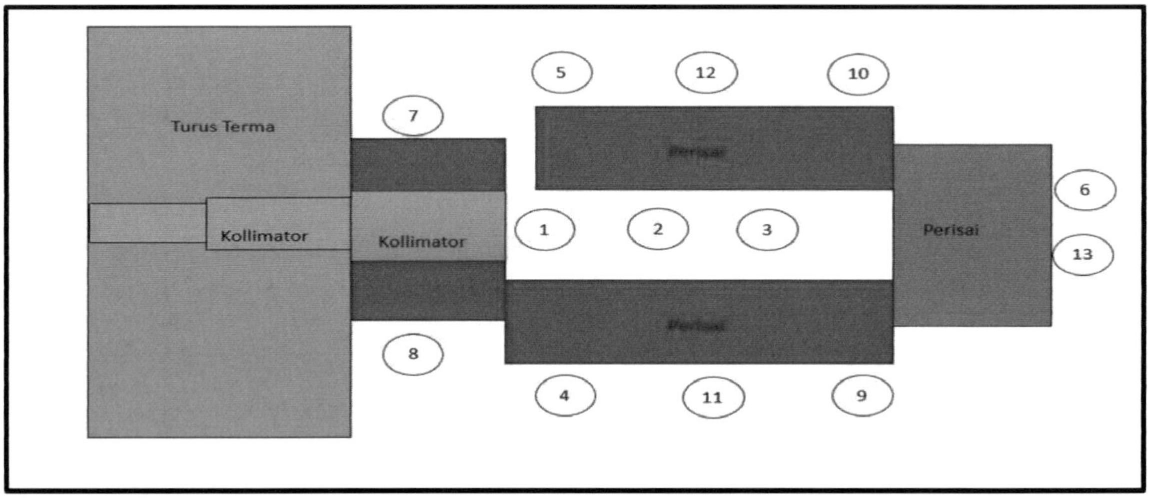

Figure 3. The position for neutron and gamma dose measurement at BNCT research facility

3. Results and discussion

3.1 Dose measurement before, during and after installation of BNCT research facility at RTP

Table 1 shows the dose measurement on beamline height before, during and after the installation of collimator at the different position around BNCT research facility and Table 2 display the dose measurement on 2-meter from the beamline high before, during and after the installation of collimator at the different position. Position B1 shows the highest beamline dose of 0.50 μSv due to a leak in the RTP thermal column door itself. For the 2-meter top position, the side door of the RTP thermal column door, namely B1, B2 and B3, also has a high dose reading due to the same leak. However, leaks that occur in the RTP thermal columns are still within the controlled radiation range set and do not endanger the safety of workers [6]. During installation in RTP thermal columns where the reactor does not operate (0 kW), dose mapping is very important because collimator installation involves the production of existing old collimators in RTP thermal columns.

When the installation of the new collimator is performed, there is a dose difference that occurs at positions B1, B2, B3, B4, B5 and B8. All of these positions indicate the dose reading increases as the beam plug is pulled out of the RTP thermal column. This is because the radioactivity of the beam plug itself is high even though this installation is done 3 days after the reactor is not operating (shutdown) [7]. Position B2 showed the highest increase from 0.05 μSv to 0.25 μSv. Dose mapping for reactors operating at full power was performed one week after collimator testing was performed at RTP thermal columns.

Upon 750 kW of reactor power, each position either A or B indicates an increase in beamline and also at the 2-meter height. At the beamline, positions B1 and B3 show the highest dose readings with 15.0 μSv while for dose records above 2-meters it shows B2 and B3 positions have the highest dose with 12.0 μSv. Through this research the maximum and minimum dose measurement can be identified and referenced to formulate and validate the irradiation parameters for BNCT in addition to being able to help regulate radiation safety in the research reactor itself. Positions B1, B2 and B3 received the highest reading points while B8 was the reading point receiving the minimum dose in this study. With this dose mapping, workers' safety against radiation will be more assured because it can plan the implementation of the way of working on BNCT research facility.

Table 1. The dose measurement on beamline height before, during and after the installation of collimator at the different position around BNCT research facility.

	Before Installation(0kW)	During Installation(0kW)	After Installation (750kW)
Position		Beamline Dose, µSv	
A1	0.02 ± 0.02	0.02 ± 0.00	9.0 ± 0.03
A2	0.01 ± 0.02	0.01 ± 0.00	1.0 ± 0.00
A3	0.07 ± 0.02	0.07 ± 0.01	7.0 ± 0.02
A4	0.03 ± 0.06	0.03 ± 0.00	3.0 ± 0.04
A5	0.25 ± 0.00	0.25 ± 0.02	2.5 ± 0.00
A6	0.4 ± 0.00	0.4 ± 0.01	4.0 ± 0.00
A7	0.15 ± 0.01	0.15 ± 0.00	1.5 ± 0.00
A8	0.10 ± 0.02	0.10 ± 0.00	1.0 ± 0.03
A9	0.17 ± 0.04	0.17 ± 0.03	1.7 ± 0.06
B1	0.50 ± 0.01	0.70 ± 0.04	15.0 ± 0.05
B2	0.05 ± 0.00	0.25 ± 0.08	2.0 ± 0.03
B3	0.06 ± 0.03	0.09 ± 0.00	15.0 ± 0.03
B4	0.07 ± 0.03	0.07 ± 0.00	2.5 ± 0.03
B5	0.02 ± 0.00	0.35 ± 0.03	5.0 ± 0.02
B6	0.07 ± 0.09	0.07 ± 0.01	2.5 ± 0.01
B7	0.03 ± 0.01	0.03 ± 0.01	1.0 ± 0.00
B8	0.03 ± 0.03	0.10 ± 0.01	1.0 ± 0.00
B9	0.03 ± 0.02	0.03 ± 0.02	1.5 ± 0.003

Table 2. The dose measurement on 2-meter from the beamline height before, during and after the installation of collimator at the different position around BNCT research facility.

	Before Installation(0kW)	During Installation(0kW)	After Installation (750kW)
Position		Dose at 2-meter height, µSv	
A1	0.01 ± 0.04	0.01 ± 0.00	7.0 ± 0.03
A2	0.06 ± 0.03	0.06 ± 0.00	1.5 ± 0.03
A3	0.04 ± 0.03	0.04 ± 0.00	3.0 ± 0.03
A4	0.03 ± 0.03	0.03 ± 0.00	3.0 ± 0.03
A5	0.05 ± 0.06	0.05 ± 0.00	4.0 ± 0.03
A6	0.04 ± 0.01	0.04 ± 0.00	3.0 ± 0.03
A7	0.05 ± 0.01	0.05 ± 0.00	1.7 ± 0.03
A8	0.04 ± 0.01	0.04 ± 0.00	1.5 ± 0.03
A9	0.05 ± 0.02	0.05 ± 0.03	1.5 ± 0.03
B1	0.45 ± 0.03	0.55 ± 0.00	10.0 ± 0.03
B2	0.55 ± 0.03	0.65 ± 0.00	12.0 ± 0.03
B3	0.55 ± 0.00	0.80 ± 0.01	12.0 ± 0.03
B4	0.01 ± 0.00	0.01 ± 0.01	3.5 ± 0.03

B5	0.15 ± 0.00	0.35 ± 0.03	2.0 ± 0.03
B6	0.06 ± 0.03	0.06 ± 0.02	1.5 ± 0.03
B7	0.09 ± 0.00	0.09 ± 0.01	0.7 ± 0.03
B8	0.08 ± 0.00	0.06 ± 0.00	0.5 ± 0.03
B9	0.08 ± 0.00	0.08 ± 0.02	0.7 ± 0.03

3.2 Neutron and gamma measurement at BNCT research facility during standard operation power

Neutron beam at 13 different positions in the RTP thermal column section were recorded to test the newly installed collimator and test the leakage that occurred for the BNCT research facility in the thermal column. Positions 1, 2 and 3 are Radiation areas for BNCT applications that require a high neutron flux rate of 1.0 x 10^ 9 neutron.cm- 2 s- 1 [8]. This application only uses a low reactor power of 250 kW. Therefore, this study is limited to the power rate for this BNCT application only.

Table 3 shows the recorded neutron beam readings for positions 1, 2 and 3 measured at a power of 50 kW due to the safety factor of the worker which may cause excessive radiation exposure. Neutron surveillance meters are different from TLD that have access to record dose in areas that are farther and safer to use than neutron survey meters. The result of neutron dose for locations 1,2 and 3 clearly indicate that the collimator in the installation works well through a dose increase of more than 200%. Positions 4 to 13show that the shielding for the BNCT research facility are safe for RTP. This test can also look at identifying leaks found in the BNCT research facility itself. Up to 250kW, the maximum dose of neutrons recorded at position 8 is 6.26 μSv while the minimum dose recorded for the same power is 0.93 μSv. From a power of 10 kW to 250 kW. The results clearly show that the collimator works well to supply neutrons for BNCT applications and the shielding used also work well and record neutron doses that are still controlled and meet the standards for radiation workers.

Table 3. Neutron dose measurement at different reactor power and position around BNCT research facility.

	Neutron Dose, μSv			
Position	10 kW	50 kW	100 kW	250 kW
1	8.83 ± 1.26	2500.00 ± 0.18	-	-
2	14.00 ± 1.00	820.00 ± 0.15	-	-
3	12.87 ± 0.81	56.00 ± 0.16	-	-
4	0.06 ± 0.04	0	0.03 ± 0.02	2.06 ± 0.03
5	0	0.13 ± 0.04	0.08 ± 0.02	1.00 ± 0.33
6	0	0.04 ± 0.08	0.09 ± 0.01	0.93 ± 0.47
7	0.06 ± 0.04	0.10 ± 0.02	0.20 ± 0.11	6.26 ± 2.21
8	0.13 ± 0.06	0.08 ± 0.00	0.20 ± 0.01	2.47 ± 0.38
9	0	0.15 ± 0.06	0.02 ± 0.03	1.37 ± 0.59
10	0	0.09 ± 0.01	0.17 ± 0.10	0.94 ± 0.32
11	0	0.04 ± 0.08	0.01 ± 0.02	1.11 ± 0.33
12	0	0.09 ± 0.01	0.09 ± 0.01	1.20 ± 0.39
13	0.08 ± 0.01	0.04 ± 0.07	0.06 ± 0.06	2.04 ± 0.73

Referring to Table 4, the gamma produced by testing show the recorded dose is greater than the neutron dose. If we look at points 1, 2 and 3 which are the current flow and also the study area of BNCT application, the gamma dose at 50 kW power is higher than the recorded dose during 250 kW power. This occurs due to the occurrence of gamma instability and the mixing of gamma dose during collimator installation. However, the gamma ray dose at the beamline is still under control and will not interfere with BNCT operations in the future [8]. Positions 7 and 8 that need to be considered because they have a high gamma ray dose rate of 28.47 μSv and 22.08 μSv. Gamma ray leaks at positions 7 and 8 need to be improved by placing lead shields that have the ability as shields for gamma beam [9]. Improving the shielding is very important to ensure the safety of employees remains at a controlled level. At a power of 250 kW, the maximum gamma ray dose is recorded at position 1 while the minimum dose recorded is at position 9. With this test, it can help employees to take optimistic measures before conducting research of BNCT on RTP in the future.

Table 4. Gamma dose measurement at different reactor power and position around BNCT research facility

	Gamma Dose, μSv			
Position	10kW	50kW	100kW	250kW
1	3706.67 ± 4.5	25300.00 ± 5.50	500.00 ± 0.98	7000.00 ± 3.80
2	92.33 ± 9.71	41366.67 ±0.03	800.00 ± 1.59	3000.00 ± 0.83
3	403.33 ± 4.16	3275.00 ± 3.90	700.00 ± 0.07	500.00 ± 4.82
4	0.05 ± 0.03	0.73 ± 0.40	0.98 ± 0.07	2.28 ± 0.65
5	0.08 ± 0.02	1.36 ± 0.15	2.88 ± 1.09	4.62 ± 1.22
6	0.06 ± 0.01	1.46 ± 0.62	1.52 ± 0.42	2.69 ± 0.14
7	0.20 ± 0.03	6.55 ± 0.55	12.53 ± 1.56	28.47 ± 1.10
8	0.05 ± 0.03	4.44 ± 1.15	8.96 ± 1.31	22.08 ± 0.32
9	0.04 ± 0.01	0.49 ± 0.08	0.66 ± 0.34	0.61 ± 0.23
10	0.04 ± 0.01	0.38 ± 0.08	0.63 ± 0.06	1.41 ± 0.39
11	0.05 ± 0.01	0.66 ± 0.32	0.35 ± 0.01	0.74 ± 0.26
12	0.06 ± 0.02	0.34 ± 0.10	0.56 ± 0.21	1.24 ± 0.51
13	0.03 ± 0.00	2.71 ± 0.76	3.70 ± 0.36	7.26 ± 0.03

4. Conclusions

Referring this research was conducted to provide a basic safety assessment of the BNCT facility installed at the PUSPATI TRIGA reactor. To accomplish, a set of radiation measurements were taken around the newly-installed BNCT facility during its commissioning. Based on the measurements, it can be safely concluded that the collimator and shielding of the BNCT were properly designed and installed

5. References

[1] Chiragkumar J G and Malleshappa N N 2015 Selective cancer treatment by Boron Neutron Capture Therapy (BNCT)–a review *Int. J. of Pharmaceut. Chem. and Analy.* **2**(3) 136-138.

[2] Munem E M E A 2007 *Neutron flux measurements with Monte Carlo verification at the thermal column of a TRIGA MARK II reactor: Feasibility study for a BNCT facility* (Universiti Sains Malaysia).

[3] Wagiran H 1997 *Prinsip Asas Sinaran dan Reaktor Nuklear* (UTM).

[4] Lee E, Lee C W, Cho G 2018 Radiation safety analysis for the A-BNCT facility in Korea *Appl. Radiat. and Isot.* **142** 92-103.
[5] Shalbi S, Salleh W N W, Idris F M Rosdi M A A Sarkawi, M S Jamsari, N L, Nasir N A M 2018. Neutron measurement at the thermal column of the Malaysian Triga Mark II reactor using gold foil activation method and TLD *IOP Proc. MSE* **298**(1) 012031.
[6] Bryce G K and Heinmiller G R 1997 Worker Radiation Health Protection*Follow-up Report* **2**.
[7] Lamarsh A J B J R 2001 Introduction to nuclear engineering 3r ed. (United States of America: Prentice-Hall).
[8] International Atomic Energy Agency 2001 Current status of neutron capture therapy *IAEA-TECDOC-1223*.
[9] Erdem M Baykara, O Doğru M and Kuluöztürk F 2010 A novel shielding material prepared from solid waste containing lead for gamma ray *Radiat. Phys. and Chem.* 79(9) 917-922.

Acknowledgement
Authors would like to extend their gratitude to Ministry of Higher Education Malaysia and Universiti Malaysia Pahang (UMP) with grant number RDU1902615 and PGRS200304.

IOP Conf. Series: Materials Science and Engineering 1106 (2021) 012023 doi:10.1088/1757-899X/1106/1/012023

Study the dose profile of low energy electron accelerator

Leo Kwee Wah[1,a),]M. Azhar[1], R. M. Chulan[1], S. A. Hashim[1], M. Mokhtar[1], Khaidawaton[1], H. Baijan[1], R. M. Sabri[1]

[1]Malaysian Nuclear Agency, Bangi 43000 Kajang. Selangor. Malaysia

[a]leo@nuclearmalaysia.gov.my

Abstract. This paper describes dose mapping method for Low Energy Electron Accelerator (LEEA). This locally designed LEEA with the former energy of 140 keV will be upgraded to 300 keV. In the setup, the electron beam is energized by the electric field with high voltage power supply, scanning and passing through the titanium foil, to irradiate the sample. The dose profile by the energetic beam mapping results have been obtained by using Cellulose Triacetate (CTA) dosimeter.

1. Introduction

A locally own developed accelerator technology, the so-called low energy electron accelerator (LEEA) [1][2][3] has been undertaken by upgrading the low energy electron accelerator from 140 keV to 300 keV, as shown in Figure 1. For the upgrading stage, several auxiliary systems were upgraded such as insulating gas system [2], scanning power supply [3], window cooling system [4], filament power supply [5] and etc. In the current stage, the beam commissioning is carried on with the dose mapping profile measurement by using Cellulose Triacetate (CTA) dosimeter, whereby the dose profile mapping method and results shall be described and discussed.

2. CTA

CTA film dosimeter is formed with a long and thin shape as shown in Figure 2. It is a colourless, transparent film with the specification as shown in Table 1 [6]. It is utilized for measuring radiation dose in gamma, electron beam or X-ray applications and industrial processes such as sterilization of medical device, crosslinking of cable and curing and coating of polymer surface and irradiation of food.

Table 1. Specification of the CTA [6].

Parameters	Figure
Film thickness	125 μm (~15 mg/cm^2)
Dimension	width: 8mm,length: 100 m/reel
Composition	cellulose triacetate, triphenyl phosphate
Analysing wavelength	280 nm

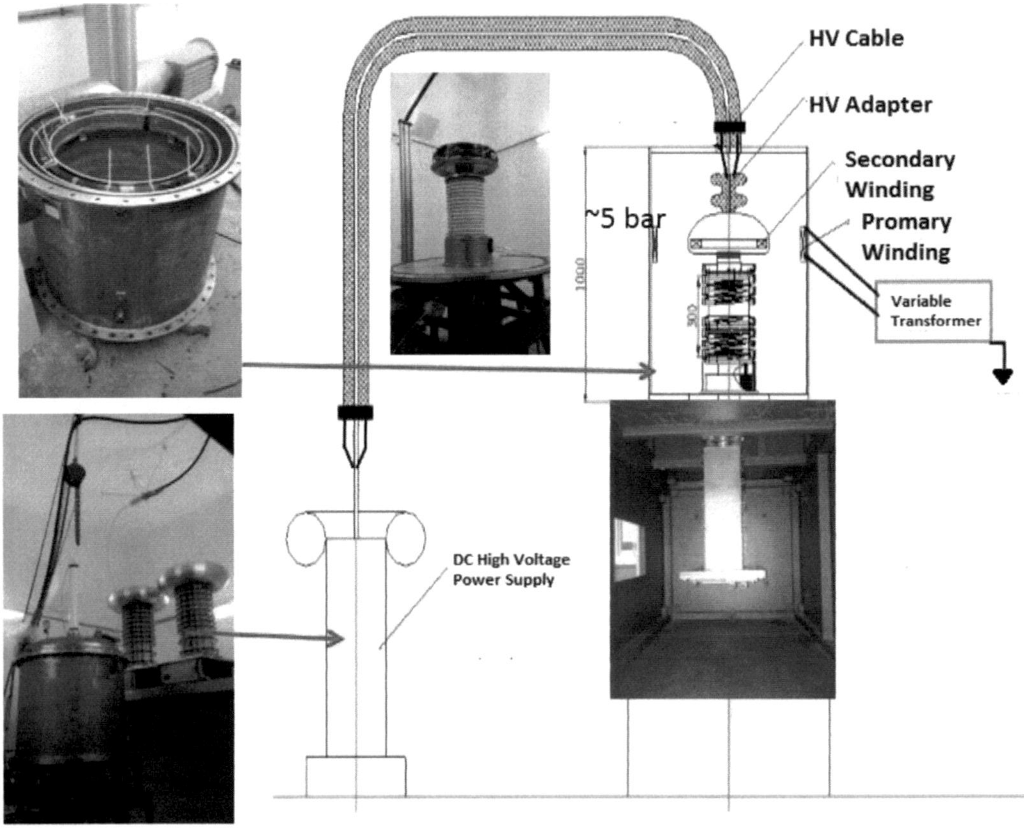

Figure 1. Low energy electron accelerator.

Figure 2. CTA film dosimeter.

The linearity dose response characteristic as shown in Figure 3 make it user friendly with simple measurement procedure [6]. In principle, the optical density of CTA is increasing by the radiation exposure. The absorbed dose is evaluated from the measured value of the increment of the optical density with the wavelength of 280 nm during irradiation. The CTA is inserted into an UV spectrophotometer and its optical density is measured by the spectrophotometer. The absorbed dose versus the length of CTA is recorded by the spectrophotometer as well.

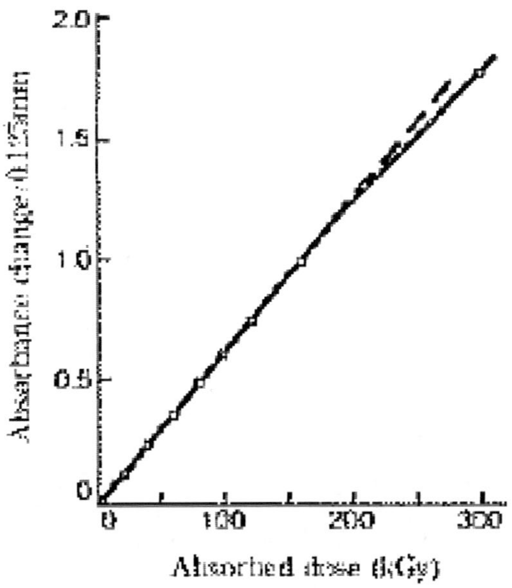

Figure 3. Dependence of absorbance change on the observed dose (measured at 280 nm).

3. Dose profile measurement

Typically, the industrial electron accelerator is equipped with a magnetic scanning system in horizontal and vertical directions [7][8]. The scanning system is required to avoid the heat at the titanium window and to get a homogeneity dose profile along the titanium window. In this experiment, a desired frequencies of 80 Hz at horizontal axis and 800 Hz at vertical axis are fixed .Whereby the excitation current of individual coils is set according to the beam energy [3]. For the dose profile measurement setup, the electron beam energy is fixed as 200 keV and excitation coil currents at horizontal axis is adjusted for ~1.0 A and ~0.5 A. Whereby the dose profile is expected to distribute uniformly along the limited length of the along titanium window [3]. A CTA film is hold on the sample holder and moved by a conveyor with fixed speed as shown in Figure 4. After irradiation, the irradiated CTA is evaluated by a UV spectrophotometer as shown in Figure 5.

Figure 4. Irradiated CTA is hold on a sample holder with a constant speed of conveyor.

Figure 5. Model GENENSYS 5 UV- spectrophotometer.

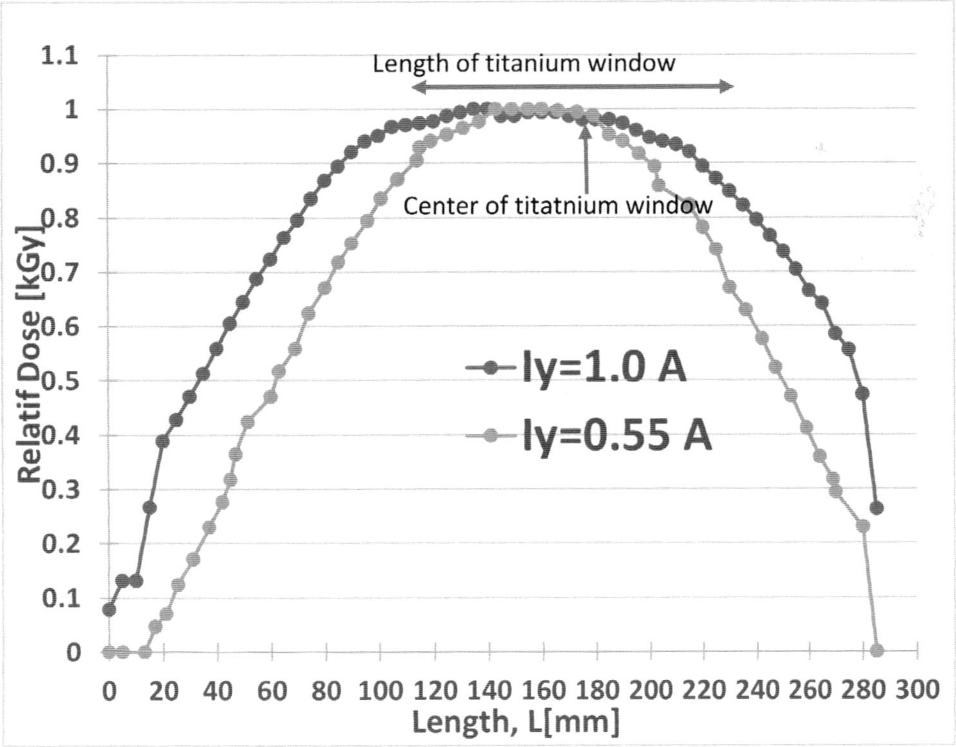

Figure 6. Dose profile dependence of y-axis excitation coil current for 1.0 A and 0.55 A.

As a result, the dose profile with dependence of horizontal excitation coils has been evaluated and studied. The result is shown in Figure 6, the dose profile length is increasing with the increment of the horizontal excitation currents. It is obvious, as recorded in our experiment; the homogeneity dose profile of with 1.0 A excitation coil current is longer than the homogeneity dose profile of 0.55 A excitation coil current. The results are within our expectation, whereby the scanning beam angle is increased with the increment of the magnetic field which induced by the excitation coil current. Although the dose profile with desired length and homogeneity along the window length as 19.25 cm [3] has been clarified. But beyond the window length, the left-right dropping slope of the dose graph is quite long ~ 5 cm that indicated the total detected dose length is ~ 30 cm. The remained dose might be caused by the secondary photon produced by the interaction between electron beam and the metal columns as shown in Figure 4, this secondary photon might cause interference in the dose profile of the CTA. This speculation should be clarified in the near future.

4. Conclusion

In this paper, preliminary dose profile of LEEA has been obtained and evaluated. In the near future, the dose profile with dependence of beam currents and conveyor speeds must be identified. In additional, dose profile by using calorimeter is planned and will be implemented as well.

5. References

[1] Ghazali A B, Rizal M M, Rosli D, Rizal M C, Lee C H, Leo K W and Zahidee M T 21 – 22 Nov 2005 Development of electron beam machine at Nuclear Malaysia *Technology and Application Accelerator Meeting VIII, P3TM-BATAN*, Yogyakarta.

[2] K W Leo *et al.* 2017 Study on the insulation gas system of the 300 keV electron accelerator *AIP Conf. Proc.* **1799** 050008-1–050008-8.

[3] K W Leo *et al.* 2016 Study on the parameters of the scanning system for the 300 keV electron accelerator *AIP Conf. Proc.* **1704** 020010-1–020010-12.

[4] K W Leo *et al.* 2018 Study on the window cooling system of 300 keV Electron Accelerator *IOP Conf. Series: Mater. Sci. and Eng.* **298** 012045.

[5] K W Leo *et al.* 2020 Development of the scanning power supply for 300 keV Electron Accelerator. *IOP Conf. Series: Mater. Sci. and Eng.* **785** 012037.

[6] Data sheet of Fuji Film Radiation Film FTR-125

[7] Fadeev S N 2013 Lecture at workshop on electron accelerator technology in Nuclear Malaysia (10th October -11st November, 2013).

[8] Ozer Z N 2017 Electron beam irradiation processing for industrial and medical applications *EPJ Web of Conferences* **154** 01019.

Acknowledgements

We would like to express our gratitude to Dr. Fadeev, from the Budker Institute of Nuclear Physics, Russia, for his guidance and willingness to join us at the Workshop on Electron Accelerator Technology during which he had shared his experience and knowledge with us, theoretically and practically.

IOP Conf. Series: Materials Science and Engineering 1106 (2021) 012024 doi:10.1088/1757-899X/1106/1/012024

Environmental radio frequency radiation (RFR) safety assessment at the vicinity of telecommunication structures in Malaysia

S R Parthasarathy[1,a)]

[1]Radiation Safety and Health Division, Malaysian Nuclear Agency, Bangi, 43000 Kajang, Selangor

[a]shamesh@nm.gov.my

Abstract. Radio Frequency Radiation (RFR) from the existing and newly erected telecommunication structures has turned out to be a major public health concern in many countries of the world including Malaysia. The public in general are not only apprehensive about the growing weight of scientific evidence associating health effects to mobile phone use, nevertheless they are also increasingly concerned about the chronic, low-level RFR exposure from the stated structures. The paper highlights the study on the RFR emanated from the antennas installed on the telecommunication structures in Malaysia. The objectives of the study are to determine the level of RFR present and to assess the potential exposure received by the general public at the vicinity of the telecommunication structures. The study encompassed on the assessment of the RFR emitted by the antennas which displayed the results in the power density unit measured in microwatts per square centimeter ($\mu W/cm^2$) at each identified locations. The results obtained from the study showed that the RFR level present in the area assessed were well below and less than 1% of the reference level for exposure stated in the Malaysian Communications and Multimedia Commission (MCMC) Mandatory Standard and International Commission on Non-Ionizing Radiation Protection (ICNIRP).

1. Introduction

Mobile phones play an essential part of modern telecommunications, whereby in many countries, over half the population use mobile phones. The usage of mobile phones has been increasing exponentially over the past decades in line with the needs of high-speed communication. As of first quarter 2020 (Q1 2020), the number of mobile subscriptions in Malaysia was at 41.74 million [1]. In conjunction, for the year 2020, the number of smart phone users in Malaysia is estimated to reach 30.41 million. In line with this, the telecommunications companies are establishing more telecommunication structures at every strategic location as the demand for mobile users are tremendously high at the stated areas and to ensure the quality of services to their customers. The number of telecommunication structures built by the Network Service Providers (NSP) and Network Facility Providers (NFP) in residential areas has raised concerns from the public about possible radio waves effects caused by these structures to them. Therefore, this study aims to determine and assess the Radio Frequency Radiation (RFR) present at the concerned areas and at the same time, analyse the RFR level based on the Malaysian Communications and Multimedia Commission (MCMC) Mandatory Standard and the International Commission on Non-Ionizing Radiation Protection (ICNIRP) guidelines.

2. Description of assessment locations

The RFR safety assessment was conducted at 25 locations at 5 selected states namely Wilayah Persekutuan Kuala Lumpur, Selangor, Penang, Malacca and Sarawak. 5 locations were selected from each states for the safety assessment. It should be noted that the RF radiation levels at the time of the measurement were from the background radiation and sources from several rectangular and point-to- point parabolic antennas installed on the telecommunication structures. The sites were named as Location 1-5, followed by the name of the state to represent the 25 locations. The locations of measurement selected was based on the worst-case situation, as an example those which were potential to present the higher radiation exposure to people living or working in that area and the areas with the direct antenna beam path. The main objective of the measurement were to determine and assess radio frequency radiation (RFR) present in all accessible places at the concerned area and to analyse the RF radiation level based on the Malaysian Communications And Multimedia Commission (MCMC) Mandatory Standard 2010 [2] and the International Commission on Non-Ionizing Radiation Protection (ICNIRP) [3] recommendation for the reference level of exposure at identified locations around the site.

3. Safety guidelines and exposure limits

The reference level or exposure limits are referred to RFR frequency involved and set differently for workers and members of the public. Public being more critical to radiation because of longer exposure time involved, unknown exposure situation and are more diverse in term of health status and age groups represented, has had the limits established at much lower levels (by 2 to 5 times lower) than the workers [2-10]. Based on the recommended guidelines for public, the Exposure limit of RF radiation is very much dependent on frequency of the transmitted radiation. The frequencies transmitted by the rectangular antennas are within the range of 2.0 GHz and above. The frequency range is based on the spectrum allocation of each service providers in Malaysia. Therefore, for this study, the lowest frequency radiation expected to be transmitted by the antennas is taken as the reference frequency point to derive the reference level or the exposure limit for public at each site. The reference level for each frequency range is summarised in Table 1. For this study, the unit of power density was stated in microwatts per square centimeter ($\mu W/cm^2$).

Table 1. Mandatory Standard Reference Levels of EMF exposure.

Type of Exposure	Frequency range (f)	Equivalent plane wave power density (W/m^2)
	10MHz-400 MHz	10
Occupational	400MHz-2000 MHz	f/40*
	2GHz-300 GHz	50
	10MHz-400 MHz	2
General Public	400MHz-2000 MHz	f/200*
	2GHz-300 GHz	10

*Note: f is the frequency of operation in MHz

4. Methodology

The RFR safety assessment was performed using standard measurement protocol and procedures developed by Malaysian Nuclear Agency based on the standards recommended by more established organizations and government agencies, such as International Telecommunication Union (ITU), International Commission on Non-Ionizing Radiation Protection (ICNIRP), Institute of Electrical and Electronics Engineers (IEEE) and Canadian Safety Code 6 guidelines [5-10]. All measurements were performed at 1.5 meters above the ground. At each point, measurements were taken for 6 minutes within which 360 data are recorded, which means that for each second, one data will be logged into the instrument.

The data will be analysed by using the software embedded in the equipment itself and transferred to the laptop for further analysis and computation. The measurement encompassed only on the assessment of the RFR generated by the antennas installed on the telecommunication structures. The measurement was arranged to include both the electric field strength and power density at identified locations around the building [2-10]. Results of radio-frequency measurement were recorded in unit of power density measured in microwatts per square centimeter (μW/cm²). The measured radiations were assessed and evaluated against the exposure limits and the recommendations of standard guidelines issued by the MCMC Mandatory Standard and the ICNIRP. The broad band frequency was measured by using Narda NBM-550 attached with isotropic probe Model EF0691(100kHz - 6GHz). For detailed spectrum analysis of radiations involved, measurements were made using a Narda-Safety Test Solution instrument Model SRM-3006 attached with Three-Axis- Antenna, which measure radio-frequency and microwave electric field strength from 27MHz up to 3 GHz.

5. Results and discussion

The RFR power density measured at all the 25 locations were ranging from 0.02 μW/cm² to 4.15 μW/cm². The highest value of RFR power density was recorded at Location 4 in the state of Selangor. This level corresponds to 0.42 % power density of the reference level or the exposure limit as stated in MCMC Mandatory Standard. The results of the 25 locations of the 5 states in Malaysia are as summarised in Table 2. The plot of the radiation level compared to the reference level or the permissible exposure limit is shown in Figure 1. Detailed investigation of the RFR present utilizing the spectrum analyser around the telecommunication structure was performed on all the selected sites. The RFR observed consists of radio, TV, mobile phones and signals from other telecommunication structures. On average for all the RFR analysed, it was recorded that around 60% of the radiation strength is from the signals from the transmitting antennas of the stated telecommunication structures, 22% of the strength is from mobile phone signals, 10% from TV signals and 8% from radio signals.

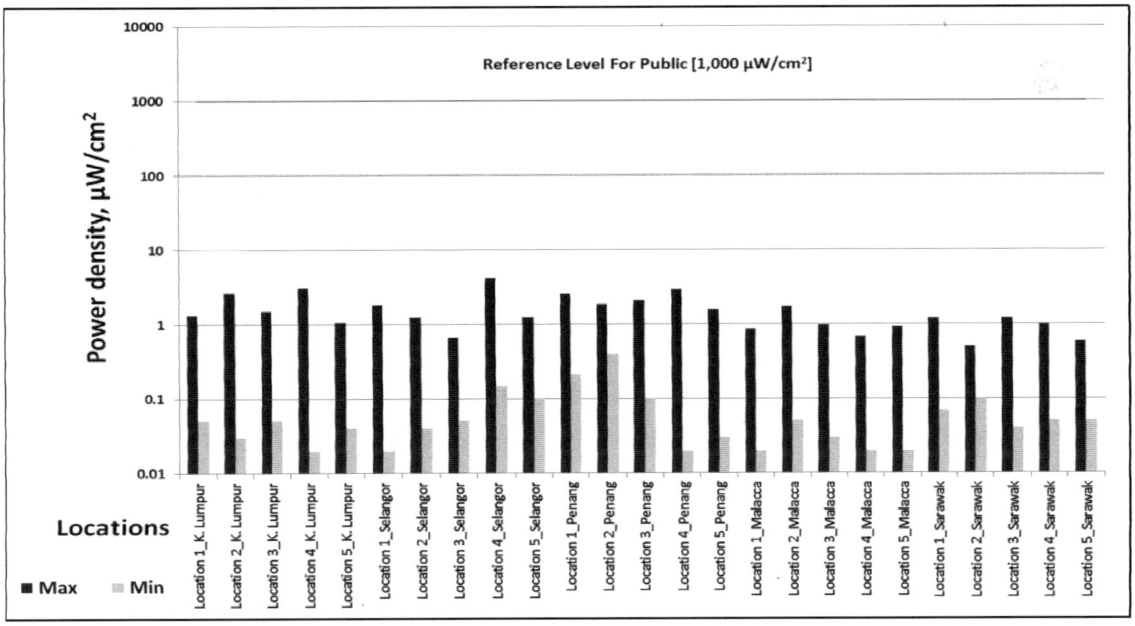

Figure 1. Plot of RFR levels in microwatts per unit area (μW/cm²) against location of the measurement and their comparison with MCMC & ICNIRP Reference Level for public.

Table 2. Summary of the RFR results of the 25 locations consisting of 5 states in Malaysia

Location	Maximum RFR Level in terms of Power density (μW/cm²)	Reference Level (μW/cm²)	Comparison with Reference Level (%)
Location 1_K. Lumpur	1.32	1,000	0.13
Location 2_ K. Lumpur	2.61	1,000	0.26
Location 3_ K. Lumpur	1.50	1,000	0.15
Location 4_ K. Lumpur	3.05	1,000	0.31
Location 5_ K. Lumpur	1.05	1,000	0.11
Location 1_Selangor	1.80	1,000	0.18
Location 2_Selangor	1.25	1,000	0.13
Location 3_Selangor	0.66	1,000	0.07
Location 4_Selangor	4.15	1,000	0.42
Location 5_Selangor	1.25	1,000	0.13
Location 1_Penang	2.52	1,000	0.25
Location 2_Penang	1.85	1,000	0.19
Location 3_Penang	2.05	1,000	0.21
Location 4_Penang	2.89	1,000	0.29
Location 5_Penang	1.58	1,000	0.16
Location 1_Malacca	0.85	1,000	0.09
Location 2_Malacca	1.70	1,000	0.17
Location 3_Malacca	0.95	1,000	0.10
Location 4_Malacca	0.67	1,000	0.07
Location 5_Malacca	0.91	1,000	0.09
Location 1_Sarawak	1.18	1,000	0.12
Location 2_Sarawak	0.50	1,000	0.05
Location 3_Sarawak	1.18	1,000	0.12
Location 4_Sarawak	0.98	1,000	0.10
Location 5_Sarawak	0.59	1,000	0.06

6. Conclusion

The RFR measured in all the sites were well below the reference level and comply with the MCMC Mandatory Standard and ICNIRP Guidelines for members of the public. Detailed analysis at each telecommunication structures shows the presence of other RF signals such as Radio, TV and Mobile Phone Signals. Based on the findings from this assessment, the presence of the RFR emitted by the antennas from the telecommunication structures would not lead to any significant radiation exposure received by the public in the area. It should be noted that RFR safety assessment must be done during the commissioning of new transmitting antenna or any modification on the existing antenna.

7. References

[1] MCMC 2020 Communication and Multimedia *Facts and Figures 1Q 2020.*
[2] MCMC 2010 Commission determination on the mandatory standard for electromagnetic field emission from radio communications infrastructure *Determination No. 1 of 2010.*
[3] International Commission on Non-Ionizing Radiation Protection (ICNIRP) 2020; Guidelines for limiting exposure to electromagnetic fields (100kHz to 300 GHz) *ICNIRP Guideline* **118**(5) 483-524.
[4] International Commission on Non-Ionizing Radiation Protection (ICNIRP) 1998 Guidelines for limiting exposure to time-varying electric, magnetic and electromagnetic fields (up to 300 GHz) Health Phys. **74** 494-522.
[5] ITU 2019 Measurement of radio frequency electromagnetic fields to determine compliance with human exposure limits when a base station is put into service *ITU-T K 100.*
[6] ITU 2018 Guidance on measurement and numerical prediction of electromagnetic fields for compliance with human exposure limits for telecommunication installations *ITU-T K 61.*
[7] ITU 2019 Guidance for assessment, evaluation and monitoring of human exposure to radio frequency electromagnetic fields *ITU-T K. 91.*
[8] ITU 2018 Guidance on complying with limits for human exposure to electromagnetic fields *ITU-T K 52.*
[9] Institute of Electrical and Electronics Engineers (IEEE) 2019 IEEE standard for safety levels with respect to human exposure to electric, magnetic, and electromagnetic fields, 0 Hz to 300 GHz.
[10] Health Canada 2015 Technical guide for interpretation and compliance assessment of health canada's radiofrequency exposure guidelines *Canadian Safety Code 6.*

Seismic hazard assessment (SHA) in Sabah using the international monitoring system (IMS) data of the Comprehensive Nuclear-Test-Ban Treaty Organization (CTBTO): The preliminary findings

M Z Zolkaffly[1, a)] and F I A Rashid[1, b)]

[1]Planning and International Relations Division, Malaysian Nuclear Agency, Bangi, 43000 Kajang, Selangor, Malaysia.

[a]zulfakar@nuclearmalaysia.gov.my
[b]faisal_izwan@nuclearmalaysia.gov.my

Abstract. Sabah is the most seismically active state in Malaysia where it has recorded higher number of moderate seismological activities for the past decades, as compared to other states in the country. The seismicity map of Sabah shows the presence of two zones of distinctive seismicity, which are Ranau in Kota Kinabalu and Lahad Datu in the southeast of Sabah. The International Monitoring System (IMS) network setup by the Comprehensive Nuclear-Test-Ban Treaty Organization (CTBTO) has successfully detected seismic events occurred in Sabah for the past decades. This paper aims at presenting the preliminary findings of seismic hazard assessment (SHA) in Sabah using the IMS data. This study utilised seismic data from the CTBTO International Data Centre and then analysed to identify the related seismicity parameters including magnitude, depth, and intensity. The Extreme Value Distribution Type-I has been applied to evaluate the maximum magnitude data, where the results of analysis have enabled the quantification of seismic hazard in Sabah in terms of recurrence periods and probabilities of occurrence of earthquake at any given magnitude. Consequently, the findings from this study could be to assess the impact of seismic events in Sabah as well as assist relevant entities in development planning and disaster management.

1. Introduction

Malaysia is situated on the southern edge of the Eurasian and considered to have low seismicity profile. It is closed to the most two seismically active plate boundaries, the inter plate boundary between the indo-Australian and Eurasian plates on the west and the inter-plate boundary between the Eurasian and Philippines plates on the east [1]. Different with Peninsular Malaysia which is classified as a seismically stable, the region of Sabah is known as prone to earthquake activities compared to other parts of Malaysia. The region has suffered several earthquakes of moderate magnitude, where of these events have caused structural damage to buildings and other infrastructures and injuries to humans [2]. In 1976, one of the worst earthquakes occurred in Sabah when a 5.8 magnitude on Richter scale earthquake shocked Lahad Datu. Later in 1991, an earthquake of magnitude 4.5 on Richter scale shook Ranau, resulting in structural damage to a school. Eastern Sabah is also exposed to tremors caused by earthquakes originated in the southern Philippines and the Straits of Macassar, Sulu Sea and Celebes Sea. Figure 1 shows the epicenter of felt earthquakes ever recorded in East Malaysia, majority in Sabah [2].

Based on the above facts, seismic hazard assessment (SHA) for Sabah is essential in order to mitigate the effects of potential large earthquake that may occur in the future. One important measure in mitigating the earthquake hazard is to design and build structures using appropriate engineering practices, so that these structures exhibit sufficient resistant against earthquake [3]. This paper presents a preliminary study regarding SHA for Sabah utilizing seismic data of the Comprehensive Nuclear-Test-Ban Treaty Organization (CTBTO). The analysis was performed using statistical theory of extreme values, where the results have enabled the quantification of seismic hazard in Sabah in terms of recurrence periods and probabilities of occurrence of earthquake at any given magnitude.

2. Earthquake data

Seismic technology is one of three waveform technologies which are part of the International Monitoring System (IMS) of the Comprehensive Nuclear Test Ban Treaty (CTBT). The IMS consists of 337 monitoring stations and laboratories that include 50 primary and 120 auxiliary seismic stations, installed world-wide and transmitting data to the International Data Centre (IDC) in Vienna [4]. Since June 1999, the IDC began routine automatic and interactive processing of seismic data, where the detected and located events are systematically included in the Reviewed Event Bulletin (REB) [5].

The analysis of seismic hazard assessment requires data recorded from earthquake events that occurred around the site of interest and observed for a specific time interval. For this study, the data of earthquake events occurred around Sabah region (bounded by 3.95°N to 7.16°N latitude and 115.4°E to 119.3°E longitude) from January 2002 to June 2020 have been taken from the CTBTO's REB. In order to study the earthquake risk, probability of occurrence and return periods, the earthquake data distributed over 19 years periods has been divided into one year time interval such as at least one event in each year duration is observed. To ensure the continuity of the data, for a few years that no events were reported for Sabah region in the REB, the event with magnitude Mw 4 has been assumed for the year of 2011, 2013 and 2018. Table 1 shows the list of earthquake events in Sabah recorded by the CTBT seismic stations from the year 2002 to 2020. In term of depth, it is observe that depth of most of the events are fixed at 0 km as the IMS is looking for explosions signature (manmade activities), which is generally shallow (depth less than 5 km) in seismology. This 0-km depth classification is telling us that the sources of these earthquakes are deep in the upper mantle to the surface [6].

3. Method of analysis

Various statistical models have been proposed for the analyses of earthquake occurrence with different degrees of success [7]. The earthquakes occurrence in space and time can be explained using mathematical models of a given physical system change in accordance with the laws of probability [8]. These models have usually incorporated the Poisson distribution, or extended to clustering of events using Markovian models of non-independent events. Calculations obtained are often unconvincing because of incompleteness in the data sets or inherent uncertainties in the distributed parameters [9].

However, the method of Extreme Value Distribution Type-I, which is also known as Gumbel's Type I, has certain clear and advantages as far as the requisite data are concerned [10]. If compared with other methods requiring the whole data set, and rarely completely reported, the Gumbel's need only part of the event data, such as, the largest earthquakes i.e. extremes. The CTBTO IMS data represent a continuous and complete set of annual maximum magnitude events, thus, it is very useful to apply such method for the calculation process of this study. The results obtained are useful to determine a variety of statistics including average recurrence periods of annual maximum magnitude earthquakes and probabilistic seismic hazard assessment of the region. Results also are informative for seismic threat and related earthquake engineering determinations that usually require estimation of return periods or probabilities of exceedance of specific levels of design load criteria or external safety conditions.

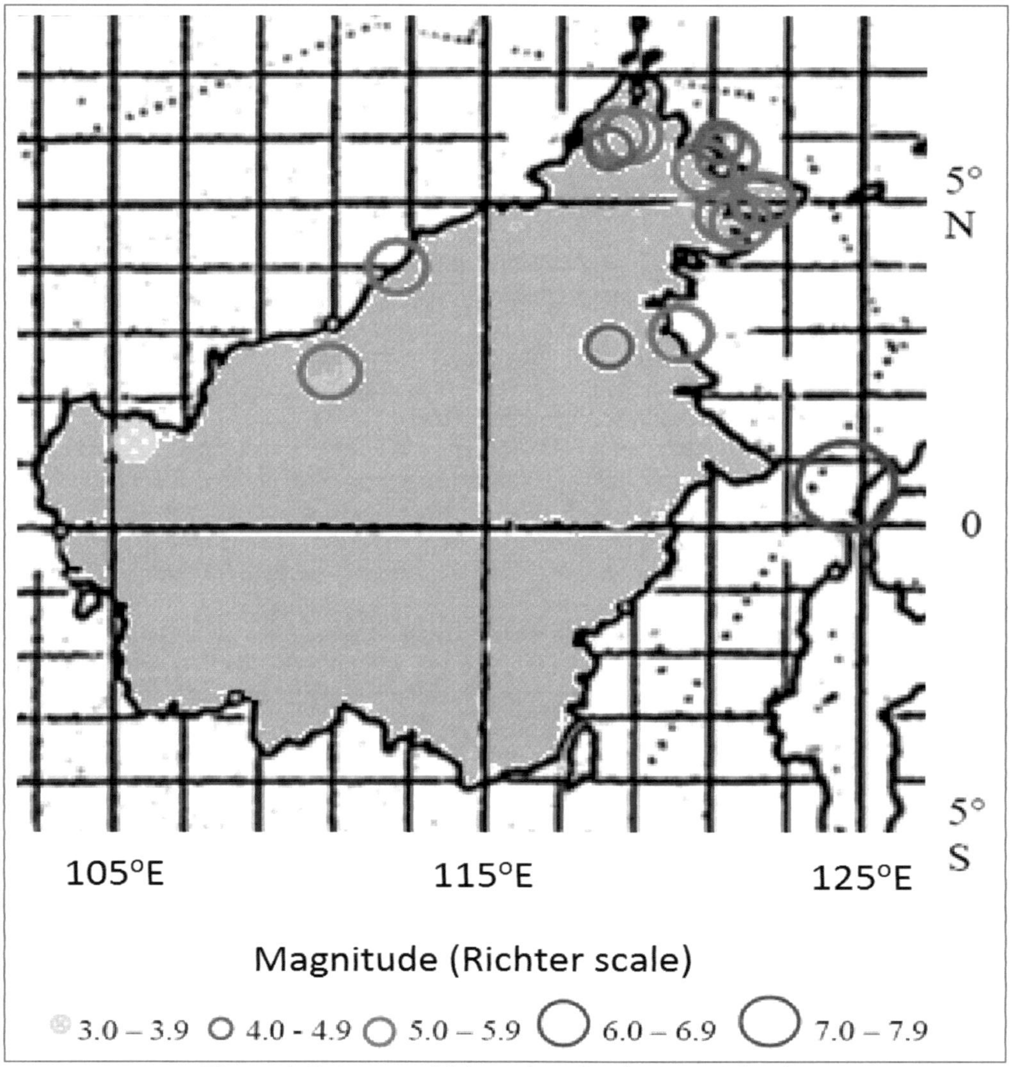

Figure 1. Epicenter of felt earthquakes in East Malaysia.

Table 1. 2002-2020 Earthquake events in the region of Sabah recorded by the IMS network.

YEAR	CTBT IMS Seismic Network						
	Event ID	Date	Time	Latitude	Longitude	Magnitude	Depth (KM)
2020	18767431	14/04/2020	20:04:22	5.0212	118.9558	mb 3.6	0
2019	17680942	03/08/2019	15:31:06	6.2831	117.2185	mb 3.6	0
2018	15530461	08/03/2018	13:06:10	5.8856	116.5306	mb 4.4	0
	15848813	27/05/2018	7:53:42	5.2794	119.0089	mb 3.8	0
	15916404	16/06/2018	7:47:58	1.0666	111.3856	mb 3.8	0
2017	14228945	26/03/2017	9:30:49	4.7966	118.7287	mb 3.7	53
	14888716	24/09/2017	0:25:23	5.9464	118.3581	mb 3.5	0
2016	13497338	26/08/2016	1:39:32	6.4172	117.3407	mb 3.6	0
2015	11783042	19/03/2015	21:56:00	5.4795	118.3777	mb 3.8	0
	11997735	14/05/2015	10:25:07	1.8141	110.5797	mb 3.6	0
	12047012	04/06/2015	23:15:42	6.039	116.6377	mb 5.2	0
	12048513	05/06/2015	13:12:14	6.6386	117.191	mb 3.7	0
	12048801	05/06/2015	15:13:32	6.0524	116.6861	mb 3.9	0
	12050340	06/06/2015	5:45:13	6.2528	116.8675	mb 4.1	0
	12068968	12/06/2015	18:25:34	6.4085	117.3888	mb 3.4	0
	12068970	12/06/2015	18:29:24	6.0878	116.8661	mb 4.4	75
	12098600	23/06/2015	9:32:34	5.9739	116.5596	mb 3.7	49
	12199515	26/07/2015	10:04.9	6.2731	117.0466	mb 3.9	0
2014	10715051	14/04/2014	20:42:03	4.5645	117.7927	mb 3.4	0
	10471451	01/02/2014	11:35:08	6.095	116.6187	mb 4.1	0
2013	10099555	19/09/2013	21:50:37	5.6921	116.223	mb 3.6	0
	9440838	29/01/2013	1:04:30	5.1358	118.7193	mb 4.0	0
2012	8726575	28/05/2012	16:44:08	4.7371	118.3378	mb 4.1	0
2011	-	-	-	-	-	-	0
2010	6607825	21/08/2010	19:43:27	5.4801	118.5899	mb 3.5	0
2009	-	-	-	-	-	-	0
2008	4536681	10/01/2008	13:18:35	4.2583	116.5442	mb 3.6	0
2007	4428540	23/10/2007	20:34:39	5.6865	119.2532	mb 4.3	0
2006	3550325	06/02/2006	14:54:07	5.1248	118.8151	mb 3.6	0
	3674353	22/04/2006	2:01:25	6.0392	117.5982	mb 3.7	0
	3859855	28/09/2006	15:11:34	6.4085	118.0916	mb 3.7	0
2005	3249883	23/05/2005	19:58:12	6.2446	117.713	mb 4.3	0
2004	-	-	-	-	-	-	0
2003	1978643	02/11/2003	8:43:19	6.2718	117.5447	mb 3.5	0
	1852956	22/08/2003	18:01:00	5.8225	119.2862	mb 3.3	0
2002	1328028	06/10/2002	21:03:24	5.9584	117.8815	mb 3.8	0

Gumbel's extreme value theory postulates that if the earthquake magnitude is unlimited, if the number of earthquakes per year decreases with their increase in size, and if individual events are unrelated, then the largest annual earthquake magnitude is distributed by cumulative distribution function G(m), where, [11]:

$$G(m; \alpha, \beta) = exp^{[-\alpha exp^{(-\beta m)}]}; \ m \geq 0 \tag{1}$$

where α is the average number of earthquakes with magnitude > 0 per year, β is the inverse of the average magnitude of earthquakes under the considered region, and *m* is the maximum annual earthquake magnitude. The probability integral transformation theorem and manipulation of equation (1) gives the relation:

$$-\ln[-\ln(p_m)] = \beta m_i - \ln(\alpha) \tag{2}$$

where p_m represents the plotting position. The mean frequency of *i*-th observation in the ordered set of extremes may be represented as:

$$p_m = \frac{i}{N+1} \tag{3}$$

where *N* is the total number of observed data. The relationship between Gumbel parameters α and β and Gutenberg-Richter parameters *a* and *b* can be given by the expression

$$b = \beta \log_{10} e \tag{4}$$

and

$$a = \log_{10} \alpha \tag{5}$$

The expected number of earthquakes, *Nm*, in a given year having magnitude exceeding *M* can be expressed by the Gutenberg- Richter seismicity relation as

$$\log_{10} N_m = a - bM \tag{6}$$

where, *aa* and *bb* are constants. From equation (6),

$$\log_{10} N_m = a - bM \tag{7}$$

the probability of at least one earthquake of magnitude ≥*M* occurring within one year is given by the Poisson process as:

$$p = 1 - e^{-N_m} = 1 - e^{-10^{a-bM}}$$
$$= 1 - e^{-e^{\ln 10^{a-bM}}} \tag{8}$$

After derivation, Equation (8) becomes,

$$M = \frac{a}{b} - \frac{1}{b \ln 10} \ln[-\ln(1-p)] \tag{9}$$

where p lies in the interval (0,1). The probability of at least one earthquake of magnitude $\geq M$ within t years can be given by the equation:

$$p = 1 - e^{-Nt} = 1 - e^{-\left(10^{a-bM}\right)*t} \tag{10}$$

The expected number of earthquakes in a given year which have magnitude exceeding m can be found using Equation (11):

$$\ln N_m = \ln \alpha - \beta m \tag{11}$$

and the return period of earthquakes having magnitude greater than m is given by:

$$T_m = {}^{1}/_{N_m} = \exp\ (\beta m)/\alpha \tag{12}$$

4. Results and Discussion

As method of Extreme Value Distribution Type-1 is applied for this study, the annual maximum magnitudes of seismic events recorded in the region of Sabah from 2002 to 2020 are need to be rearranged in rank order so that a variety of statistics can be determined using the Equations (1), (2) and (3) as described above. These 19 years of annual maximum magnitudes and the subsequent statistics are shown in Table 2. The values of cumulative frequency probability are calculated using Equation (3), while the Gumbel's extreme Type I reduced variant is calculated as per Equation (2). Figure 2 shows the variation of maximum magnitude with years, which indicates that the maximum magnitude increases with time in the considered region. Figure 3 demonstrates the mean Line of Expected Extreme (LEE) to study the probability of largest extreme magnitude in Sabah.

The values of estimated Gumbel's parameters α and β are then estimated from a least-square fit to the Reduced Linear Variation Equation, as described in Figure 4. These values of α and β are summarized in Table 3.

Table 2. Calculations of Gumbel's Annual Maximum Distributions

Extremes	Rank (j)	Plotting Position G(m)	Reduced Statistics ln (-ln(G(m)))
3.5	1	0.05000	1.09719
3.5	2	0.10000	0.83403
3.6	3	0.15000	0.64034
3.6	4	0.20000	0.47588
3.6	5	0.25000	0.32663
3.7	6	0.30000	0.18563
3.7	7	0.35000	0.04862
3.7	8	0.40000	-0.08742
3.8	9	0.45000	-0.22501
4	10	0.50000	-0.36651
4	11	0.55000	-0.51444
4	12	0.60000	-0.67173
4	13	0.65000	-0.84215
4.1	14	0.70000	-1.03093
4.1	15	0.75000	-1.24590
4.3	16	0.80000	-1.49994
4.4	17	0.85000	-1.81696
4.4	18	0.90000	-2.25037
5.2	19	0.95000	-2.97020

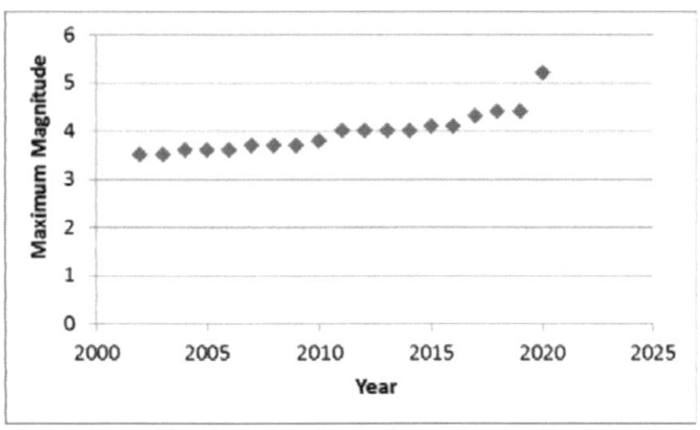

Figure 2. Variation of maximum magnitude with year

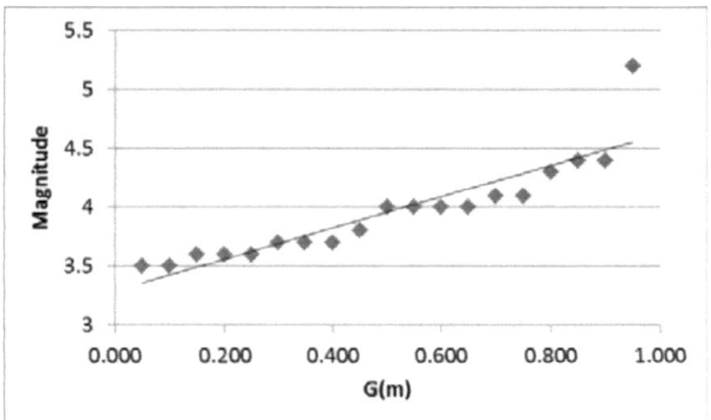

Figure 3. Variation of extreme magnitude with probability

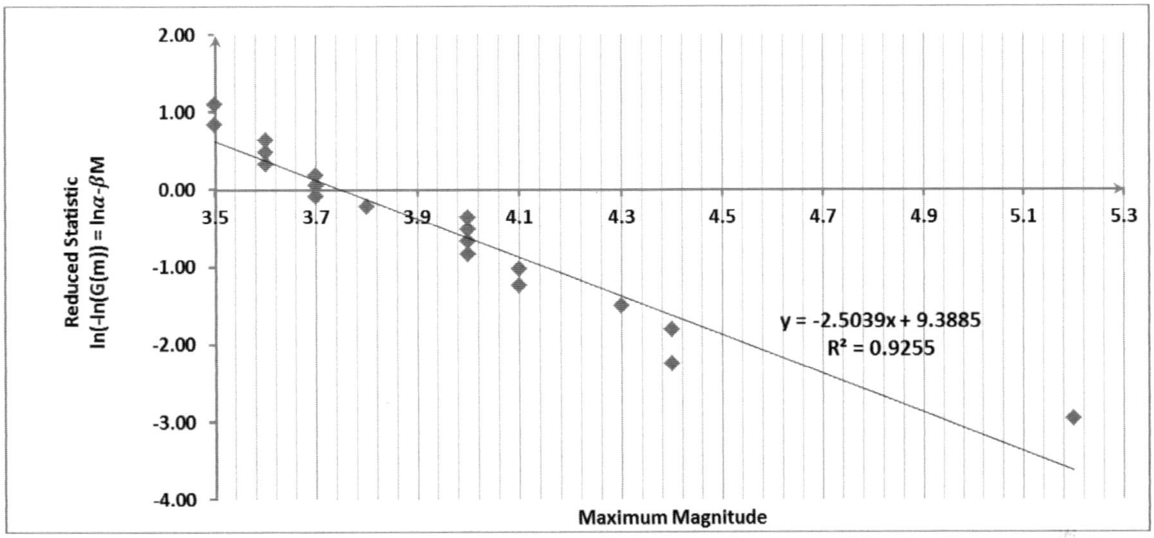

Figure 4. Display Reduced Variant with extreme magnitudes to estimate α and β using linear regression of data

Table 3. Estimated Gumbel's Parameters α and β

Statistics	Value
Slope(-β)	-2.5039
β	2.504
Intercept(ln(α))	9.3885
α	11950.16

Risk analysis that involves a detailed consideration, includes of hazards and uncertainties is important from the economic and safety point of view, prior to any decision made to embark on development and investment in the region. Seismic threat and related earthquake engineering determinations usually require estimation of return periods, probabilities of exceedance of specific levels of design load criteria, and external safety conditions. Assessment of seismic hazard has become one of major problems in the field of earthquake engineering.

The earthquake yearly numbers and their return period for different magnitude expected in Sabah region are calculated and summarized in Table 4. These indicate that as return period increases, frequency of earthquake occurrences decreases. For quantifying the earthquake hazard *Ht (m)* in Sabah, which is the probability of occurrence of earthquake of magnitude m within a period of t year, the following equation is considered:

$$H_t(m) = 1 - \exp\left(-\alpha t e^{-\beta m}\right) \tag{13}$$

From the histories of maximum earthquake magnitude recorded, earthquake hazard probabilities for different magnitudes with time are calculated for 10, 20, 30, 50, 75 and 100 years periods. The results are summarized in Table 5. The relation between of these yearly numbers of earthquakes, their return periods and the calculated earthquake hazard is further illustrated in Figure 5 and Figure 6. The observations suggest that within hundred year period the probability of occurrence of larger magnitude earthquakes decrease with time. The general interpretation of this curve reveals that the probability of an earthquake of magnitude 4.1 occurring in the considered region with 20-years period is estimated to be 0.999. This means that at least one earthquake of magnitude 4.1 is predicted to occur within that period of time. However, probability of such particular event is never absolutely certain.

Table 4. Predicted yearly number of earthquakes and their return periods.

Magnitude	N_m	T_m
3.5	1.867965642	0.535341752
3.5	1.867965642	0.535341752
3.6	1.454205854	0.687660552
3.6	1.454205854	0.687660552
3.6	1.454205854	0.687660552
3.7	1.132095054	0.883318054
3.7	1.132095054	0.883318054
3.7	1.132095054	0.883318054
3.8	0.881332728	1.134645257
4	0.53413853	1.872173497
4	0.53413853	1.872173497
4	0.53413853	1.872173497
4	0.53413853	1.872173497
4.1	0.415825302	2.404856065
4.1	0.415825302	2.404856065
4.3	0.252014147	3.968031205
4.4	0.196192285	5.097040379
4.4	0.196192285	5.097040379
5.2	0.026469026	37.78000723
6	***0.003571034***	***280.0309278***
6.5	***0.001021125***	***979.311774***
7	***0.000291987***	***3424.805818***
7.5	***8.34928E-05***	***11977.07941***
8	***2.38745E-05***	***41885.71234***

Table 5. Most probable largest earthquake hazard H_t (m) for Different Magnitudes and Time Periods (t=10, 20, 30, 50, 75 and 100 years).

Magnitude (m)	H_{10}(m)	H_{20}(m)	H_{30}(m)	H_{50}(m)	H_{75}(m)	H_{100}(m)
3.5	0.999999992	1	1	1	1	1
3.5	0.999999992	1	1	1	1	1
3.6	0.999999516	1	1	1	1	1
3.6	0.999999516	1	1	1	1	1
3.6	0.999999516	1	1	1	1	1
3.7	0.999987884	1	1	1	1	1
3.7	0.999987884	1	1	1	1	1
3.7	0.999987884	1	1	1	1	1
3.8	0.999851262	0.999999978	1	1	1	1
4	0.995210768	0.999977063	1	1	1	1
4	0.995210768	0.999977063	1	1	1	1
4	0.995210768	0.999977063	1	1	1	1
4	0.995210768	0.999977063	1	1	1	1
4.1	0.984365152	0.999755552	0.999996	1	1	1
4.1	0.984365152	0.999755552	0.999996	1	1	1
4.3	0.919551775	0.993528083	0.999479	0.999997	1	1
4.4	0.859412169	0.980235062	0.997221	0.999945	1	1
4.4	0.859412169	0.980235062	0.997221	0.999945	1	1
5.2	0.23255638	0.41103029	0.547999	0.733785	0.862644	0.92913

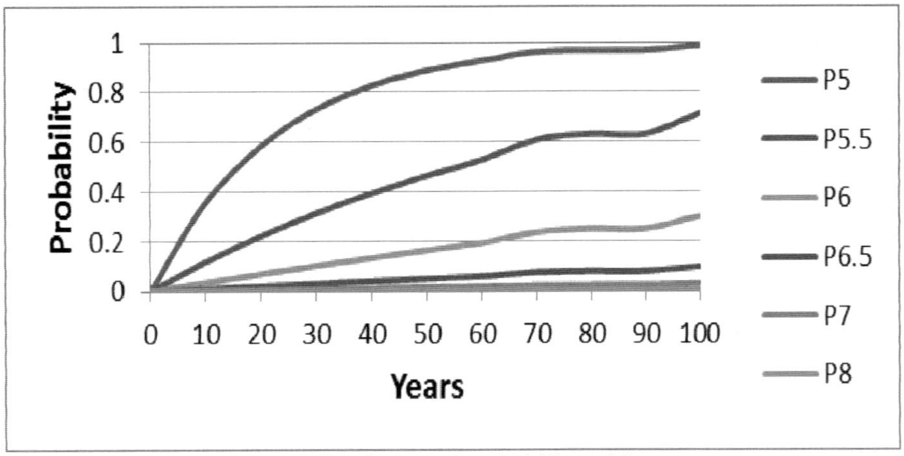

Figure 5. Variation of probability with year.

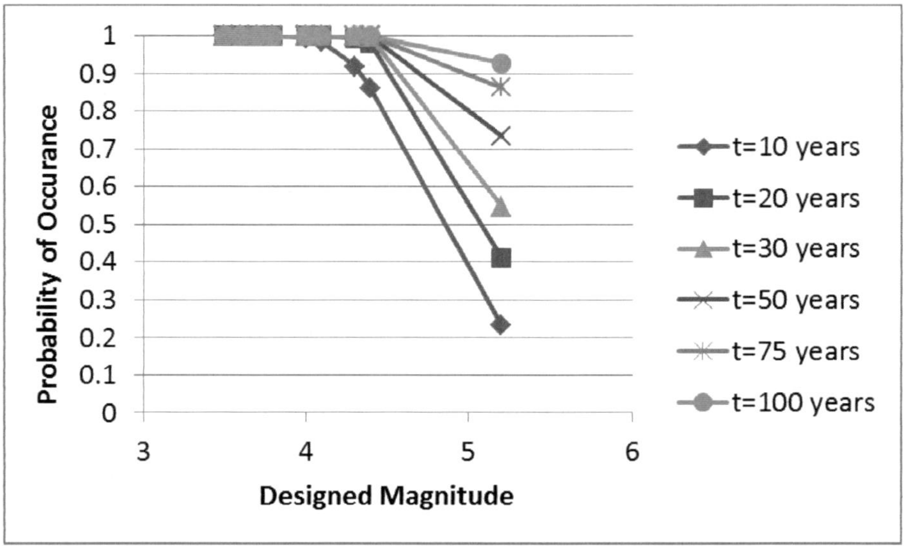

Figure 6. Earthquake hazard in Sabah region for different period.

As for probabilistic seismic hazard assessment in the region, the probability P (t ≥ T) that the recurrence period of the design earthquake of magnitude m exceeds a random recurrence period of T years is given by [12]:

$$P\ (t \geq T) = \exp\ (-T/T_m) \tag{14}$$

Therefore, the probability of recurrence period being less than random time T is given by:

$$P\ (t \leq T) = 1 - \exp\ (-T/T_m) \tag{15}$$

where, one can deduce that period T within which at least one earthquake of magnitude m will occur with probability P express as:

$$T = -T_m \ln\ (1-P) \tag{16}$$

In this regard, one can estimate expected period within which at least one earthquake of any given magnitude will occur with any specified probability. For example, recurrence period for at least one earthquake of magnitude m within a probability of 90% is given by:

$$T_{90} = -T_m \ln\ (1-90) \tag{17}$$

From Table 4, the design earthquake recurrence period with 90% probability is calculated and the values are presented in Table 6. The 90% probability recurrence period could be understood that in Sabah region, there is 90% probability in 29 years period that at least one earthquake of magnitude 4.5 or greater will occur and conversely that 10% probability an earthquake of the same magnitude or more will not occur.

Table 6. Design earthquake Recurrence Period with 90% probability.

Magnitude (m)	Return Period (years)	Recurrence Period (years)
3	0.153	0.69
3.5	0.535	2.40
4	1.872	8.40
4.5	6.547	29.39
5	22.897	102.78
5.5	80.074	359.42
6	280.031	1256.96
6.5	979.312	4395.77
7	3424.806	15372.71
8	41885.712	188009.73

Subsequently, the findings from this study could be useful in order to assess the impact of seismic events in Sabah, which is one of the important elements that could assist the relevant entities in the planning for infrastructural development and activities in the region. As the study is useful for engineering investigations at particular site, the predicted values are also useful and important for decision making and planning with regards to disaster management of the site of interest.

5. Conclusion

The study has managed to produce the preliminary findings of seismic hazard assessment in Sabah using the IMS data of the CTBTO. The results of analysis have enabled the quantification of seismic hazard in terms of recurrence periods and probabilities of occurrence of earthquake at any given magnitude. Genarally, it can be summarized that the methodology of the Gumbel's Type I extreme distribution could provide future seismic and hazard status of the region, and the availability of CTBTO's IMS complete earthquake data sets leads to reliable results that the study aimed for. The results demonstrate a clear consistency with the stability postulate from which distributions of extremes are deduced and provides further confidence in the application of such method.

6. References

[1] Rashid F I A and Zolkaffly M Z 2019 Analysis of Seismic events in malaysia using international monitoring system (IMS) data of the Comprehensive Nuclear-Test-Ban Treaty Organization (CTBTO) *IOP Conf. Ser.: Mater. Sci. Eng.* **785** 012019 (doi: 10.1088/1757-899X/785/1/012019).

[2] Che Abas M R 2001 Earthquake monitoring in Malaysia *The Seismic Risk Seminar (25 September 2001).*

[3] Hu Y X 1996 *Earthquake Eng.* (London: E & FN Spon).

[4] Rashid F I A and Zolkaffly M Z 2018 The Comprehensive Nuclear-Test-Ban Treaty (CTBT): seismic monitoring *IOP Conf. Ser.* **555** (doi: 10.1088/1757-899X/555/1/012010) 012010.

[5] Singh M, Shanker D and ALI S M 2015 Seismic hazard assessment in Hindukush-Pamir Himalaya using IMS network *Geosciences* **5**(3) p-ISSN:2163-1697 e-ISSN:2163-1719 (doi: 10.5923/j.geo.20150503.0) 81-85.
[6] Andriamampandry J, Andrianasolo R and Rakotoarisoa T 2020 Unusual seismic activities in northwestern Madagascar using Spot Them All *J. of Geol. Resource and Eng.* **8** (doi:10.17265/2328-2193/2020.03.002) 80-89.
[7] Gayskiy V N and Katok A P 1965 Primeneniye teorii extremal "nykh znacheniy dlya ocenki povtoryayemosti sil" nykh zemletryaseniy, sbornik Dinamika zemnoy kory *AN SSSR* **9** (Moskva: Nauka).
[8] Chandra U 1992 Seismotectonics of Himalaya *Current Sci.* **62**(1-2) 40-71.
[9] Shanker D and Shubham 2018 Earthquake hazard update in central Himalaya (doi: 10.5923/j.geo.20180801.01).
[10] Shanker D, Ali S, Singh M 2017 Earthquake hazard and engineering determinations for Indonesian region using IMS network data *Geosci.* (doi: 10.5923/j.geo.20170705.02) 150-155.
[11] Gumbel E J 1958 Statistics of extremes (New York: Columbia University Press) 375.
[12] Lomnitz C 1974 Global tectonics and earthquake risk (Amsterdam: Elsevier) 320.

IOP Conf. Series: Materials Science and Engineering 1106 (2021) 012026 doi:10.1088/1757-899X/1106/1/012026

Big data analytics nuclear security framework

Maizura Ibrahim[1,a)], Siti Nurbahyah Hamdan[1], Saa'idi Ismail[1], Mohd Fauzi Haris[1], Mohamad Safuan Sulaiman[1], Sufian Norazam Mohamed Aris[1], Mohd Hasnor Bin Hasan[1], Mohd Dzul Aiman Aslan[1], Nur Fatini Abdul Ghani[1], and Amy Hamijah Ab Hamid[1]

[1]Information Technology & Application Centre, Technical Support Division, Malaysian Nuclear Agency, Bangi, 43000 Kajang, Selangor Darul Ehsan, Malaysia.

[a]maizura@nm.gov.my

Abstract. Nuclear security is defined as the prevention and detection of, and response to, theft, sabotage, unauthorized access, illegal transfer, or other malicious acts involving nuclear material, other radioactive substances, or their associated facilities. Whereas, big data analytics is denoted as a new way of collecting, analyzing large amounts of data, finding the appropriate patterns to support decision making, hence improve the action taken in solving the problems. The problem with the nuclear security system in most of the nuclear organizations and nuclear facilities in the world today is that the unintegrated data management in nuclear security systems is causing sub-optimal efficiency in the detection of malicious acts involving nuclear/radioactive materials. We argue that a big data analytics framework for nuclear security should be created and utilized to integrate all data in nuclear security systems as well as serving for needs of intelligence especially involving predictive analytics, both within organizations and inter-organization. Therefore, in this paper, we present a conceptual framework of big data analytics for nuclear security. The framework is formulated based on a holistic methodology with the aims to integrate all the data of the nuclear security systems at the organization/facilities level and national level so that the data could be analyzed to derive the appropriate patterns, hence increase the accuracy of decision making that could lead to increased efficiency of detection process and response to the nuclear security event.

1. Introduction

Information technology is the key driver for the development strategy and progress in most areas. The nuclear security area is among the affected areas. Development in information technology contributes to enhancing the quality of security systems, speeding up getting data for security monitoring purposes, effective security management, emerging novel forms of security system, improving communication and interaction between security personnel, and providing access to a wide range of information for further security improvement. The fast evolution of the IT field and its rapid obsolescence stimulate demand for new research, development, and application in these areas. Thus, the evolvement of big data analytics technologies is leading to innovative changes in strategy for nuclear security. A concept for the development of a combination of nuclear security strategy with big data analytics technology could be formulated to provide the solution to problems facing by most of the nuclear organization and nuclear facilities in the world, i.e. a 'silo' data management practice for a nuclear security system that is causing a sub-optimal nuclear security system [1].

Based on IAEA Nuclear Security Series, nuclear security could be defined as the prevention and detection of, and response to, theft, sabotage, unauthorized access, illegal transfer, or other malicious acts involving nuclear material, other radioactive substances, or their associated facilities [2]. Thus, any system that is created and implemented to achieve nuclear security, whether it is a computer-based system or a manual system, is regarded as nuclear security systems. Apart from that, big data is denoted as high volume, high velocity, high variety, and high veracity, i.e., the 4Vs of data (information assets) that require new forms of processing, to enable enhanced decision making, insight discovery, and process optimization [3]. Therefore, this paper reveals a new conceptual framework adopting big data analytics technology with various other sub-technological solutions, concepts, and principles as a strategic IT-ecosystem adopting big data analytics for enhancing nuclear security.

2. Related works

This work is in the preceding of two reports published by the World Institute of Nuclear Security in 2015[1][4]. This is also an extended version of our unpublished works that started from 2017 until 2020[5][6][7]. Based on our literature review, there exist several research initiatives in applying big data analytics in the nuclear sector [8] adopting big data for an integrated disaster management system that comprises of flood management and nuclear disaster management. Research on applying big data analytics for the operation and maintenance of nuclear power plant has been conducted by [9][10][11]. It is also found by Roh [12] that adopting the big data analytics approaches specifically using sentiment analysis of social media, to get an insight into the public acceptance of nuclear power in South Korea [12]. However, up to now, we have yet to find evidence that there exists research in big data analytics specifically for nuclear security [13]. Therefore, a new conceptual framework of big data analytics for nuclear security has to be developed from scratch. The fundamental elements of our framework are based on the existing International Legal Framework for Nuclear Security, which we constituted as a non-big data analytics nuclear security framework. The International Legal Nuclear Security Framework has been illustrated by the International Atomic Energy Agency (IAEA) in multiple publications of the IAEA Nuclear Security Series, staggered in four different levels of grouped materials, i.e., the Nuclear Security Fundamentals [14], Recommendations, Implementing Guides, and Technical Guidance, Reference Manuals, Training Materials.

3. The proposed conceptual framework

This conceptual framework is developed in the context of a data management point of view. This concept illustrated how data from the nuclear security system should be managed and integrated so that real-time and autonomous analysis could be carried out to improve or optimize the required situational awareness mentioned in [1] and [4]. In this context, we follow the term 'Integrated Data Analytics' introduced by [1] which describes the approach in managing organizations that are led from the top and, where all information and data are used to generate value and support the success of the organization. This framework is also tailored to the statement that security must be viewed and operated as an enterprise-wide activity and be fully integrated into other business processes and objectives [1]. Furthermore, for autonomous analytics, the related data from nuclear facilities/operators must be fed in real-time into the state-level data management system.

Here, we propose a conceptual framework that is developed on two levels. The first level is dedicated to the implementation at the nuclear facilities/operators' level. The second level is proposed for a state-level implementation by the related state regulatory. However, at this stage, the focus of this framework is only limited to physical security [14][15], personnel security [16][17], nuclear/radioactive materials accounting and control system [18]. Additional elements namely, data visualization, data acquisition, data cleaning, and data exploration are adapted into the framework from the big data analytics concept. The next section will explain the framework that is dedicated to being implemented at the nuclear facilities/operators' level.

IOP Conf. Series: Materials Science and Engineering 1106 (2021) 012026 doi:10.1088/1757-899X/1106/1/012026

3.1. Level 1: Framework for nuclear facilities/operators

The first level of the proposed framework is depicted in Figure 1. This framework is named the integrated and intelligent nuclear security, I2NS framework. To realize the big data analytics for nuclear security at the state level, the security system that falls into the regime of nuclear security at facilities/operators needed to be fully digitalized. All manual systems should be upgraded into a computerized based system so that the related records and files could be stored electronically in a digitalized form utilizing existing database technology. Depicted in the lowest level of the framework illustrated in Fig. 1 are four components that are critically important for digitalization, i.e. the physical security system, the nuclear/radioactive materials accounting and control system, the personnel trustworthiness evaluation system, and the data visualization component.

Apart from that, all components are linked to the integrated online alert system so that trigger could be initiated from each of the components, whenever the abnormality is detected, as depicted in the second level of the framework. For each of the components, the related back-end databases must be also integrated to form the data pond or data lake, depending on the needed da2ta storage capacity for the respective facility/operator.

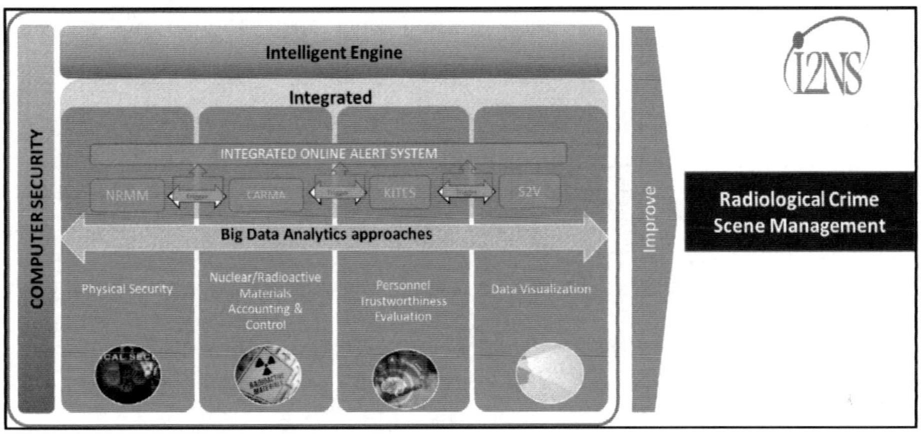

Figure 1. A big data analytics nuclear security framework for facilities /operators.

Figure 2 presents the recommended integrated database architecture that consists of the respected data components of physical security, personnel security, and nuclear/radioactive materials accounting and control. The Apache Hadoop [20] framework is recommended here because instead of relying on expensive, and different systems to store and process data, Hadoop is open-source software that enables distributed parallel processing of huge amounts of data across inexpensive, industry-standard servers that both store and process the data. It is important to state here that before the data could be used to form the data lake/data pond/data warehouse, the data need to be cleaned to ensure the correctness of the data, maintain the high-quality level of the data, and to ensure the data integrity. After that, the data must be explored to develop the analysis model that could lead to the development of data visualization and analytics product, that is usually displayed as a dashboard system.

In this framework, we have identified three types of data that must be dealt with to develop the big data analytics nuclear security system at the facilities/operators' level, which is, structured data, semi-structured data, and unstructured data. Structured data is data that is extracted from databases that may consist of a multi-platform database management system such as Microsoft SQL Server, MySQL, MariaDB, Oracle, or MongoDB. The semi-structured data may come from the radiation portal, sensors, or text files related to the nuclear security system. Lastly, unstructured data is the most difficult data type to be handled. This kind of data may come from the CCTV monitoring system, or surveillance system, which consists of images and video stream frames.

As depicted in Fig.2, the personnel trustworthiness evaluation is included as a component that utilized the sentiment analysis of the media social approach. The result of the sentiment analysis is then injected into the personnel profile databases for the evaluation of personnel trustworthiness. The trustworthiness evaluation is important for the detection of insider threats.

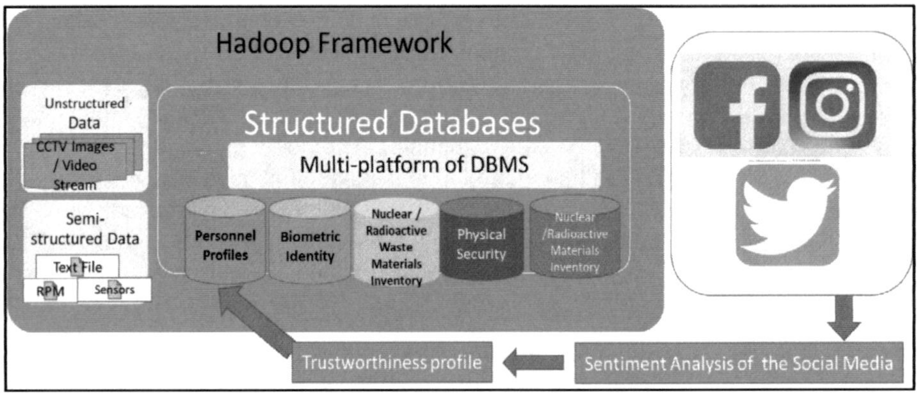

Figure 2. The integrated database architecture.

3.2. Level 2: State-level framework

Figure 3 illustrates the state-level framework. The elements in the framework are designated based on the nuclear security elements stated in the International Legal Nuclear Security framework and the need to perform autonomous analysis of the data for the realization of the enhancement situational awareness. It is recommended that the state-designated regulatory bodies that are appointed to safeguards nuclear/radioactive materials to take charge of owning the respective system based on this framework.

In this framework, we had specifically identified the external data that is needed to be acquired from external parties such as nuclear facilities/operators and GIS/GPS data providers. The data that is required from the related regulatory bodies is also stated in the framework. Once the data acquisition is done, data cleansing and exploration needed to be performed to develop the state-level analysis model that could result in the finding of a data product that could give an insight of what, where, when, and how the improvement shall be made with regards to the state-level nuclear security

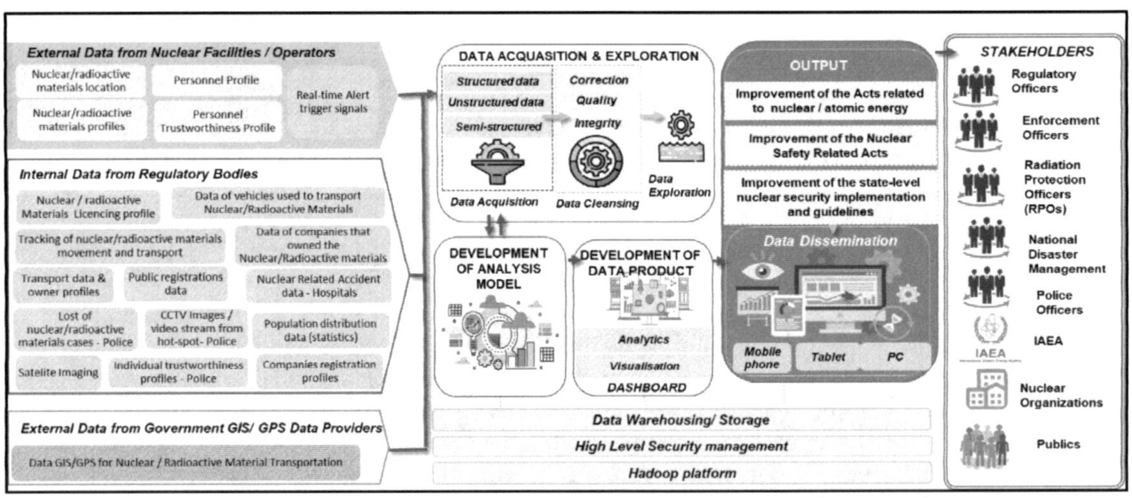

Figure 3. State-level big data analytics nuclear security framework.

The output could be an improvement of the Act, guidelines, or other related regulations tailored to the security, safety, and safeguards of the nuclear/radioactive materials which will be beneficial to the public, nuclear organizations, and other stakeholders.

4. Conclusion and future works

Nuclear security is the process of prevention and detection of, and response to unauthorized removal, sabotage, unauthorized access, illegal transfer, or other malicious acts involving nuclear or radiological material or their associated facilities. The possibility of nuclear material or other radioactive material that could be used for criminal purposes or intentionally used in an unauthorized manner cannot be ruled out in the current global situation. Currently, most information and data in nuclear security applications are managed in silos both within and between organizations which resulted in many current data management systems for the nuclear security system that are suboptimal hence cannot provide support to improve security decision and control.

In this paper, we had proposed a conceptual big data analytics framework for nuclear security. The framework is developed in the context of a data management point of view. It consists of two levels. The first level is dedicated to the implementation of nuclear facilities/operators. The second level is for the state-level implementation. This framework has yet to be evaluated. Therefore, our future works are to evaluate this framework by applying the empirical methodology. We are currently in the process of developing survey questionnaires. We had reviewed some related techniques for the evaluation process and choose to implement the Rasch Measurement Model for the evaluation process.

References

[1] World Institute of Nuclear Security, WINS 2015 Data analytics for nuclear security, how real-time integrated data management could support nuclear security *WINS Special Report Series,* (Vienna, Austria).

[2] International Atomic Energy Agency IAEA 2008 Security in transport of radioactive material, implementing guide. *IAEA Nuclear Security Series.***9** (Vienna, Austria).

[3] Beyer M A and Laney D 2012 The importance of big data: A definition. (Stamford: CT: Gartner).

[4] World Institute of Nuclear Security WINS, 2015 Big data in motion, how real-time integrated data management could support nuclear security *WINS Special Report Series* (Vienna, Austria).

[5] Ibrahim M, Ismail S, Haris M F, Hamdan S N, Sulaiman M S, Hamid A H A, Aris S N M, Aslan M D A, Ghani N F A and Hassan H 2017 A proposal on adaptation of big data analytics framework in nuclear security toward development of integrated, real-time,and intelligent nuclear security system *Nuclear R & D Seminar 2017* (Selangor, Malaysia).

[6] Ibrahim M, Ismail S, Haris M F, Hamdan S N, Sulaiman M S, Hamid A H A, Aris S N M, Aslan M D A, Ghani N F A and Hassan H 2018 The integrated intelligent nuclear security (I2NS) system: A progress report for 2018 *Technical Report No.Nuklearmalaysia/L/2018/173* (Kajang, Selangor, Malaysia: Malaysian Nuclear Agency).

[7] Ibrahim M, Ismail S, Haris M F, Hamdan S N, Sulaiman M S, Hamid A H A, Aris S N M, Aslan M D A, Ghani N F A and Hassan H 2020 Malaysian Nuclear Agency integrated and intelligent nuclear security (I2NS) system: A progress report 2019 *Technical Report No.Nuklearmalaysia/L/2020/10* (Kajang, Selangor, Malaysia: Malaysian Nuclear Agency).

[8] Pu C and Kitsuregawa M 2013 Big data and disaster management a report from the JST/NSF joint workshop *Technical Report No. GIT-CERCS-13-09;* (USA: Georgia Institute of Technology, CERCS).

[9] Jae-Min C, Junguk S, Choong-Sub Y 2015 A Review on applicability of big data technology in nuclear power plant: focused on O&M phases *Trans. of the Korean Nucl. Soc. Spring Meeting, May 7-8* (Jeju, Korea).

[10] Moon-Ghu P, Seong-Ki K, Jangbom C, Ji-Hun K 2017 Development of a big data framework

for operation support of nuclear power plants *Proc. of the KNS 2017 Spring Meeting.*

[11] Liu K 2019 Big data analytics solutions to improve nuclear power plant efficiency: online monitoring, visualization, prognosis, and maintenance decision making *Adv. Sensors and Instr. Ann. Webinar, Nov. 6, 2019* (University of Wisconsin-Madison, USA).

[12] Roh S 2017 Big data analysis of public acceptance of nuclear power in Korea J. of *Nucl. Eng. and Technol.* **49** 850-854.

[13] Ibrahim M, Ismail S, Haris M F, Hamdan S N, Sulaiman M S, Hamid A H A, Aris S N M, Aslan M D A, Ghani N F A and Hassan H 2020 big data analytics for nuclear security: a literature review *Nuclear R & D Seminar 2020, 16 -19 Nov. 2020 (*Kajang, Selangor, Malaysia).

[14] International Atomic Energy Agency IAEA 2013 Objective and essential elements of a state's nuclear security regime, nuclear security fundamentals. *IAEA Nuclear Security Series* **20** (Vienna, Austria).

[15] International Atomic Energy Agency IAEA 2013 nuclear security systems and measures for the detection of nuclear and other radioactive material out of regulatory control, implementing guide *IAEA Nuclear Security Series* **21** (Vienna, Austria).

[16] International Atomic Energy Agency IAEA 2015 Risk-informed approach for nuclear security measures for nuclear and other radioactive material out of regulatory control, implementing guide. *IAEA Nuclear Security Series*.**24-G** (Vienna, Austria).

[17] International Atomic Energy Agency IAEA 2008 Preventive and protective measures against insider threats,implementing guide *IAEA Nuclear Security Series* **8** (Vienna, Austria).

[18] International Atomic Energy Agency IAEA 2020 Preventive and protective measures against insider threats, implementing guide. *IAEA Nuclear Security Series (Rev. 1)* **8-G** (Vienna, Austria).

[19] International Atomic Energy Agency IAEA 2015 Use of nuclear material accounting and control for nuclear security purposes at facilities, implementing guide *IAEA Nuclear Security Series* **25-G** (Vienna, Austria).

[20] https://hadoop.apache.org accessed on 13th November 2020.

Acknowledgments

This study is supported by the International Atomic Energy Agency (IAEA) research grant under the IAEA Coordinated Research Project CRP J02011: Improving the Security of Radioactive Material throughout its Lifecycle, Associated Facilities, and Associated Activities (Contract No. 22165).

IOP Conf. Series: Materials Science and Engineering 1106 (2021) 012027 doi:10.1088/1757-899X/1106/1/012027

Laboratory facilities for nuclear forensic investigation in Malaysian Nuclear Agency

S. Muhd Sarowi[1,a)], Z. Laili[2], A.Ahmad[1] and K. Samuding[2]

[1]Radiation Safety & Health Division
[2] Waste & Environmental Technology Division
Malaysian Nuclear Agency,43000 Kajang, Selangor.

[a]suzie@nm.gov.my

Abstract. Nuclear forensic is a new field and a key element in nuclear security in any country. The purpose of nuclear forensic investigation is to establish the history of nuclear material of unknown origin by analysing the nuclear material in the laboratories. Hence, the results of the nuclear forensic examination will provide the evidence to support law enforcement or nuclear security investigation in the country. The typical analysis used in nuclear forensic investigation is radiological assessment, physical characterization, isotope analysis, elemental/chemical analysis and traditional forensic analysis. This paper will discuss the capability of laboratories in the Malaysian Nuclear Agency and internal arrangement in handling nuclear forensic analysis.

1. Introduction

The use of radioactive sources or nuclear material all over the world is increased, some country also introducing the nuclear power plant then nuclear security aspect should be strengthened. The nuclear security aims in preventing and detecting of and response to theft, unauthorized access, illegal transfer or other malicious acts involving radioactive or nuclear material. It was reported that as of 31 December 2019, a total of 3,686 incidents involving unauthorized activities and events of nuclear and other radioactive material, including incidents of trafficking and malicious use [1]. Nuclear forensic is define as "the examination of nuclear or other radioactive materials or of evidence contaminated with radionuclides in the context of international or national law or nuclear security" [2]. From the nuclear forensic analysis these such questions; what the materials are, how, when, and where the materials were made, and what their intended uses were can be answered. Therefore, nuclear forensics is important tool in the nuclear security [3].

In this regard, countries like Malaysia recognizing to develop the capabilities in this forensic field. An International Atomic Energy Agency (IAEA) expert mission was carried out in 2016 to assess Malaysia's capability in nuclear forensics. Thus, Malaysian Nuclear Agency (Nuclear Malaysia) as the promoting agency in the country in using nuclear technology and equipped with several facilities and instruments learned that this nuclear forensic could be developed in the agency. Nuclear Malaysia found that nuclear forensic is integration and cooperation among several laboratories and standard operating procedures (SOP) should be documented [4]. The document is purposely to provide guidance for nuclear forensic investigation in Nuclear Malaysia laboratories which technical report from the analysis will help the designated authority in decision making once dealing with the nuclear forensic case in the country.

2. Methodology

The nuclear forensic analysis consists of categorization, characterization, interpretation and reconstruction. The term 'material signature' is commonly used in this field. There are several analysis or techniques could be useful to obtain the material signature for example by physical dimensions measurement, analytical techniques, isotopic composition, trace elemental impurities, age determination and of course we cannot left behind the traditional forensic [5]. As mentioned earlier, the development of nuclear forensic in the agency is involvement of several division. Sharing knowledge gained from the training abroad was conducted which lastly the process flow as in Figure 1 was proposed to be used internally in handling the nuclear forensic investigation.

The designated authorities have to send the evidence to the Nuclear Malaysia. Then, the description, classification, categorization of the evidence will be performed. Laboratories will be identified based on the requirement and need from designated authorities before the distribution of evidence for further analysis. Finally, the full report will be written based on finding from the analysis in laboratories and will be submitted to the designated authorities.

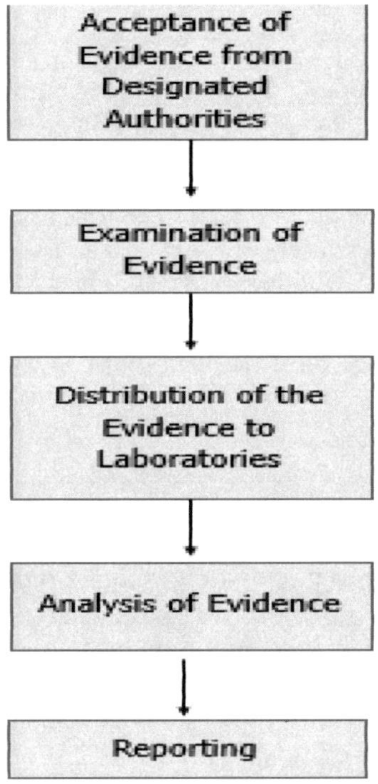

Figure 1. Proposed process flow for handling nuclear forensic case in Nuclear Malaysia.

3. Result and discussion

The results shows that 5 divisions in the agency could be involved in the nuclear forensic facilities. These division equipped with instrumentations which capable to obtain the "material signature". Laboratories will be integrated as in the process flow shown in Figure 1 and this development need a secretariat in order to coordinate, communicate and as the central of administrative. The identified laboratories under each division with their assigned task for nuclear forensic facilities is shown in the Table 1.

Table 1. Identified laboratories under each division with their assigned task for nuclear forensic facilities.

No.	Laboratory(Division)	Task
1.	Health Physic Group (Radiation Safety & Health)	Secretariat: receive evidence, conducting physical radiological assessment and physical characterization, distribution of sample, report writing & communicate to designated authorities
2.	Radiochemistry and Environmental Laboratory Analytical Chemistry Application Laboratory (Waste and Environmental Technology Division)	Radio analytical measurement & chemical analysis
3.	Material Technology Laboratory Micro Computed Tomography Laboratory (MicroCT) (Industrial Technology Division)	Material characterization analysis
4.	Reactor (Technical Support Division)	Operating Analytical Chemistry Lab
5.	Transmission Electron Microscope Laboratory (Radiation Technology Division)	More advanced analysis : nano-scale microscope

The Health Physics Group under the Radiation Safety and Health Division responsible to plan, prepare and implement health and environmental safety (industrial, radiological and nuclear) programs and activities in accordance with procedures and legal provisions enforced by regulatory bodies. Hence, it seems ideal to be the secretariat in this nuclear forensic facilities. This group equipped with instrumentations used for radiological assessment (dose rate, surface contamination, total activity) which is very important for radiological hazard precaution througout the various stages during the investigation.

A radio analytical chemistry procedure is to identify the radionuclide of interest and determining its amount in a sample. Chemical analysis will determine the physical properties or chemical composition of samples. Both of this analysis can be carried out by Radiochemistry and Environment Laboratory (RAS) and Analytical Chemistry Laboratory (ACA). RAS can identify the types of radionuclides (isotopes) present in the material within 24 hours Gamma Spectrometry System while ACA can determine Uranium (U)/Thorium (Th) using Neutron Activated Analysis (NAA) and Inductively Couple Plasma-Mass Spectrometer ICP-MS technique which is very important in nuclear forensic investigation. In addition, these laboratory adopt high quality control and quality assurance practices in order to provide reliable analytical test results, which meet international standards. RAS Laboratory also had been accredited with MS ISO/IEC 17025:2017 since December 2005 by Department of Standards Malaysia for the analysis of gamma emitting radionuclides.

Material characterization analysis is a further method for physical characterization to achieve the goal of nuclear forensic investigation. Material Technology Laboratory which equipped with Field Emission Scanning Electron Microscopy (FE-SEM) techniques, Scanning Electron Microscopy-Energy Dispersive X-Ray Spectroscopy (SEM-EDX), Energy Dispersive X-Ray Fluorescence Spectroscopy (EDXRF), Wavelength Dispersive X-Ray Fluorescence Spectroscopy (WDXRF), and X-Ray Diffraction (XRD), and Micro Computed Tomography Laboratory (MicroCT) under Industrial Technology Division together with

Transmission Electron Microscope Laboratory (Radiation Technology Division) can perform this analysis.

Above all, the secretariat plays a role in distributing the sample of evidence for analysis to identified lab based on the requirement and need from designated authorities. Beside that, traditional forensic (fingerprints, fiber, DNA) still have to be carried out. Those tecniques describe above is commonly used in nuclear forensic investigation [5]. Hereby, major elements of the IAEA's recommended pieces of the nuclear forensics 'model action plan' are in place in Nuclear Malaysia [4].

4. Conclusion

Currently, there are 10 SOP's was compiled and will be used in nuclear forensic investigation in Nuclear Malaysia. As nuclear forensic facilities is integration of several division in the agency, practices and familiarization of the process flow is a must. This development could be serve as a reference and guidelines for nuclear forensic analysis not only in Nuclear Malaysia but also could be used in supporting nuclear security system in Malaysia.

5. References

[1] IAEA 2020 Incident and trafficking database (ITDB), incidents of nuclear and other radioactive material out of regulatory control *Fact Sheet.*

[2] IAEA 2015 Nuclear forensic in support of investigations *Nuclear Security Ser. No. 2-G Rev. 1.*

[3] Michael J K, Amy M G, Naomi M, Kim K, William S C and Ian D H 2016 Nuclear Forensic Science:Analysis of Nuclear Material Out of Regulatory Control *Annu. Rev. Earth Planet. Sci.* **44** 555–79.

[4] Smith D K 2016 IAEA Expert mission report on assessment of malaysia capability in nuclear forensics.

[5] Mayer K *et al.* 2011 Nuclear forensics: a methodology applicable to nuclear security and to non – proliferation, *J. Phys. Conf. Ser* **312**.

Acknowledgements

Special thanks to the various laboratories's staff in Malaysian Nuclear Agency who involved in this work directly or indirectly for their support.

Two dimensional array of MPPC and CsI(Tl) for radiation monitoring prototype

A A Jasni[1], YS Yap[1,a)], I H.Hashim[1], N E Ahmad[1], N Ramlee[1]

[1]Department of Physics, Faculty of Science, Universiti Teknologi Malaysia, 81310, Johor Bahru, Johor, Malaysia.

[a]yungszen@utm.my

Abstract. An effcient radiation and position monitoring system is useful for preliminary security inspection and radiation emergency preparedness and response. A high sensitivity and effective monitoring system is needed for these applications. In this study, two position-sensitive prototypes were proposed by using of 8 Multi Pixel Photon Counters (MPPC) and coupled to the cesium iodide with Thallium (CsI(Tl)) scintillator. This inorganic scintillator was chosen for its output wavelength compatibility with MPPC, slightly hygroscopic, high photon transmission, and input energy. For both prototypes, eight MPPCs with di_erent con_gurations were used to obtain active areas of 2.0 × 2.4 cm^2 and 2.5 × 4.5cm^2, respectively. The prototype has been tested with radioactive (Tl-204 and Co-60). The intensity received by each MPPC's readout will be recorded using Extended Analogue SiPM Integrated ReadOut Chip (EASIROC) module. Our test showed that both prototypes were sensitive to both beta and gamma sources and were able to reliably determine the sources' positions. It was shown that the first prototype had greater resolution and performance than the second prototype. This studies may help to improve radiation detection system which is suitable for homeland security, medical field and scanning system.

1. Introduction

Radiation monitoring is important as a preventive measure before an accident occurs as well as to respond to a radiation incident. Current systems rely on the use of sensors capable of detecting radiation, such as photomultiplier (PMT) tubes or avalanche photodiodes (APD).

PMT for radiation monitoring is a photon detector containing of a scintillator, photocathode and electron multiplier anode. Incident ionising photons are transformed into light by the scintillator, which in turn ejects electrons by photoelectric effects at the photocathode. The electron multiplier anode then amplifies the weak signal into a measurable electrical signal.

An APD is a highly sensitive photodiode semiconductor that has the same feature as the photocathode in the PMT. To transform the incident ionising photons, the APD still needs a scintillator. The APD has higher detection efficiency (400 nm to 1100 nm wavelength range) [1], smaller and does not require high voltage since it does not use an electron multiplier anode compare to the PMT.

There are several disadvantages that make them undesirable for PMT and APD detectors. PMTs need a stable high voltage to operate and cannot be used in a strong magnetic field [2]. In addition, PMTs are mechanically fragile, very costly despite having lower quantum efficiency and large in size [2, 1]. The APDs are linear devices and have moderate gain (50-100) and unable to measure low energy (<50 keV)

[3]. As such, a suitable type of detector is needed for an efficient position-sensitive radiation monitoring system.

In this study, PMT and APD are replaced with a semiconductor detector with excellent photon counting capability known as Multi-Pixel Photon Counter (MPPC). It is a solid state detector using multiple combined APDs as a pixel operating in Geiger mode. The MPPC is sensitive enough to detect single photons for photo counting at low light [4]. MPPC is a type of silicon photomultiplier. The main features of MPPC are high resolution for single photon detection, small in size, insensitive to a magnetic field and low operating voltages.

The MPPC is a type of photomultiplier of silicon. High resolution, small in size, magnetic field resistant and low operating voltages are the key characteristics of the MPPC for single photon detection. The MPPC operates at low bias voltage operation (typically 70 V) with high gain almost comparable to conventional PMT [5][6]. MPPC is also good in signal multiplication and has higher photon detection efficiencies [7] despite having higher dark count rates compared to PMT (that the detection of weak scintillation light signals is difficult due to the many contamination of dark counts) [6]. Although the active area of MPPC is much smaller compared to PMT and APD, scintillation light signals still can be observed separately. It can also be used under a strong magnetic field range (1-12 Tesla) [8]. MPPC is also affordable and durable compared to PMT [6] while having better energy threshold than APD because of MPPC have better timing resolution and higher gain [1].

Besides a sensitive detector, a good scintillation material is needed to get a high detection efficiency. There are many types of scintillation materials which are primarily solid crystals, but can also be made of liquid or gas [9]. For solid scintillators, there are two common types: organic and inorganic scintillators. For high detection performance and low energy X-ray or gamma-rays, inorganic scintillators are suitable. The organic scintillators, however, are fragile, have low efficiency of detection and poor resolution [10].

In order to maximise the generation of scintillation light, an inorganic CsI(Tl) crystal was selected as the scintillation material to provide excellent photo-fluorescence. The CsI(Tl) offers a high effective atomic number and the consequently large cross-section of the radiation photo absorption [11]. Although, sodium iodide with thallium activator, NaI(Tl) gives a better light transmission, but NaI(Tl) is hygrocopic [12] and easily tampered by humidity especially in Malaysia. CsI(Tl) is only slightly hygroscopic than NaI(Tl) [10] and has good radiation hardness (103 rad) that suitable applied in high-energy physics experiment [13].

Additional information such as the position of the radiation source is imperative and can be determined using position-sensitive detection. In general, a position-sensitive detection is a measurement technique in 2 or 3 dimensions from a reference point, in order to determine the origin of the signal. This technique is useful for many applications such as vibration, heat and light measurement. Thus, in this research, a position-sensitive radiation detection device was built by using MPPC and CsI(Tl). In order to assess the precision and resolution of the prototypes, two proposed systems, each with different array configurations, were tested with different sources.

2. Experimental setup

In this study, a new system position-sensitive monitoring was developed to identify the position of the radiation. There were two type of prototype were tested by using EASIROC-NIM for the performance of prototype of MPPC array. Therefore, each MPPC (S10362-11-50C) will be coupled to CsI(Tl) scintillator. Each pixel of them MPPC has a photosensitive area of 1.0×1.0 mm^2 and is composed of Geiger-mode APDs arranged at a pixel pitch of 100 μm. The surface of CsI(Tl) scintillator will be polished using sandpaper to make sure the surfaces were smooth and clear. Then, the CsI(Tl) scintillator was cleaned using 2-propanol. The silicone oil was used to coupling between CsI(Tl) scintillator and MPPC so it can allow propagation of photon signal from scintillator to the MPPC. The refractive index of silicone oil and MPPC are 1.405 and 1.41 respectively [4][14]. The density of CsI(Tl) is 6.01 g/cm^3.

The prototype 1 housing is as shown in Figure 1. The prototype 1 outer dimension was around $(3.0 \times 7.0 \times 6.6)$ cm^3 where 8 CsI(Tl) filled up the insides of the housing with a total volume of $(2.4 \times 2.0 \times 7.0)$ cm^3.

There are contains four CsI(TI) for each layer. Then each MPPC was coupled to CsI(Tl) crystal and all the MPPC were in contact with the surface of the CsI(Tl) crystal by using silicon oil. A radiation source was placed directly to prototype 1. The scintillation photons signal information was recored by each MPPC that coupled to a single CsI(Tl) scintillator.

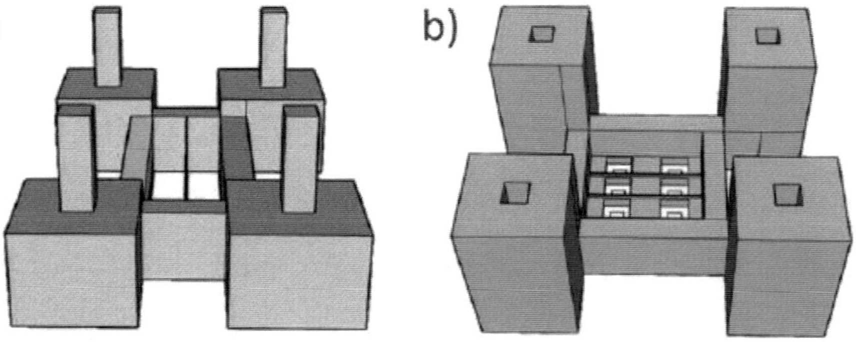

Figure 1. The design of Prototype 1 housing for (a) female and (b) male part.

Figure 2 shows the design of prototype 2 for female part and male part for prototype 2. The prototype 2 outer dimension was around (8.9 × 2.0 × 5.9) cm^3. where 4 CsI(Tl) filled up the insides of the housing with a total volume of (4.5 ×2.0 × 2.5) cm^3. Similar to prototype 1, all of the MPPC were coupled to the CsI(Tl) crystals with silicon oil. A radiation sources were placed in the middle of the CsI(Tl) scintillator. Then, the scintillation photons signal information were towards to both ends of MPPCs and recoreded the position information.

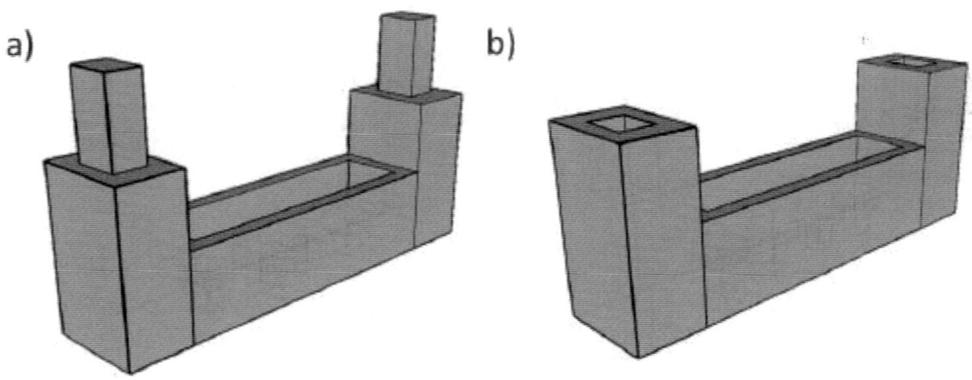

Figure 2. The design of Prototype 2 housing for (a) female and (b) male part.

The distance of prototype from radiation source is keep constant but was varied the position of source to the prototype in order to observe the photons detected by all MPPCs. The schematic experiment setup is as shown as in Figure 3. The photoelectron threshold was set from 0.5 p.e. A software that is interfacing with EASIROC module had been developed by Omega [3]. EASIROC-NIM module is connected to MPPCs using rainbow cable as stated in the user guide. The output of MPPC is a high-speed pulse signal and need to delay the signal for detection by 50 ns. In this system, the optimum delay was 50 ns and 5 meter of lemo coaxial cable 50 were used according RS component guide [15]. Operating voltage for each MPPC was varied within its range given by supplier. If the operating voltage applied to the MPPC is significantly higher than the recommended operating voltage, noise components such as afterpulses and crosstalk will increase and make accurate measurement impossible.

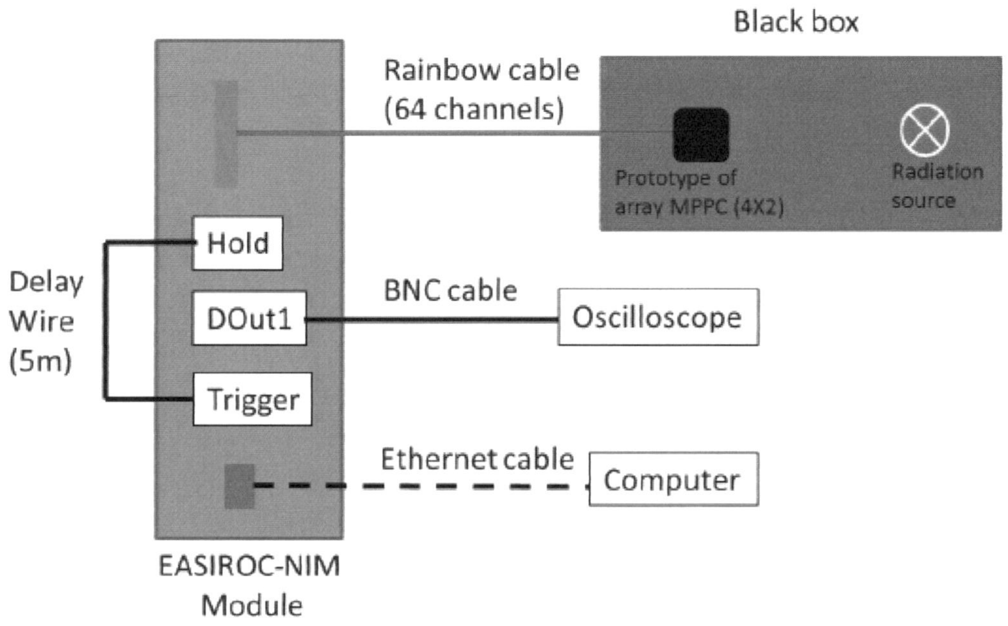

Figure 3. The schematic diagram for the experiment..

3. Result

In order to observe the amount of counts events, the data was analyzed using ROOT software. The number of photon distribution detected by MPPs was calculated based on area under of histogram obtained.

The detector response function gives spectrum of signal amplitudes by a particle of defined energy. Photons with definite energy may interact with the detector which results into a broad spectrum of deposited energies due to the subsequent interaction of the recoil electrons. Charged particles with definite energies which are stopped within the detector material will follow Gaussian distribution of signal amplitudes [16].

The output charge radiation distribution was calculated using full energy peak (FEP) event formula [17]:

$$FEP = \tilde{x} \times \sigma \times \sqrt{2\pi} \tag{1}$$

where the $\tilde{x}$ is the highest peak for 1 p.e. and σ is the FWHM.

In both prototype, scintillation photon will travel within the CsI(Tl) and MPPS located at the fo the scintillator. However, loss of the scintillation light can occur in two ways which are escape through scintillator boundaries and absorption due to the scintillator materials itself [16].

Only scintillation photon that reaches the MPPC will be converted into signals. As such, the light intensity observed by each MPPC corresponds to a fraction of the light emitted by the scintillation process due to attenuation along the scintillator. The light intensity as a function of length can be described as:

$$I(x) = I_0 e^{\frac{-x}{l}} \tag{2}$$

where *l* is the attenuation length, x the path length travelled by the light and I_0 is the initial light intensity.

The scintillation light that are converted into electrical signals are then counted to form the histogram. Once the histrogram was plotted, the number of photon counts can be processed to form spatial images.

For prototype 1, this information can be used directly since each MPPC was connected to a single CsI (Tl) scintillator (refer in Figure 4) and a spatial image with intensity (photon count) can be plotted.

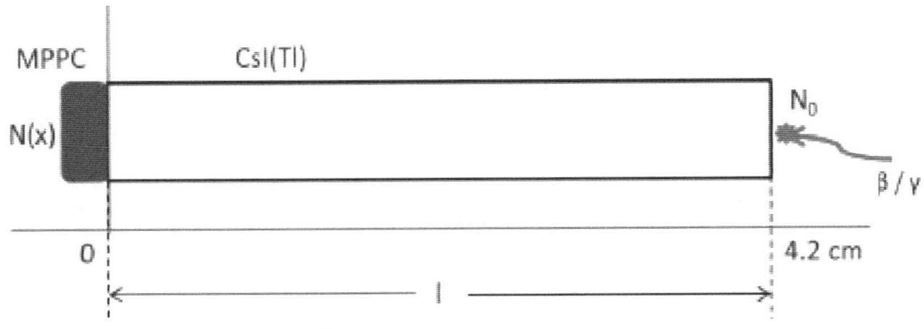

Figure 4. Number of scintillation photon detected by MPPC for prototype 1.

However, for the prototype 2, some processing is required to extract information to form the spatial image. In this design, the scintillation photon will move within the CsI(Tl) to both MPPCs, located at both ends of the scintillator. According to the Figure 5, the number of photons signal N1 and N2 of the left and right MPPCs located at the respective ends of the scintillator are [18]:

$$I(x) = I_0 e^{\frac{-x}{l}} \tag{3}$$

Thus, the number of photons in each CsI(Tl) scintillator was calculated by taking root of the product the number of photons signal N1 and N2 [18] is,

$$N = N_1 \times N_2 = \sqrt{\frac{4N_1N_2}{e^{(-1)}}} \tag{4}$$

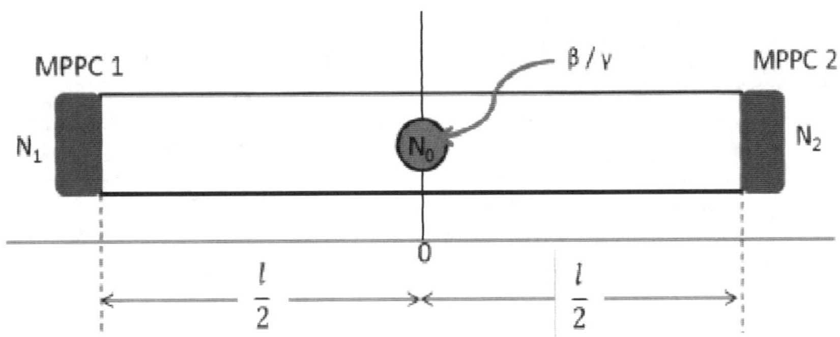

Figure 5. Number of scintillation photon detected by MPPC for prototype 2.

IOP Conf. Series: Materials Science and Engineering 1106 (2021) 012028 doi:10.1088/1757-899X/1106/1/012028

In order to evaluate the 2-D plane position monitoring, the mapping of photons count has been done by using Gnuplot software as shown in Figure 6. The contour plot of these 2-D images will represent different scintillation light intensities. These heat maps or contours are expected to be in agreement with the areas of high incident radiation photons where a colour map can represent areas of higher or lower radiation photon counts. As such, the position and spatial resolution of both prototypes can be determined. These parameters are important to determine the quality of the position-sensitive radiation monitoring system.

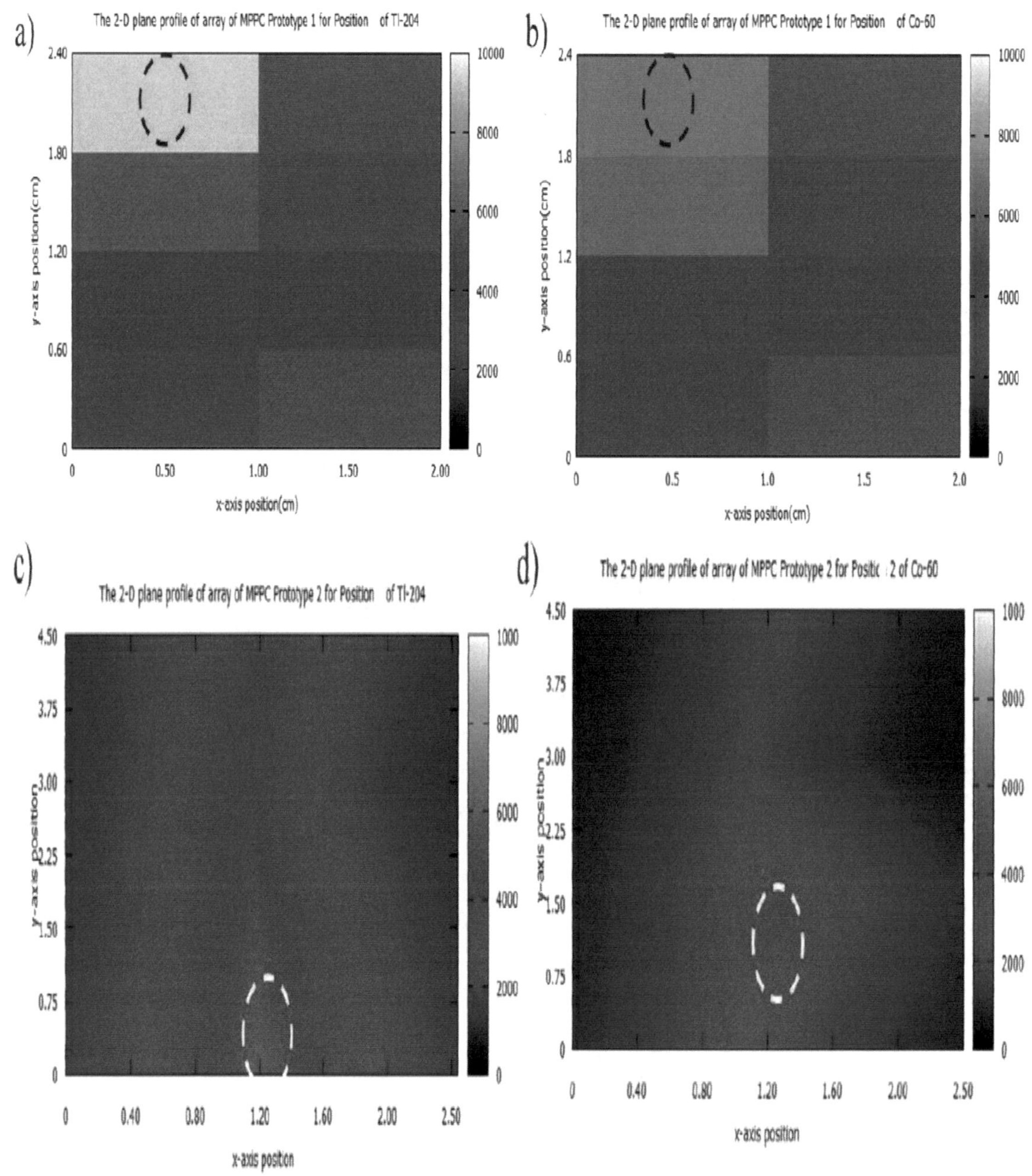

Figure 6. The 2-D plane profile monitoring for prototype 1 and 2.

4. Conclusion

In this study, the 2-D plane location monitoring indicates that both prototype arrays are feasible in the radiation position seen for both sources. The 2-D plane position monitoring is capable of separating high and low radiation counts for both prototypes. On both prototypes, the intensity photons of Tl-204 are higher compare to Co-60. This is because the energy deposited by the beta source is greater than the source of gamma. Plus, beta radiation has high ionizing compared to gamma radiation. In this case, the particle will decelerate when the ionising phase happens as beta particles travel into the scintillator. While the gamma emission brings the photon down to a more stable energetic state (lower state) and produce little ionization.

In this research, even change the position of source, it gets the same outcome depending on the position that has been placed. This proves that, in beta radiation, the MPPC has high-energy sensitivity and specifically separates between high and low radiation energy. The non-uniform photons distribution could be caused by the background noise and ambient fluctuation. The pattern for the highest count is the same from this study for positioning the source and dependent on photon distribution. The pattern found from this study is the same as the previous Wee *et al.* study [19] study the flexible temperature sensor array with different heat sources from different shape.

In terms of photon distribution, the position sensitivity response was calculated in order to obtain a 2-D plane profile. The number of photon events reported by the MPPC was specified on each channel by colours. The stability between the EASIROCNIM module MPPC and CsI(Tl) was observed by understanding the actual position of the MPPC detector that detects multiple source positions. The differences between high and low intensity radiation can be easily seen from the plane profile. Testing at different locations found that good agreement between the position allocated and detected position contributed to good position monitoring.

In this research, the absolute efficiency of the detector was calculated [20]. These measurements are the percentage of radiation that detect overall yield that emitted from the source to the detector. The total efficiency of prototype 1 with beta and gamma source is 8.86 % and 18.51 % while prototype 2 is 1.65 % and 2.07 %, respectively. Therefore, in terms of detection efficiency, prototype 1 was greater than prototype 2, since prototype 1 directly detected the distribution of scintillation photons compared to prototype 2.

5. Reference

[1] Lawrence W G, Varadi G, Entine G, Podniesinski E and Wallace P K 2008 A comparison of avalanche photodiode and photomultiplier tube detectors for flow cytometry *Imaging, Manipulation, and Analysis of Biomolecules, Cells, and Tissues VI.* Conf. Ser. **6859** (Soc. of Photo-Optical Instr. Eng. (SPIE)) doi:10.1117/12.758958 68590M.

[2] Polyakov S V 2013 Single-Photon Generation and Detection: Chapter 3. *Photomultiplier Tubes* **45** (Elsevier Inc.).

[3] Nassalski A, Moszynski M, Syntfeld-Kazuch A, Szczesniak T, Siderski L, Wolski D, Batsch T and Baszak J 2010 Multi pixel photon counters (MPPC) as an alternative to APD in PET applications *IEEE Trans. on Nucl. Sci.* **57**(3) 1008-1014.

[4] Hamamatsu Photonic 2015 *What is the MPPC? Technical Report* (Solid State Division).

[5] MPPC 2014 *MPPC Module. Technical Report.* (Solid State Division).

[6] Yamamoto K, Yamamura K, Sato K, Ota T, Suzuki H and Ohsuka S 2006 Development of multi-pixel photon counter (MPPC) *2006 IEEE Nucl. Sci. Symp. Conf. Rec.* (IEEE) **2**. 1094-1097.

[7] Hosomi F 2016 *Characterization of Multi-Pixel Photon Counters for a new neutrino detector* (Ph.D. thes.).

[8] Roncali E and Cherry S R 2011 Application of silicon photomultipliers to positron emission tomography *Ann. of Biomed. Eng.* **39**(4) 1358-1377.

[9] Prekeges J 2012 *Nucl. Med. Instr. 2nd ed.* (Jones and Bartlett Publishers). ISBN 1449652883.

[10] Saha G B 2012 *Physics and Radiobiology of Nuclear Medicine* (Springer Science and Business Media).

[11] Yanagida T 2018 Inorganic scintillating materials and scintillation detectors *Proc. of the Japan Acad. Ser. B* **94**(2) 75-97.

[12] Saha G B 2006 Scintillation and semiconductor detectors *Phys. and Radiobiol. of Nucl. Med.* (Springer) 81-107.
[13] Kobayashi, M. and Sakuragi, S. Radiation damage of CsI (Tl) crystals above 103 rad. Nuclear Instruments and Methods in Physics Research Section A: Accelerators, Spectrometers, Detectors and Associated Equipment, 1987. 254(2): 275-280.
[14] Hamamatsu Photonics 2008 *K MPPC Modules Technical Report.*
[15] *R C Tecnical Data Assembled MFB Models Cables Technical Report 2019.*
[16] Leo W R 2012 Techniques for nuclear and particle physics experiments: a how-to approach (Springer Science and Business Media).
[17] Gultekin A, Kaynak G and Gurler O 2006 Determination of full energy peak efficiency of HpGe detector from 59.5 to 1332.5 keV.
[18] Hashim I H 2010 *Development of Novel Particle Detector with Multi Pixel Photon Counter Readout* (Universiti Teknologi Malaysia: Ph.D. thes.).
[19] Shih W-P, Tsao L-C, Lee C-W, Cheng M-Y, Chang C, Yang Y J and Fan K C 2010 Flexible temperature sensor array based on a graphite-Polydimethylsiloxane composite *Sensors* **10**(4): 3597-3610.
[20] Crystals S G 2004 Efficiency calculations for selected scintillators.

Acknowledgments
This work was financially supported by the Universiti Teknologi Malaysia under the Research University Grant; Q.J130000.2526.12H74 and Q.J130000.2526.20H07.

IOP Conf. Series: Materials Science and Engineering 1106 (2021) 012029 doi:10.1088/1757-899X/1106/1/012029

A multi-year study of natural radionuclides Beryllium-7 (Be-7) activity concentrations in surface air in Tanah Rata, Cameron Highlands, Pahang

F I A Rashid[1, a)] and M Z Zolkaffly [1, b)]

[1] Planning and International Relations Division, Malaysian Nuclear Agency, 43000 Kajang, Selangor, Malaysia.

[a]faisal_izwan@nuclearmalaysia.gov.my
[b]zulfakar@nuclearmalaysia.gov.my

Abstract. The Comprehensive Nuclear-Test-Ban Treaty (CTBT) is a multilateral treaty that bans any nuclear explosions, either for military or civil purposes. The treaty is supported by an international network of monitoring stations, which inter-alia include radionuclide monitoring stations. Malaysia is a Member State to the CTBT and hosts a radionuclide monitoring station (Site code: RN42) in Tanah Rata, Cameron Highlands, Pahang. The continuous daily measurement at RN42 has created an invaluable resource for understanding variations in natural background radioactivity, including natural radionuclide, Beryllium-7 (Be-7). This study aims at assessing seasonal variations of Be-7 in surface air in Tanah Rata, Cameron Highlands, Pahang and determining its correlation with local meteorological parameters using statistical analysis. The results have been found that Be-7 activity concentration at the RN42 showed a seasonal variation and strong inverse correlation with precipitation data. The findings from this study could serve as basis for further studies on local and regional radiological impact assessment as well as for exploring the potential application of Be-7 as atmospheric tracer for civil and scientific purposes.

1. Introduction

The Comprehensive Nuclear-Test-Ban-Treaty (CTBT) is a multilateral treaty which bans all nuclear explosions – everywhere and by anyone. Under the provisions of the treaty, a global monitoring system based on waveform and radionuclide technologies known as the International Monitoring System (IMS) is currently being set up to verify country's compliance to the treaty [1]. The IMS is one of the key components of the verification regime of the CTBT [2]. Malaysia is among the CTBT contracting states through its commitment in signing the treaty on 23 July 1998 and ratifying it on 17 January 2008 [3]. Under the CTBT, Malaysia has hosted an IMS Radionuclide Monitoring Station RN42 in Tanah Rata, Cameron Highlands, Pahang since 2010. The continuous daily measurement of atmospheric radionuclides indicative of nuclear explosion at RN42 has created an invaluable resource for understanding variations in atmospheric radioactivity.

Among the atmospheric radionuclides regularly measured at the station is Beryllium-7 (Be-7). Be-7 is a sampled particle-bound natural radionuclide which is mainly produced in the upper troposphere and lower stratosphere by spallation of nitrogen and oxygen nuclei of atmospheric molecules induced by galactic cosmic rays [4]. It decays with a 53.3 days half-life which make it suitable tracer for atmospheric dynamics [5]. Over two third of Be-7 is produced in the lower stratosphere (around 20 km altitude) while the remaining is produced in the upper troposphere [6]. The Be-7 production rate in the air is associated with solar activity, seasonal variation and latitude dependence [7]. Once formed in the atmosphere, it attached to submicron-size aerosol particles and transport with air masses. The aerosols are transported by winds and redistributed vertically to earth's surface via atmospheric vertical mixing and gravitational sedimentation [8][9]. Moreover, Be-7 is removed from the atmosphere by wet and dry deposition in addition to its radioactive decay. Studies have showed that worldwide Be-7 concentrations on ground level follow an annual cycle whereby the transport of the radionuclide in the atmosphere is strongly governed by seasonal atmospheric processes [7].

In general, Be-7 has been used for numerous applications. Accordingly, several studies have proposed the radionuclide as natural atmospheric tracer [10][11][12]. Additionally, there are also studies on connection between air mass transport and ground level Be-7 concentration, such as seasonal variations caused by stratosphere-troposphere exchange, vertical air mass transport and transport from mid-latitudes to polar regions [13][14][15]. Hence, the variation of Be-7 deposited near the earth's surface can be used as atmospheric tracer to study the behaviour of radionuclides in aerosols and atmospheric processes including tropospheric dynamics, stratosphere-troposphere coupling and cosmic ray flux variations due to solar activity. This understanding could serve as basis for further studies on local and regional radiological impact assessment as well as for exploring the potential application of Be-7 as atmospheric tracer for civil and scientific purposes.

This study aims at assessing seasonal variations of Be-7 in surface air in Cameron Highlands, Pahang and determining its correlation with local meteorological parameters using statistical analysis. The study takes advantage of the detection capabilities of the IMS Radionuclide Monitoring Station RN42, wherein Be-7 is regularly measured in the air particulate samples collected by station.

2. Methods

2.1. Site description

Be-7 data used in this study was obtained from the IMS CTBTO Radionuclide Monitoring Station RN42 (4.48°N, 101.37°E) as illustrated in Figure 1. RN42 is a particulate radionuclide monitoring station co-located with the Meteorological Station in Tanah Rata, Cameron Highlands, Pahang. The station is situated 1545 m above sea level. The station is one of the 80 radionuclide monitoring stations under the IMS that monitor atmospheric radionuclides particles indicative of a nuclear explosion.

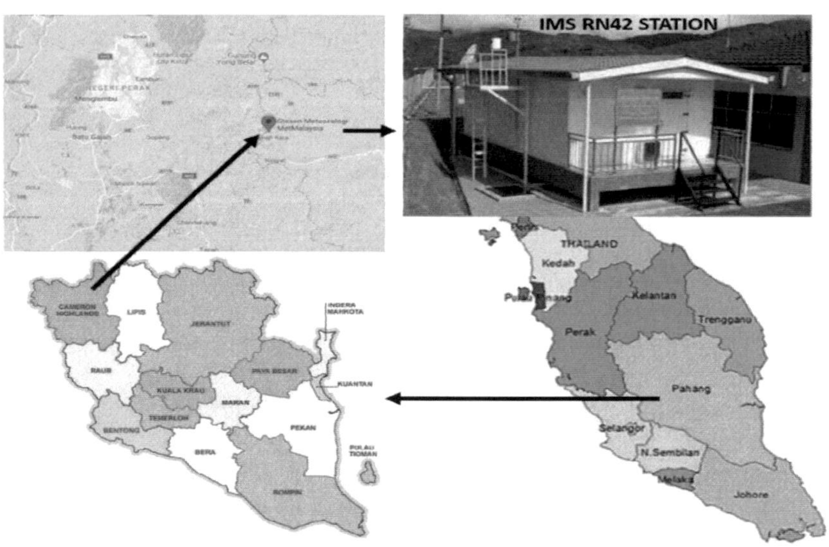

Figure 1. The location of IMS CTBTO Radionuclide Monitoring Station RN42.

Tanah Rata is located in the district of Cameron Highlands in the state of Pahang, Malaysia. It is situated in the main mountain range of Banjaran Titiwangsa, with elevations ranging from 100 m at the east to 2031 m at the west [16]. The climate in Tanah Rata's vicinity is between a tropical rainforest climate and a subtropical highland climate. It has recorded a temperature ranged from 15.0 °C to 25.4 °C with an average temperature of 19.4 °C [17]. Moreover, the historical meteorological record showed that the mean rainfall is 95.8mm with the maximum of 365.1 mm and minimum of 8.9 mm [18].

2.2. *Sample collection*

The RN42 is equipped with a high-volume air sampler, detection equipment, computers and a communication set-up. At high-volume air sampler, the air is forced through a 57 cm × 46 cm polypropylene air filter for 24 hours with a sampling rate of 300 - 900 m^3/h. Next, the air filter was compressed to a 5 cm diameter disc using a hydraulic press and placed on a plastic sample holder before being allowed for decay for 24 hours. Then, the filter is analysed for another 24 hours using a high-purity Germanium (HPGe) gamma spectrometer with a relative efficiency of 40%. The generated gamma radiation spectra are sent to the International Data Centre (IDC) in Vienna, Austria for further analysis and sample categorisation.

2.3. *Statistical analysis*

In this study, the daily activity concentrations of Be-7 from 1 January 2011 to 31 December 2018 were obtained from the CTBTO Concentration Reporting Tool (CRTool) [19]. Additionally, the daily precipitations in Tanah Rata, Cameron Highlands, Pahang for 1 January 2011 to 31 December 2018 were acquired from the Malaysian Meteorological Department. For both data, monthly average activity concentrations and precipitation for each year were plotted against acquisition month using Microsoft Excel. The generated graphs and tables were then examined for seasonal variation in activity concentrations of Be-7 and its correlation with precipitation data.

In order to determine the influence of meteorological conditions on the deposition of atmospheric radionuclide on earth's surface, this study has attempted to correlate the Be-7 activity concentration data with meteorological data such as precipitation. In this context, monthly average activity concentrations of Be-7 were overlay against monthly average rainfall measured at the RN42 using Microsoft Excel. The correlation was determined using simple liner regression, with value of correlation (R^2) nearest to 1 indicates strong liner relationship, and vice versa. In addition, for confirmation and verification purposes, the correlation between Be-7 activity concentrations with precipitation was also assessed using Pearson correlation coefficient (ρ), with values between ± 0.50 to ± 1.00 considered as having strong correlation, between ± 0.30 to ± 0.49 as moderate, and between 0 to ± 0.29 as having weak correlation.

3. Results andDiscussion

3.1. *Seasonal variation and frequency distribution of Be-7*

Figure 2 shows the time series plot of average monthly Be-7 activity concentrations against month of sample collection measured at RN42 from 2011 to 2018. Table 1 summarises the average monthly Be-7 activity concentrations measured at RN42 from 2011 to 2018 and its standard deviations by months. Over the period of 2011 to 2018, the average monthly Be-7 activity concentrations ranged from 1.13 mBq/m^3 to 2.85 mBq/m^3 as shown in Table 1. Be-7 activity concentrations show two noticeable annual peaks, with the highest maximum activity concentration in March and second maximum in June as illustrated in Figure 2. Conversely, two annual minimums occurred throughout the year, with the lowest occurring in November and second lowest in May. The elevated Be-7 concentrations observed during the months of March and June as shown in Figure 2 can be explained by the low amount of precipitation in the vicinity of Tanah Rata recorded during these months. Additionally, the finding also suggest that the study area could experiencing stratosphere-troposphere exchange of air masses in the mid-latitudes caused by an intense vertical jet stream which lead to an increase in Be-7 depositional fluxes to the Earth's surface [20]. The finding is also consistent with the fact that the variation of near surface Be-7 activity concentrations is contributed by several factors including wet scavenging, vertical mixing and horizontal atmospheric transport [21].

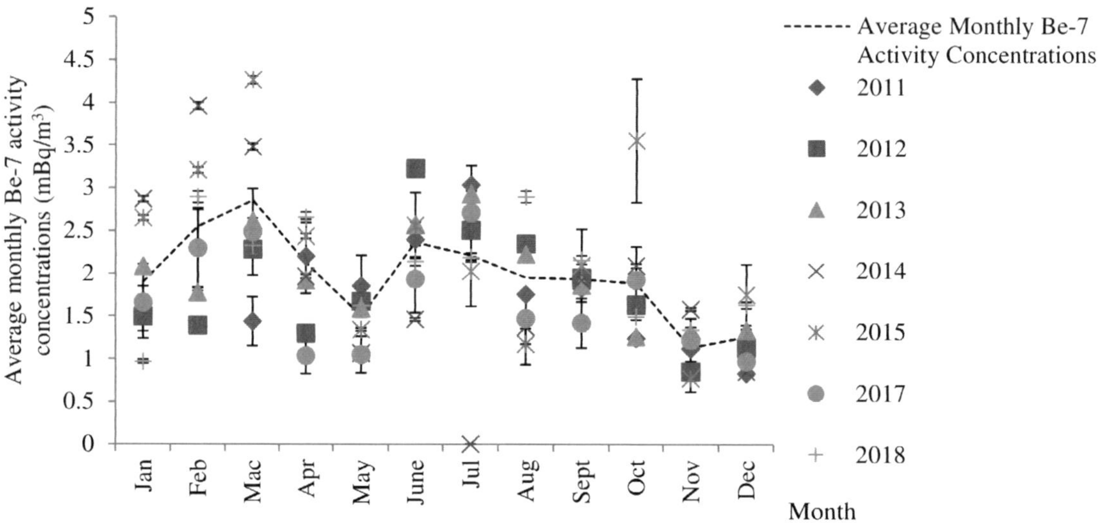

Figure 2. Average monthly Be-7 activity concentrations measured at RN42 from 2011 to 2018

Table 1. Summary of average monthly Be-7 activity concentrations measured at RN42 from 2011 to 2018

Months (2011-2018)	Average monthly Be-7 activity concentration (mBq/m^3)	Standard deviation, ρ
January	1.90	0.63
February	2.55	0.81
March	2.85	0.95
April	2.13	0.78
May	1.50	0.31
June	2.37	0.53
July	2.21	0.96
August	1.95	0.60
September	1.94	0.26
October	1.88	0.74
November	1.13	0.28
December	1.26	0.37

3.2. Correlation of Be-7 activity concentrations with precipitation data

The relationship of precipitation and Be-7 activity concentrations is showed in Figure 3 and Figure 4. Figure 3 shows the average monthly radionuclide activity concentrations and average monthly precipitation at RN42 from 2011 to 2018 plotted against month of sample collection. The interpretation of Be-7 activity concentrations trendline is already discussed as Figure 2 in previous section. On the other hand, the trendline of precipitation showed two annual peaks which indicate the high amount of precipitation occurred over the study area, with the highest in November and the second highest in April. The first peak in April is largely caused by the South-West monsoon winds which usually starts in the last week of April and northward motion of the Intertropical Convergence Zone (ITCZ) which passes over Peninsula Malaysia between April and May [22]. Additionally, the second peak in November is attributed by the southward motion of the ITCZ which crosses Peninsula Malaysia around November [22].

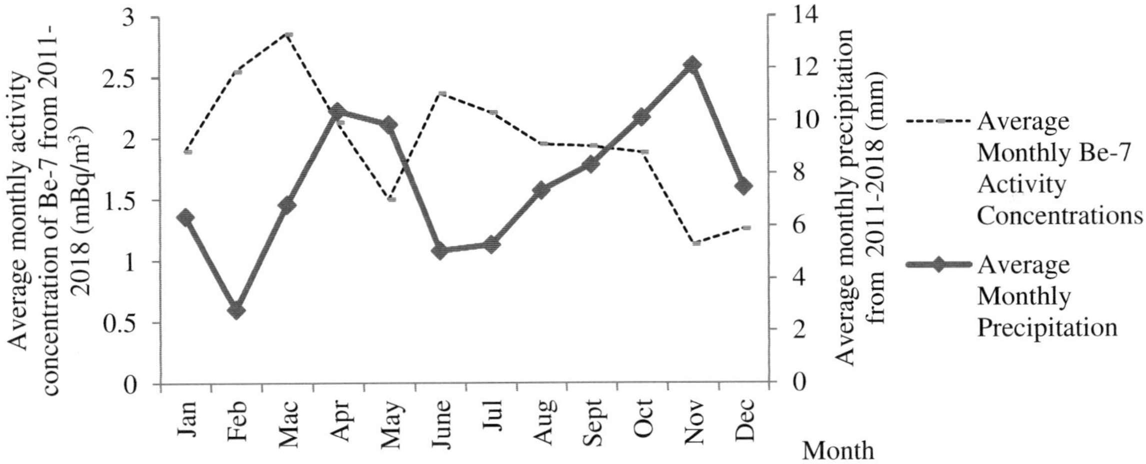

Figure 3. Correlation of average monthly Be-7 activity concentrations with average monthly precipitation.

Figure 4 represents the result of simple linear regression analysis using Be-7 activity concentrations and precipitation. It could be seen that the trendline is sloping downward which indicates inverse correlation. The R^2 value of 0.2277 specifies moderate linear correlation. Furthermore, Pearson correlation coefficient also showed the value of -0.65 which validate that both parameters have strong inverse correlation. This draw the conclusion that the increase of precipitation will decrease the Be-7 activity concentrations, and vice versa. This conclusion is well-concurred with the findings from other studies which found that wet scavenging as the dominant mechanism for the removal of Be-7 in surface air [23].

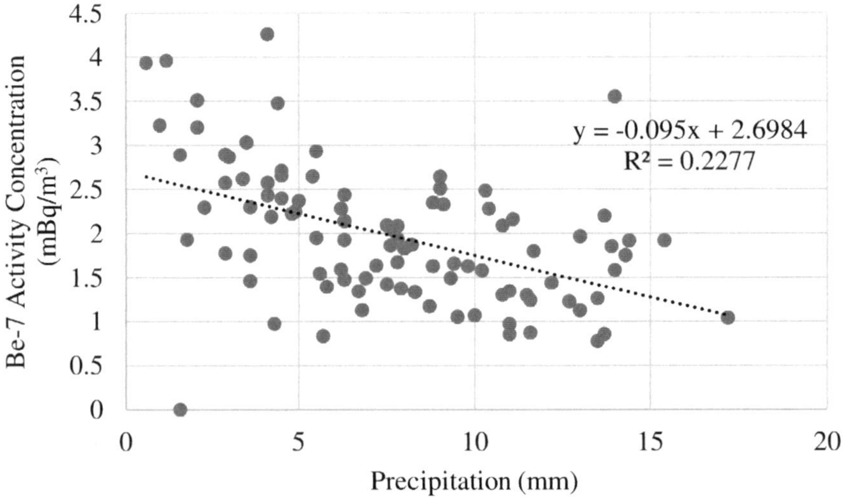

Figure 4. Simple linear regression of Be-7 activity concentrations and precipitation.

4. Conclusion

The study presents the seasonal variation of natural atmospheric radionuclide Be-7 activity concentrations in Tanah Rata, Cameron Highlands, Pahang from 2011 to 2018. The mean monthly activity concentration of this radionuclide ranged between 1.13 mBq/m^3 to 2.85 mBq/m^3. Over the period of this study, there were two annual peaks of Be-7 activity concentrations throughout the year, with the highest maximum activity concentration in March and second maximum in June. These elevated concentrations are due to low amount of precipitation in the vicinity of Tanah Rata as well as strong vertical atmospheric jet stream which lead to an increase in Be-7 depositional fluxes.

Based on correlation and simple linear regression of radionuclide data with precipitation data, a strong inverse correlation was found. This observation concurred with the findings from other studies that recognize wet scavenging as the main mechanism ofBe-7 removal in surface air. Overall, this study concludes that the seasonal variation of Be-7 activity concentration in Tanah Rata, Cameron Highlands, Pahang are dependent on atmospheric processes and conditions affecting the study area. This study could serve as basis for further studies on local and regional radiological impact assessment as well as for exploring the potential application of Be-7 as atmospheric tracer for civil and scientific purposes.

5. References

[1] Gerhard W, Lars-Erik D G, Philippe D, Martin K, Harri T, Real D, Franco D, Jean-Pierre I, Matthias L, Petra S, Andreas F, Craig S and Hiromi Y 2003 Atmospheric transport modelling in support of CTBT verification—overview and basic concepts *Atmospheric Environ.* **37** (18) 2529-37.

[2] Auer M, Kumberg T, Sartorius H, Wernsperger B and Schlosser C 2010 Ten years of development of equipment for measurement of atmospheric radioactive xenon for the verification of the CTBT *Pure Appl. Geophys.* **167** 471–86.

[3] Rashid F I A and Zolkaffly M Z 2018 Stakeholder engagement for promoting the Comprehensive Nuclear Test-Ban Treaty (CTBT): Malaysia's experience *IOP Conf. Ser. Mater. Sci. Eng* **298** 012054.

[4] Yoshimori M 2005 Production and behavior of beryllium 7 radionuclide in the upper atmosphere *Adv. in Space Res.* **36** 922–926.

[5] Leppanen A P, A Pacini A, Usoskin I G, Aldahan A, Echer E, Evangelista H, Klemola S, Kovaltsov G A, Mursula K and Possnert G 2010 Cosmogenic 7Be in air: A complex mixture

of production and transport *Journal of Atmospheric and Solar Terrestrial Physics* **72** pp 1036–43.
[6] Lal D and Peters B 1967 Cosmic ray produced radioactivity on the earth ed. Sittle K *Handbuchder Physik* **46**(2) pp 551–612.
[7] Lal D and Peters B1962 Cosmic ray produced radioactivity on the earth *Kosmische Strahlung II/Cosmic Rays II* **9/46/2** 551-612.
[8] Ananthakrishnan R and Soman M K 1991The onset of the southwest monsoon in 1990 *Curr. Sci.* **61** 447–53.
[9] L Terzi and M Kalinowski 2017 World-wide seasonal variation of 7Be related to large-scale atmospheric circulation dynamics *J. of Environ. Radioact.* **178-179** 1-15.
[10] Vecchi R and Valli G 1997 7Be in surface air: a natural atmospheric tracer *J. Aerosol Sci.* **28**(5) 895–900.
[11] Brost R A, Feichter J and Heimann M 1991Three-dimensional simulation of 7Be in a global climate model *J. Geophys. Res* **96** 22423–45
[12] Lal D and Peters B 1962 Cosmic ray produced isotopes and their application to problems in geophysics ed. Wilson J, Wouthuysen S *Prog. in Element. Particle and Cosmic Ray Phys.* **6** 77–243
[13] Feely H W, Larsen R J and Sanderson C G 1989 Factors that cause seasonal variations in beryllium-7 concentrations in surface air *J. Environ. Radioact.* **9** 223–49
[14] Aldahan A, Possnert G and Vintersved I 2001Atmospheric interactions at northern high-latitudes from weekly Be-isotopes in surface air *Appl. Radiat. Isotop.* **54** 345–53
[15] Usoskin I G, Field C V, Schmidt G A, Leppanen A P, Aldahan A, Kovaltsov G A, Possnert G and Ungar R K 2009a Short-term production and synoptic influences on atmospheric 7Be concentrations *J. Geophys. Res.* **114** D06108.
[16] Muhammad B G, Ismail S, Ekhwan T, Joy J P, Mazlin M and Md P A 2009 Integrated water resource management and pollution sources in Cameron Highlands, Pahang, Malaysia, *American-Eurasian J. Agric. and Environ. Sci.* **5** (6) pp 725-32.
[17] Malaysian Metrological Department 2014 *Monthly Weather Bulletin 2014* 45
[18] Malaysian Metrological Department 2009 *Accumulated Rainfall Observation Data 2009* 32
[19] Preparatory Commission for the Comprehensive Nuclear Test Ban Treaty 2020 Concentration Reporting Tool (CRTool) (https://crtool.ctbto.org Accessed on 1 February 2020).
[20] Du J, Baskaran M, Bi Q, Huang D and Jiang Y 2015 Temporal variations of atmospheric depositional fluxes of 7Be and 210Pb over 8 years (2006–2013) at Shanghai, China, and synthesis of global fallout data *J. Geophys. Res. Atmos.* **120** 4323–39.
[21] Gerasopoulos E, Zanis P, Stohl A, Zerefos C S, Papastefanou C, Ringer W, Tobler L, Hubener S, Gaggeler H W, Kanter H J, Tositti L and Sandrini S A climatology of 7Be at four high altitude stations at the Alpsand the Northern Apennines *Atmos. Environ.* **35** 6347–60.
[22] Alejandro L C, Mohd N S, Ahmad S A and Mohd R O 1998 Monthly distribution of precipitation in Peninsular Malaysia *Pertanika J. Sci. and Technl.* **6**(1) 59 - 70
[23] Kuśmierczyk-Michulec J and Bourgouin P 2018 Influence of mineral dust on changes of 7Be concentrations in air asmeasured by CTBTO global monitoring system *J. Environ. Radioact.* **192** 454–66

IOP Conf. Series: Materials Science and Engineering 1106 (2021) 012030 doi:10.1088/1757-899X/1106/1/012030

The elemental distribution study of heavy metals (Fe, Cu, Cd, As, Cr, Ni and Hg) in various fish species at local market in Muar, Johor using SEM-EDX

N S Mat Beyi[1], Z Embong[1,a)], N M Sunar[2], M A B Abdul Latiff [1], F Abdullah[3] and Z Suhaili[4]

[1]Faculty of Applied Science and Technology, Universiti Tun Hussein Onn Malaysia, Pagoh Education Hub, Panchor, Pagoh, 84600, Muar, Johor, Malaysia.
[2]Faculty of Engineering Technology, Universiti Tun Hussein Onn Malaysia, Pagoh Education Hub, Panchor, Pagoh, 84600, Muar, Johor, Malaysia.
[3]Department of Chemistry, Faculty of Science, Universiti Teknologi Malaysia, 81310 Skudai, Johor, Malaysia.
[4]School of Animal Sciences, Faculty of Bioresources and Food Industry, Universiti Sultan Zainal Abidin, Besut Campus, 22200 Besut, Terengganu, Malaysia.

[a]zaidi@uthm.edu.my

Abstract. The polluted marine environment causes bioaccumulation of heavy metals particularly by fish, which is consumed by humans as sources of nutrients for humans, particularly as a source of protein. Thus, the investigation of the toxic heavy metals concentration levels in fish is essential. This study aimed to evaluate the heavy metals (Cd, Cr, As, Ni, Fe, Hg and Cu) contents in five different fish species, namely *Rastelligar kanagurta, Selaroides leptolepis, Megalaspis cordyla, Decapterus ruselli* and *Euthynnus affinis*. Each sample has been dried and grinds to 100 microns of powder form. The elemental composition of the heavy metals was determined using the rapid and versatile analytical technique, which is Scanning Electron Microscopy with Energy Dispersive X-ray analysis (SEM-EDX) and the results were expressed as an atomic percentage. The percentage concentrations of Cd, As, Cr and Ni were found to be higher in *R. kanagurta* which are 0.01%, 0.02%, 0.03% and 0.02%, while Cu was found to be higher in *S. leptolepis* which is 0.29%. For Hg element analysis, it was found that this element only exists in *D. ruselli* species with atomic percentage of 0.02%. The atomic percentage concentration of As in *R. kanagurta* and *S. leptolepis* are 0.02% and 0.01%. The element of Cr, Fe and Ni were detected in *M. cordyla* with atomic percentage of 0.01%, 0.02% and 0.02%. In *E. affinis,* Fe was the highest concentration of metals detected with 0.02%, while other elements such as Cr, Cu and Ni recorded similar the atomic percentages concentration at 0.01%. This study revealed that *R. kanagurta* species contributed most of the elements compared to other species. The accumulation of these toxic heavy metals from this fish species could contribute to the toxicity of heavy metals in the human body. However, since all heavy metal concentration measurements in this study are in atomic percentages level, hence it can be considered that the measurements are below the acceptable limit that suggested by WHO, FAO and UNSCEAR

1. Introduction

Heavy metals pollution is a major concern worldwide due to toxicity, persistent nature, non-biodegradable and accumulative behaviors [1]. It is caused by industrialization, urbanization, agriculture, and other human activities especially in developing countries [2]. Heavy metals from these activities will be discharged into the sea or rivers, accumulated in water, sediments and aquatic food chain resulting in sublethal effects or death in the fish population [3]. Heavy metals enter the aquatic food chain through the direct consumption of water and food or through the muscle or gills [4]. Fish are at the top of the aquatic food chain and considered as one of the bioindicators of pollutants [5][6]. Besides that, fish is the major source of protein for humans. Thus, the human body is easily susceptible to heavy metals concentration in fish [7]. Therefore, analysis of the heavy metal in fish will reflect the level of sediment and water of the aquatic environment from which they are sourced [8].

Accumulation of heavy metals in fish are influenced by their nutritional habits, ecological requirements, concentrations of heavy metals in water and sediment, fish's life, season and physicochemical properties of water [9][10][11][12]. Among the heavy metals, cadmium (Cd), lead (Pb), mercury (Hg), arsenic (As) and copper (Cu) have a higher potential to accumulate in fish above the standard limit [13][14][15]. Some trace elements such as zinc, iron, chromium, manganese, cobalt, nickel and copper are required by marine organisms. For the aquatic organism, these elements are required in a trace amount for normal growth and development. However, elements such as arsenic, mercury and cadmium do not give any biological importance. All the heavy metals are potentially toxic to most organisms at some level of exposure and absorption [16][17]. An excessive intake of heavy metals will cause serious threats such as renal failure, liver damage, cardiovascular disease and even death to human [18].

Fishes can hold toxic materials in a small concentration to high concentrations in their body without harming them, but the toxicity will affect the humans who feed on them. A small amount of heavy metals can be held either temporarily or permanently in their body. Thus, fish in the polluted water will accumulate high concentration of heavy metals along the time [19]. For example, a high concentration of Hg, As and Cd could be harmful to the living cells, and prolonged exposure to the body can cause illness or even death [20]. The high Cd content could damage the kidney, liver, causing diarrhea and nausea [21]. Besides that, metals such as Cd, Hg, As, Ni, As and Cr that are commonly found in human diets have been reported to be carcinogenic [22][23]. The distribution of heavy metals in fish differs due to habitat and feeding habits [19]. Information on the distribution of heavy metals in fish resources from local market is still lacked in Malaysia. Hence, in this work the investigation of heavy metals composition quantification in several fish species from local market in Muar, Johor has been conducted using rapid and versatile analysis technique, SEM-EDX.

2. Materials & methods

2.1. Sample collection

The sample of fish species were obtained from wet market at Bukit Gambir, Muar. Five kind fish samples have been collected consist of *Selaroides leptolepis, Rastelligar kanagurta, Megalaspis cordyla, Decapterus ruselli and Euthynnus affinis* as shown in Figure 1. The collected samples were preserved in ice container and kept frozen in -20°C before analysis. The total length (cm) and weights (gm) for each fish samples were measured before dissection as tabulated in Table 1.

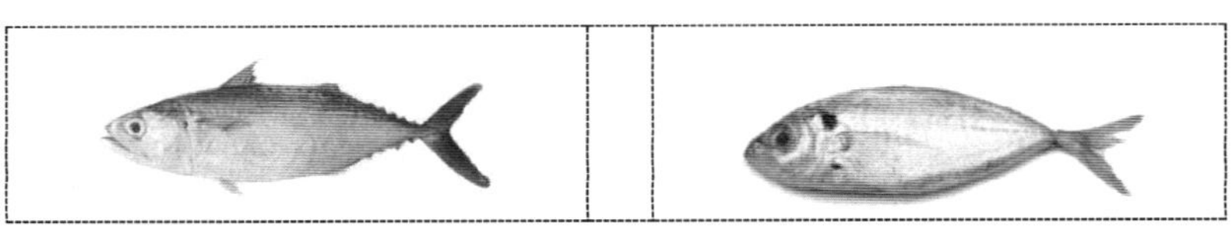

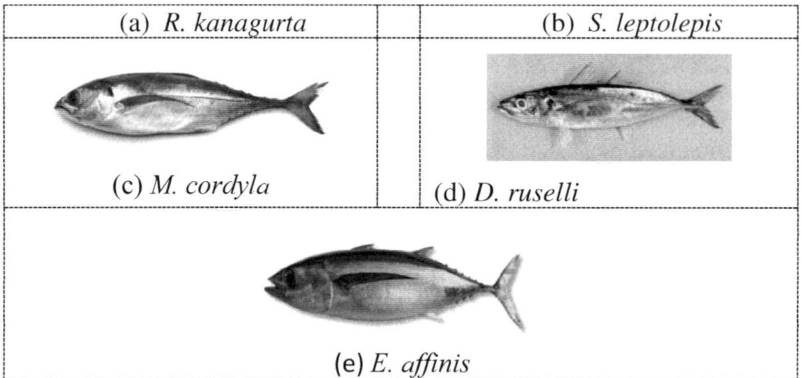

Figure 1. The appearance of all fish species, a) *R. kanagurta*, b) *S. leptolepis*, c) *M. cordyla*, d) *D. ruselli*, e) *E. affinis*.

Table 1. The description for all fish sample species.

Scientific name	Local name	Common name	Length (cm) (min- max)	Weight (g) (min- max)
Selaroides leptolepis	Selar	Yellowstripe scad	20-23	106-145
Rastelligar kanagurta	Kembung	Indian mackerel	20-25	110-160
Megalaspis cordyla	Cencaru	Torpedo Scad	23-30	110-300
Decapterus ruselli	Selayang	Slander scad	17-22	61-127
Euthynnus affinis	Tongkol	Mackerel tuna	30-41	350-925

2.2. Sample preparation
All fish samples were washed using distilled water to remove the skin surface dirt. Head, viscera, gills and flesh segments were cut into identical segments. The samples are freeze dried for 48 hours to remove the moisture content. Then, the dried samples were ground into 100 microns of powder form. The fish samples need to be in the homogenous powder form for elemental composition analysis purpose. Before the grinding process, it is important to make sure that the fish samples were completely dried.

2.3. Elemental composition analysis
In this study, Scanning Electron Microscopy with Energy Dispersive X-ray analysis (SEM-EDX) (EM-30AX, Coxem, Korea) was used to quantify the metal concentrations. The powder form samples were mounted in the stubs with conductive carbon tape. Then, the samples were sputter-coated with gold to avoid charging during the analysis. EDX system are attached to SEM instruments where the imaging capability of the microscope is used for morphology analysis. In this research, the elemental composition of the sample was analyzed by three measurements in order to achieve a good statistical analysis. Three fine spot areas were selected, and the average is determined. Data were collected at 20 kV accelerating voltage and 12 cm working distance. Figure 2 shows a typical EDX spectrum where each energy represents different element and therefore the analysis of elemental compositions was calculated by the percentage of the atomic elements, at %.

IOP Conf. Series: Materials Science and Engineering 1106 (2021) 012030 doi:10.1088/1757-899X/1106/1/012030

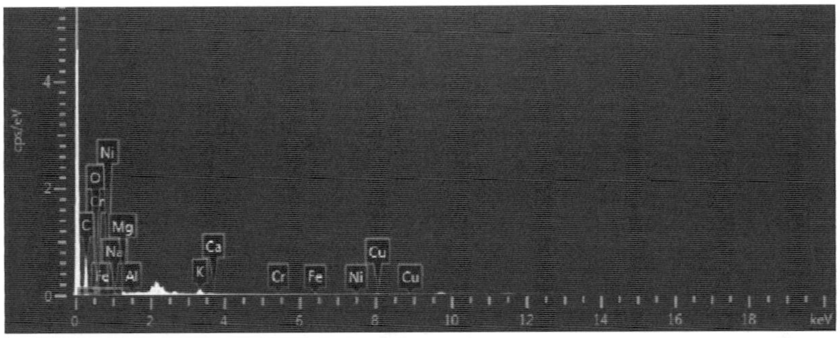

Figure 2. EDX spectrum for fish samples from SEM-EDX

3. Results and discussion

3.1. Morphology analysis for every fish species

The morphology structure obtained for all fish species were shown in Figure 3(a)-(e). From this morphology analysis, it is seen that the particle size for all types of fish species was different and the distribution of each particle is uneven. The morphology structure for *S. leptolepis* exhibited a smoother surface compared to other, while *D. ruselli* and *E. affinis* have a similar particle size.

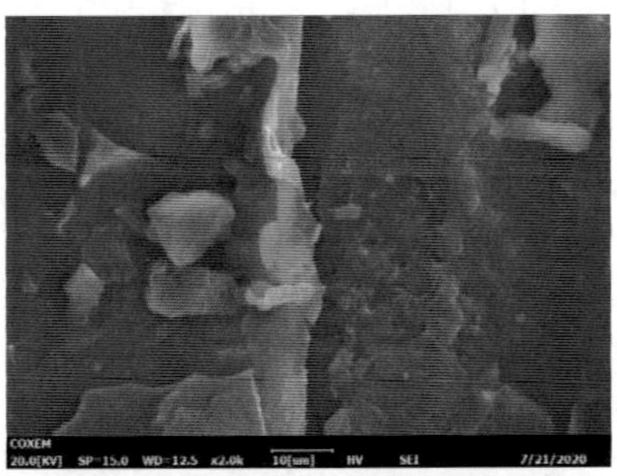

(a) *R. kanagurta*

(b) *S. leptolepis*

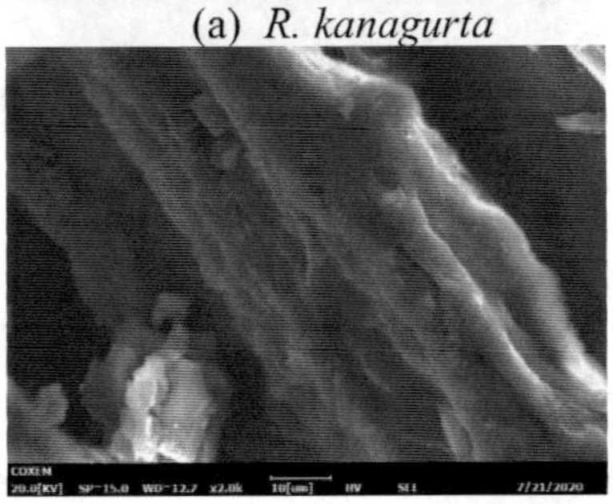

(c) *M. cordyla*

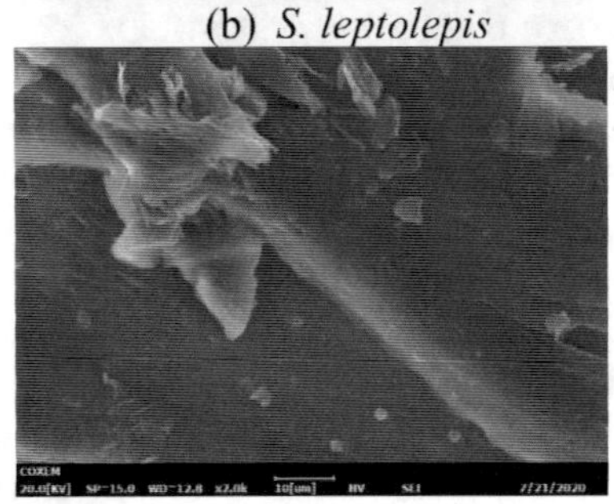

(d) *D. ruselli*

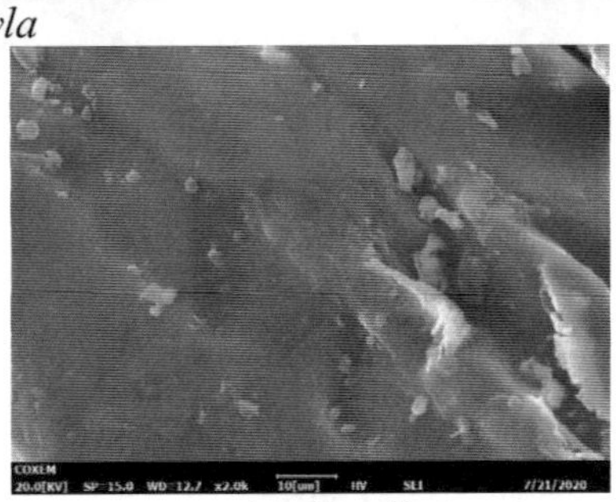

(e) E. affinis

Figure 3. Morphology image of all fish species (a) *R. kanagurta*, (b) *S. leptolepis*, (c) *M. cordyla*, (d) *D. ruselli*, (e) *E. affinis.*

3.2. Elemental percentage concentration for all fish species

The quantitative analysis of heavy metals was given in atomic percentage (%). The result in Table 2 shows that the level of Cd, Hg, As, Cu, Fe, Ni and Cr vary according to the samples. The concentrations of Cd, Hg, As, Cu, Fe, Ni and Cr were classified according to different fish species. The same applies to the study done by the previous researchers where the heavy metals distributions were varied among different fish species due to metabolism and feeding patterns of the fishes [24].

Table 2. The elemental concentration for all fish samples in atomic percentage, %.

Elemental compositions (%)	Fish species					Mean Concentrations (%)
	Rastelligar kanagurta	*Selaroides leptolepis*	*Megalaspis cordyla*	*Decapterus Ruselli*	*Euthynnus affinis*	
Cadmium	0.010±0.001	0.010±0.001	ND	ND	ND	0.010
Mercury	ND	ND	ND	0.020±0.007	ND	0.020
Arsenic	0.020±0.007	0.010±0.001	ND	ND	ND	0.015
Chromium	0.030±0.020	0.010±0.004	0.010±0.001	0.010±0.001	0.010±0.001	0.014
Copper	0.220±0.014	0.290±0.077	ND	ND	0.010±0.001	0.173
Iron	0.040±0.035	0.010±0.001	0.020±0.007	0.020±0.007	0.020±0.007	0.022
Nickel	0.020±0.007	0.020±0.007	0.020±0.007	0.010±0.001	0.010±0.001	0.016

ND=Not detected

From the observation, the trends for the concentrations of heavy metals for all fish samples were in descending order which are Cu > Fe > Hg > Ni > As > Cr > Cd. The mean concentrations of the heavy metals found in fish species were 0.173 (Cu), 0.022 (Fe), 0.020 (Hg), 0.016 (Ni), 0.015 (As), 0.014 (Cr) and 0.010 (Cd).

R. kanagurta was found to have a greater atomic percentage concentration of heavy metals as compared to the other fish species. Most of the heavy metals detected in *R. kanagurta* is mainly consist of Cd, As, Cr, Cu, Fe and Ni with atomic percentage concentration of 0.010±0.001, 0.020±0.007, 0.030±0.020, 0.220±0.014, 0.040±0.035 and 0.020±0.007 respectively. Additionally, Figure 4 shows that Cu was found in higher concentrations than the other heavy metals. The distribution of heavy metals in descending orders for *R. kanagurta* was Cu > Fe > Cr > As > Ni >Cd. This trend was similar to the outcome obtained from [25], where Cu was detected to be the highest among the other metals. The finding obtained from *R. kanagurta* suggests that the cause of high Cu reading is due to the high Cu concentration in the fish's catching area environment. This opinion is supported by [26] where Cu is highest detected in fish due to the fish species taken from waters enriched in this element. It was proposed in [24] that Cu is essential in all organism in trace amounts and it serves as a constituent of respiratory enzyme complex in the human body and due to its role in facilitating iron uptake, deficiency of Cu can cause impaired growth, anaemia-like symptoms, bone abnormalities and vulnerable to infections.

Meanwhile, Cd, As, Cr, Cu, Fe and Ni were detected in *S. leptolepis* with percentage concentration of 0.010±0.001, 0.010±0.001, 0.010±0.004, 0.290±0.077, 0.010±0.001 and 0.020±0.007 respectively. Figure 5 showed that metals Cu and Ni were found at higher percentage concentrations than the other heavy metals. The level of accumulation of these heavy metals for *S. leptolepis* species in descending order was Cu > Ni > Cd > As > Cr > Fe. In the present study, arsenic (As) and Cd were only detected in *R. Kanagurta* and *S. leptolepis*. The existence of arsenic (As) in this particular fish may be resulting in poisoning to consumers as reported by national environmental protection agencies [28]. The observed percentage concentration of heavy metals detected in *M. cordyla* is shown in Figure 6. The heavy metals compound in this particular fish species is mainly consist of Cr, Fe and Ni with a percentage concentration of 0.010±0.001, 0.020±0.007 and 0.020±0.007 respectively. The general observed trends for the metals for *M. cordyla* was Fe > Ni > Cr. The lowest percentage concentration of Cr (0.010±0.001) was recorded while Fe and Ni showed the same percentage concentration which was 0.020±0.007.

As for *D. Ruselli* species, the heavy metals concentration observed were Hg, Cr, Fe and Ni with percentage concentrations of 0.020±0.007, 0.010±0.001, 0.020±0.007 and 0.010±0.001 respectively. The trends for the metals for *D. ruselli* in descending order were Hg > Fe > Cr > Ni. The result shows Hg and Fe recorded the same concentration as shown in Figure 7. Cd, As and Cu were not detected in *D. ruselli.* Mercury is carcinogenic and produces adverse effects during developmental stages as a result of acute and chronic exposure [29]. In the present study, the mercury had only been detected in *D. ruselli.*

While for the percentage concentration of heavy metals identified in *E. affinis* is shown in Figure 8. The detected heavy metals were Cr, Cu, Fe and Ni with percentage concentration of 0.010±0.001, 0.010±0.001, 0.020±0.007 and 0.010±0.001 respectively. The heavy metals in *E. affinis* followed an order of Fe > Cr > Cu > Ni. In *E. affinis,* Fe was the highest concentration of metals detected with 0.020±0.007 while Cr, Cu and Ni record the same concentration which is 0.010±0.001. Ahmed & Bat (2015) found a similar finding, with Fe being the highest concentration found in fish. In the current analysis, chromium was found at low levels in all fish samples, close to those examined in Turkey [27] and Malacca [25].

In general, all fish species had Fe, Ni and Cr in them. The maximum concentration of Cu was observed in *R. Kanagurta* and *S. leptolepis* and was not detected in *M. cordyla*, *D. Ruselli* and *E. affinis*. From the observation, it indicated that different levels of metals were found in different fish species. The differences in the heavy metals content are due to the accumulation of each metals in different fish species that related to its ecology surrounding such as habitat and nutrient in the sea water [19, 30]. All heavy metals concentration is considered below the acceptable limits suggested by WHO, FAO and UNSCEAR [31][32][33].

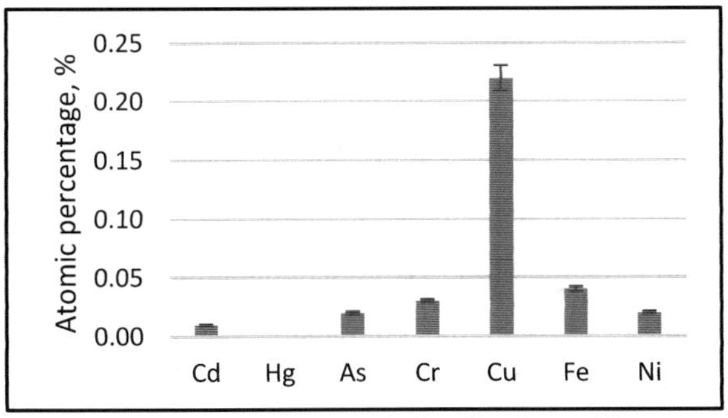

Figure 4. The elemental distribution of heavy metals in *R. Kanagurta*

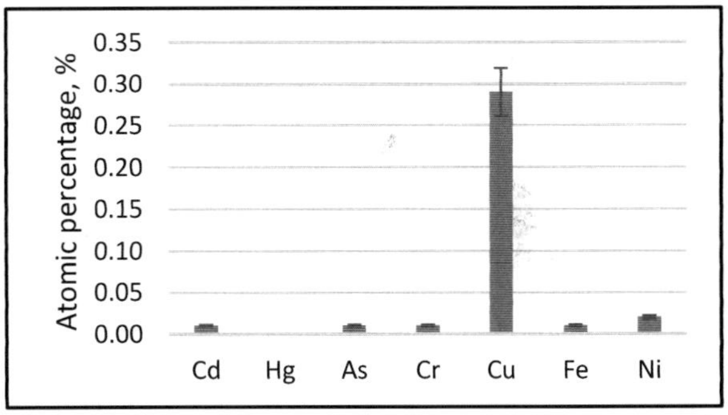

Figure 5. The elemental distribution of heavy metals in *S. leptolepis*

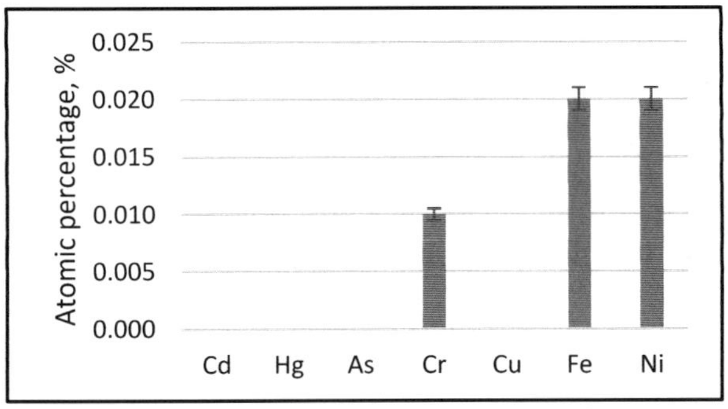

Figure 6. The elemental distribution of heavy metals in *M. cordyla*

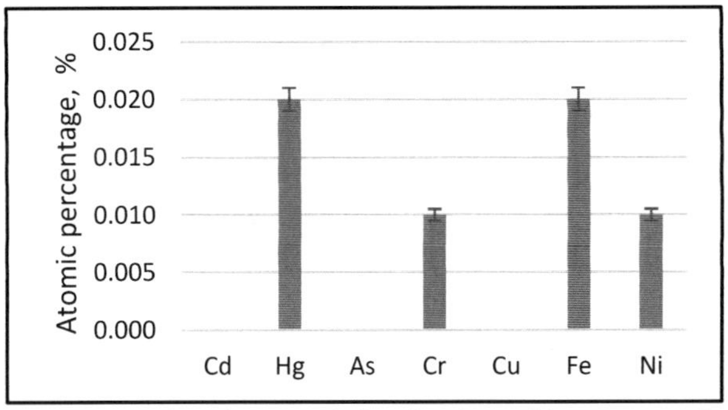

Figure 7. The elemental distribution of heavy metals in *D. Ruselli*

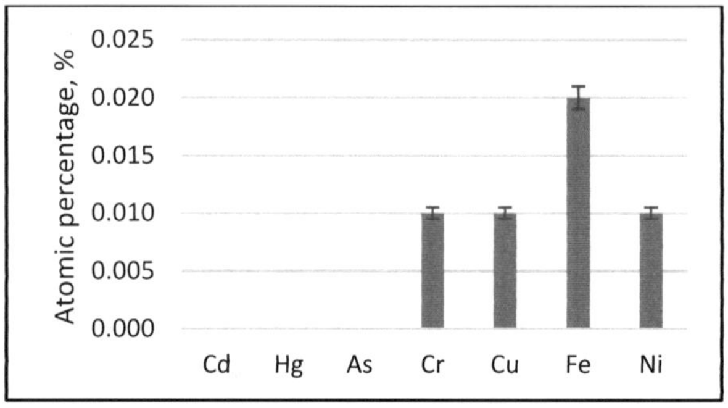

Figure 8. The elemental distribution of heavy metals in *E. affinis*

4. Conclusion

In this research, the existence of heavy metals in fish samples was successfully investigated. The distribution of heavy metals accumulation in all fish species were established and the descending order of heavy metals concentration are as the following order of Cu > Fe > Hg > Ni > As > Cr > Cd. The atomic percentage concentration of Cd, As, Cr and Fe in *R. kanagurta* was found to be the highest compared to the other fish species. Whilst, Cu was the highest in *S. leptolepis.* On the other hand, some amount of Hg was found in *D. Ruselli,* while, Cr, Fe and Ni are mainly observed in all fish samples. Overall, it was found that most of the heavy metals elements is found in *R. kanagurta,* while *M. cordyla* contributes the lowest heavy metals concentration. As *R. kanagurta* has the highest heavy metals content, hence it could give a risk of disease to human beings. However, heavy metals content that accumulated in this particular fish can be considered very low concentration and still safe to be consumed.

5. References

[1] Islam M S, Hossain M B, Matin A, Sarker M S I 2018 Assessment of heavy metal pollution, distribution and source apportionment in the sediment from Feni River estuary, Bangladesh *Chemosphere* **202** 25-32.

[2] Hossain M B, Ahmed A S S, Sarker M S I 2018 Human health risks of Hg, As, Mn, and Cr through consumption of fish, Ticto barb (Puntius ticto) from a tropical river, Bangladesh *Env. Sci. and Pollu. Res.* **25**(31) 31727-31736.

[3] Xu S, Tao S 2004 Coregionalization analysis of heavy metals in the surface soil of Inner Mongolia *Sci. of the Tot. Environ.* **320**(1), 73-87.

[4] Ribeiro C O, Vollaire Y, Sanchez-Chardi A, Roche H É L È N E 2005 Bioaccumulation and the effects of organochlorine pesticides, PAH and heavy metals in the eel (*Anguilla anguilla*) at the Camargue Nature Reserve, France *Aquatic Toxicol.* **74**(1) 53-69.

[5] Idriss A A, Ahmad A K 2015 Heavy metal concentrations in fishes from Juru River, estimation of the health risk *Bullet. of Environ. Contam. and Toxicol.* **94**(2), 204-208.

[6] Authman M M, Zaki M S, Khallaf E A, Abbas H H 2015 Use of fish as bio-indicator of the effects of heavy metals pollution *J. of Aquacul. Res. And Dev.* 6(4) 1-13.

[7] Ali H, Khan E 2018 Bioaccumulation of non-essential hazardous heavy metals and metalloids in freshwater fish. Risk to human health *Environ. Chem. Lett.* **16**(3) 903-917.

[8] Nhiwatiwa T, Barson M, Harrison A P, Utete B, Cooper R G 2011 Metal concentrations in water, sediment and sharptooth catfish Clarias gariepinus from three peri-urban rivers in the upper Manyame catchment, Zimbabwe *African J. of Aquatic Sci.* **36**(3) 243-252.

[9] Fakhri Y, Mohseni-Bandpei A, Conti G O, Ferrante M, Cristaldi A, Jeihooni A K, Sarkhosh M 2018 Systematic review and health risk assessment of arsenic and lead in the fished shrimps from the Persian gulf *Food and Chem. Toxicol.* **113** 278-286.

[10] Conte F, CopatC, Longo S, Conti G O, Grasso A, Arena G, Ferrante M 2015 First data on trace

elements in Haliotis tuberculata (Linnaeus, 1758) from southern Italy: Safety issues *Food and Chem. Toxicol.* **81** 143-150.

[11] Dadar M, Adel M, Ferrante M, Saravi H N, Copat C, Conti G Oliveri 2016 Potential risk assessment of trace metals accumulation in food, water and edible tissue of rainbow trout (Oncorhynchus mykiss) farmed in Haraz River, northern Iran *Toxin Rev.* **35**(3-4) 141-146.

[12] Başyiğit B, Tekin-Özan S 2013 Concentrations of some heavy metals in water, sediment, and tissues of pikeperch (Sander lucioperca) from Karataş Lake related to physico-chemical parameters, fish size, and seasons *Polish J. of Environ. Stud.* **22**(3) 633-644.

[13] Bosch A C, O'Neill B, Sigge G O, Kerwath S E, Hoffman L C 2016 Heavy metals in marine fish meat and consumer health: a review *J. of the Sci. of Food and Agricul.* **96**(1) 32-48.

[14] Bosch A C, O'Neill B, Sigge G O, Kerwath S E, Hoffman L C 2016 Heavy metal accumulation and toxicity in smoothhound (*Mustelus mustelus*) shark from Langebaan Lagoon, South Africa *Food Chem.* **190** 871-878.

[15] Vandermeersch G, Lourenço H M, Alvarez-Muñoz D, Cunha S, Diogène J, Cano-Sancho G, Bekaert K 2015 Environmental contaminants of emerging concern in seafood–European database on contaminant levels *Environ. Res.* **143** 29-45.

[16] Fakhri Y, Saha N, Ghanbar, S, Rasouli M, Miri A, Avazpour M, Morad 2018 Carcinogenic and non-carcinogenic health risks of metal (oid) s in tap water from Ilam city, Iran *Food and Chem. Toxicol.* **118** 204-211.

[17] Shahsavani A, Fakhri Y, Ferrante M, Keramati H, Zandsalimi Y, Bay A, Khaneghah M 2017 Risk assessment of heavy metals bioaccumulation: fished shrimps from the Persian Gulf.*Toxin Rev.* **36**(4) 322-330.

[18] Al-Busaidi M, Yesudhason P, Al-Mughairi S, Al-Rahbi W A K, Al-Harthy, K S, Al-Mazrooei, N A, Al-Habsi S. H 2011 Toxic metals in commercial marine fish in Oman with reference to national and international standards *Chemosphere* **85**(1) 67-73.

[19] Koleleni Y A, Haji O O 2014 Determination of concentration of heavy metals in fish from sea port of Zanzibar by energy dispersive x-ray fluorescence (EDXRF) *Tanzania J. of Sci.* **40**(1), 79-89.

[20] Azaman F, Juahir H, Yunus K, Azid A, Kamarudin M K A, Toriman M E, Saudi A S M 2015 Heavy metal in fish: Analysis and human health-a review *J. Teknologi* **77**(1).

[21] Faroon O, Ashizawa A, Wright S, Tucker P, Jenkins K, Ingerman L, Rudisill C 2012 *Toxicol. Profile for Cadmium* (US: Agency for Toxic Substances and Disease Registry).

[22] Ng J C, Wang J, Shraim A 2003 A global health problem caused by arsenic from natural sources *Chemosphere* **52**(9) 1353–1359.

[23] Ahmed M K, Baki M A, Islam M S, Kundu G K, Habibullah-Al-Mamun M, Sarkar S K, Hossain M M 2015 Human health risk assessment of heavy metals in tropical fish and shellfish collected from the river Buriganga, Bangladesh *Environ. Sci. and Pollut. Res.* **22**(20) 15880-15890.

[24] Mukherjee D P, Bhupander K 2011Assessment of arsenic, cadmium and mercury level in commonly consumed coastal fishes from Bay of Bengal, India *Food Sci. Qual. Manag.* **2**.

[25] Khandaker M U, Asaduzzaman K, Nawi S M, Usman A R, Amin Y M, Daar E Okhunov A A 2015 Assessment of radiation and heavy metals risk due to the dietary intake of marine fishes (Rastrelliger kanagurta) from the straits of Malacca *PLoS One* **10**(6) e0128790.

[26] Rejomon G, Nair M, Joseph T 2010 Trace metal dynamics in fishes from the southwest coast of India *Environ. Monitor. and Assess.* **167**(1-4) 243-255.

[27] Görür K, Keser F, Akçay R, Dizman S 2012 Radioactivity and heavy metal concentrations of some commercial fish species consumed in the Black Sea Region of Turkey *Chemosphere* **87**(4) 356–361.

[28] Roychowdhury T, Tokunaga H, Ando M 2003 Survey of arsenic and other heavy metals in food composites and drinking water and estimation of dietary intake by the villagers from an arsenic-affected area of West Bengal, India *Sci. of the Total Environ.* **308**(1-3), 15-35.

[29] Bose-O'Reilly S, McCarty K M, Steckling N, Lettmeier B 2010 Mercury exposure and children's health *Current Prob. in Pediat. and Adoles. Health Care 40*(8) 186-215.
[30] Lakshmanan R, Kesavan K, Vijayanand P, Rajaram V, Rajagopal S 2009 Heavy metals accumulation in five commercially important fishes of Parangipettai, Southeast Coast of India *Adv. J. of Food Sci. and Technol.* **1**(1) 63-65.
[31] United Nations Scientific Committee on the Effects of Atomic Radiation UNSCEAR (2000) Sources, effects and risks of ionizing radiation (New York: United Nations).
[32] World Health Organization 2011 *Training Package for the Health Sector: Adverse Health Effects of Heavy Metals in Children* (World Health Organization) [Retrieved from http://www.who.int/ceh/capacity/heavy_metals.pdf].
[33] FAO/WHO Expert Committee on Food Additives: Arsenic 2005 [Available from http://www.inchem.org/documents/jecfa/jeceval/jec_159.htm].

Acknowledgements
The authors would like to thank the financial support by GPPS (Vot: 497) provided by University Tun Hussein Onn Malaysia.

Residence time distribution (RTD) of water flooding activity in oil recovery using vertical sand column: Radiotracer approach

N Othman[1,*], W H B Wan Kamal[2], N S Jasni[3], N R Rosli[3], R Yahya[1], N H Yusoff[1], M A S Mohd Yunos[1], L N Shah Dahing[1], M R Shari[1], H Hassan[1], A AMahmood[1], A M Terry[1], N Abdullah[1]

[1]Plant Assessment Technology (PAT), Industrial Technology Division, Malaysian Nuclear Agency, 43000 Kajang, Selangor, Malaysia.
[2]Medical Technology Division, Malaysian Nuclear Agency, 43000 Kajang, Selangor, Malaysia.
[3]Faculty of Chemical Engineering, Universiti Teknologi MARA (UiTM), 40450, Shah Alam, Selangor, Malaysia.

*noraishah@nm.gov.my

Abstract. Waterflooding is a common secondary recovery oil extraction method that can enhance oil recovery up to 45 percent overall recovery factor. Waterflooding is commonly used due to its availability, cost effectiveness and simplicity. Radioactive tracer was introduced to get the optimum Residence Time Distribution (RTD) model which indicates the mechanism of the system. Radiotracer Gallium-68 (Ga-68) is a positron emitting radioactive material, which is eluted from parent Germanium-68 (Ge-68) generator to study the water flooding activity inside the 15 μm grain size of sand pack column oil recovery study. The sand was cleaned, occupying a Perspex sand pack column of 30 cm length and 5 cm diameter arranged in vertical position. The water inlet and tracer were injected from the bottom of the column, assisted by a syringe pump, with the flow rate optimized at 3.5 ml/min in order to push the saturated oil inside the column. Four Sodium Iodide (NaI) scintillation detectors were installed accordingly and attached to 12-channel data analyzer (DAS) to acquire signals of the emitted positron from the Ga-68 solution and monitored by a laptop. The activity utilized was 766 μCi/3ml and the half-life of Ga-68 was 68 minutes with penetrative energy of 511 keV. Initial experience of using Ga-68 was eluting it with 0.05 M HCl for 90μCi/3ml and directly inject it using pulse injection at the upstream of the column. Nonetheless, there was no indication of tracer arrival at the output after several hours which indicated the sole Ga-68 behaves as sand tracer thus, some modification was made to the respective Ga-68 solution. Chelating with DOTA-NHS solution was the second experience with Ga-68, which produced satisfactory result. A very vivid signal was retrieved at the output and RTD model analysis was carried out. The results indicated that the vertical sand pack column behaved as a Perfect Mixer with Parallel Model.

1. Introduction

Water flooding is a form of oil recovery in which continuous water injection mechanism is adopted and the overall recovery factor is about 35 to 45 % original oil in place [1]. The induced pressure in the presence of water is required to displace oil from the reservoir rock into the producing well. The extraction of residual oil requires modification in reservoir characteristics in which water flooding is most used for enhanced oil recovery [2]. Water flooding within Malaysia fields such as Tapis and Guntong had been proven to increase the overall oil recovery [3]. Common accessibility of water, ease of connection with water, retention of hydraulic head in injection well, water that able to move through an oil-bearing formation, and efficiency of water in sweeping the oil can be counted as the factors of popularity for waterflooding.

To evaluate the effectiveness of the waterflooding, sweep efficiency evaluation need to be done. Sweep efficiency is important to indicate whether trapped oil from primary recovery is displaced by the injected water or not. If the sweep efficiency for waterflooding is low, it means that the water injected to reservoir is insufficient to drive the oil towards production well. The knowledge of the total/swept efficiency is not a simple task because the reservoirs are underground formations and therefore remote to man and conventional tools. Moreover, the petroleum potential in Malaysia is mainly fractured basement rock reservoir located offshore and characterized by a fracture system of very high heterogeneity in porosity and permeability. In addition, fluid flow in most reservoirs is anisotropic. The reservoir structures are usually layered and frequently contained significant heterogeneities leading to directional variations in the extent of flow [4]. Hence, the effective fluid movement can be difficult to predict. The complexity of the reservoir structure introduces a considerable uncertainty in the reservoir model and fluid flow simulation.

Thus, in order to investigate, understand and analyze the mechanisms involved during water-flooding, radiotracer technology can be utilized [5]. Tracer is a tool that accesses and runs across the reservoir, contacting portions of interest and giving information about the total/swept efficiency [6]. Fluid flow and transport behavior inside the reservoir can be specifically described using analysis of RTD Model [7][8]. This RTD models provide the behavior of the reservoir such as the presence of anomalies like dead zone and channeling which will be addressed by the reservoir engineer in order to increase sweep efficiency that leads to increased oil recovery.

2. Methodology

2.1. Ga-68 preparation

Ge-68/Ga-68 generator is typically used in nuclear medicine to produce positron-emitting radionuclide Ga-68.The parent isotope Ge-68 has a half-life of 270.95 days and can be easily delivered to hospitals as a generator, where it can be used as the source of Ga-68 for at least 1 year. Gallium-68 (with a half-life of only 67.71 minutes, difficult to transport) can be easily eluted from the generator any time at the site of application and can be used for different purposes. In this study, Ga-68 was eluted with 0.05 M HCl (solvent) to produce GaCl3 and the activity was 90 µCi/3ml when it was directly injected inside the sand pack column.

2.2. Ga-68-DOTA-NHS Preparation

Since there was no indication of tracer arrival after 5 half-lives of Ga-68 in the first experiment, modification has been made to the eluted Ga-68 (0.05 M HCl). Firstly, 3 ml of eluted Ga-68 with activity of 9.28 mCi was obtained from the Ge-68 generator. Then, 2 ml of the respective Ga-68 was mixed with HEPES (4-(2-Hydroxyethyl)-1-Piperazineethanesulfonic acid), a buffering agent to prevent a rapid change in pH when acid is added to the solution, using vortex (shaker) to ensure the mixing is homogenous. The activity of the dissolved solution was 6.03 mCi. DOTA (1,4,7,10-Tetraazacyclododecane-1,4,7,10-Tetraacetic acid) was then introduced to chelate with dissolved Ga-68-HEPES. Prior to that, DOTA was heated to 95°C for 10 minutes in order to activate the chelating process of DOTA with Ga-68. About 500 µl of Ga-68 and 100 µl DOTA-NHS were added together and mixed using vortex to ensure homogenous solution in order to produce 68Ga-DOTA. The pH of mixed solution was regulated to 3.9 acidic medium. The activity of chelated DOTA with Ga-68 was 1.142 mCi (600 µl).

2.3. *Preparation of radiotracer set up*

The sandstone with grain size of 150 μm was pressed manually using surficial weight inside the cylindrical column that was arranged in vertical order. Formation water (original water inside the reservoir) was prepared earlier to be saturated inside the column. About 0.5 g/ml (500 ppm) NaCl solution (saline) was prepared for water flooding activity. Moreover, 180 ml of oil was saturated inside the column and aged for 24 hours to ensure that the oil settled well inside the column. On the next day, about 28.3 MBq/3ml (766 μCi) of Ga-68 was prepared and tapped at the common line together with water line as shown in figure 1. Thus, during water flooding operation, the valve of Ga-68 and water were opened sequentially. Hence, about 850 ml of water was pumped with constant flow rate 3.5 ml/min using syringe pump continuously to purge out the saturated oil from the column. Prior to that, four NaI scintillation detectors were arranged accordingly and attached to data acquisition system (DAS). The DAS was then connected to laptop for data storage and tracer monitoring. The parameters for each experiment are shown in Table 1.

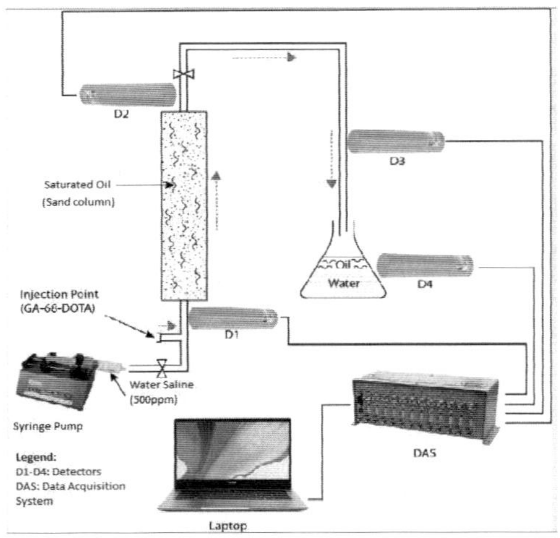

Figure 1. Schematic diagram of the experimental setup.

Table 1. Parameters of each experimental set up.

Parameters	Ga-68	Ga-68-DOTA-NHS
Q (ml/min)	3.5	3.5
V_f (ml): formation water	329	238
V_w (ml) water flooding	850	780
Activity of Ga-68 (μCi)/3ml	90	766
Saturated oil, S_{or}	180	238

The mathematical expression for the First Moment (M1) in discrete form or better known as MRT was calculated using Equation (1). The mathematical expression for RTD is shown in Equation (2). RTD is a fundamental parameter in reactor design which can give information on how long the substrate has been in the reactor and the RTD analysis can help in characterizing the extent of their deviation from ideal behavior.

2.4. Residence time distribution (RTD) model

The mathematical expression for the First Moment (M1) in discrete form or better known as MRT was calculated using Equation (1). The mathematical expression for RTD is shown in Equation (2). RTD is a fundamental parameter in reactor design which can give information on how long the substrate has been in the reactor and the RTD analysis can help in characterizing the extent of their deviation from ideal behavior.

$$Mi = \frac{\int_0^{\infty} t\, C_i(t)\, dt}{\int_0^{\infty} C_i(t)\, dt} \tag{1}$$

$$E(t) = \frac{C(t)}{\int_0^{\infty} C(t) dt} \tag{2}$$

where $C(t)$ is the concentration of radiotracer monitored by NaI scintillation detectors in counts per second (cps) as numerator and denominator is the area under the curve of plotted $C(t)$. The detected signal was normalized by dividing it by the area under the curve. The RTD models can provide theoretical information or describe the hydrodynamic behavior of the reactor.

3. Result and discussions

The results will cover the Ga-68 analysis, oil recovery and RTD Models respectively.

3.1. Ga-68-DOTA Analysis

The purity of solution was checked using thin layer radiochromatograph (TLC) scanner which produced about RCP 99.24% that indicated only 0.76% of Ga-68 did not bind with DOTA as shown in figure 2. In this study, the radioactive solution was diluted and 766 μCi/3ml to be injected inside the sand pack column for water flooding activity. Table 2 shows the RCP of chelated Ga-68 and DOTA.

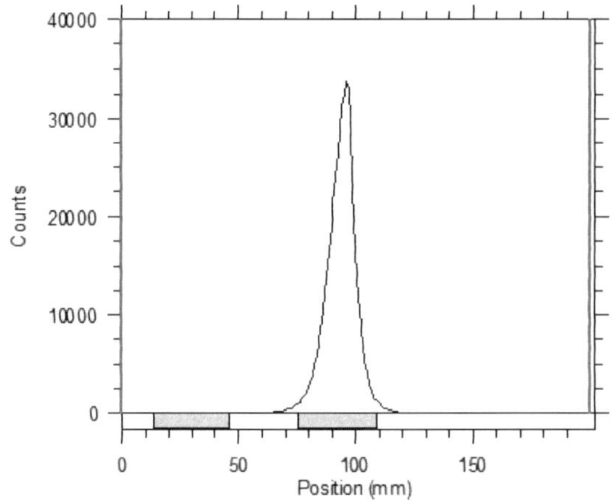

Figure 2. TLC Scanner results.

Table 2. RCP of chelated Ga-68 and DOTA-NHS.

Region	Start (mm)	Stop (mm)	Centroid (mm)	RF	Region Counts	Region CPM	% of Total	% of ROI
Region1	13.5	46.7	31.1	0.156	3337.0	6674.0	0.68	0.70
Region 2	75.5	109.7	94.0	0.470	475981.0	951962.0	96.28	99.30
2 peaks					479318.0	958636.0	96.96	100.00

No increment of radiation signal was found at the outlet of the sand column using D2. Only background reading was received that indicated Ga-68 was adsorbed onto the sand beads. To confirm the theory, the survey meter was run along the column and the counts reading were escalated. Thus, conclusion has been made that Ga-68 is suitable for sand tracking and NOT for fluid tracking and monitoring. Therefore, the modification was made to the existing Ga-68 which was best chelated with DOTA and the data was outlined accordingly.

3.2. *Oil recovery*

Table 3 shows the percentage of oil recovery from two respective radiotracers. The Ga-68 extracted more oil than chelated DOTA. This result is puzzling since the expected results should be similar. This is because radiotracer will not interfere with any chemical process since it acts as the tracking fluid for easy monitoring during water flooding. Investigation has been carried out and the results were due to negligence in saturating the water inside the column. The water saturation should use formation water and NOT the 500ppm saline water.

Table 3. Results of oil recovery.

Parameters	**Ga-68**	**Ga-68-DOTA**
Recovered oil (ml)	120	91.8
% oil recovery	66.7	38.6

3.3. *Residence time distribution (RTD) models*

Figure 3 illustrates the results of the output that has significant information about the process inside the column. It was observed that the data required necessary treatment to ensure only radioactive signals are present. Thus, data treatment as outlined by previous researchers was referred [9]. The RTD curve was then obtained after undergoing the data treatment as well as data transformation as described in Equation 1 [7]. Figure 4(a) to (d) shows the results of RTD optimization of RTD Model by adopting Software by IAEA [1].

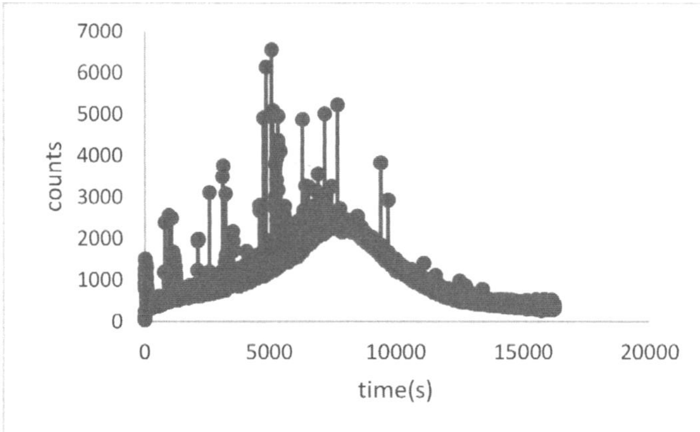

Figure 3. Raw data of Ga-68-DOTA-NHS experiment.

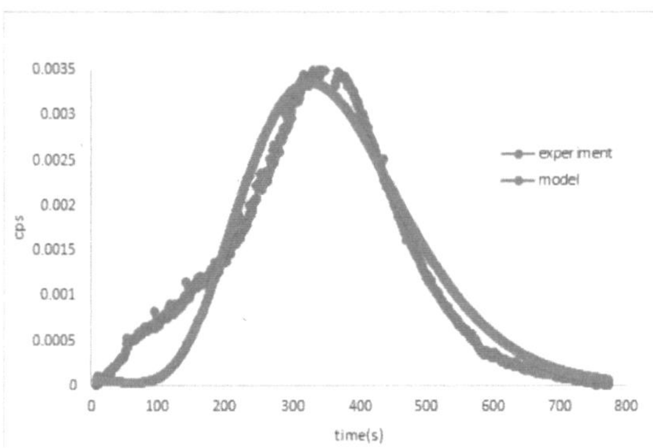

(a) Perfect Mixer in Series Model

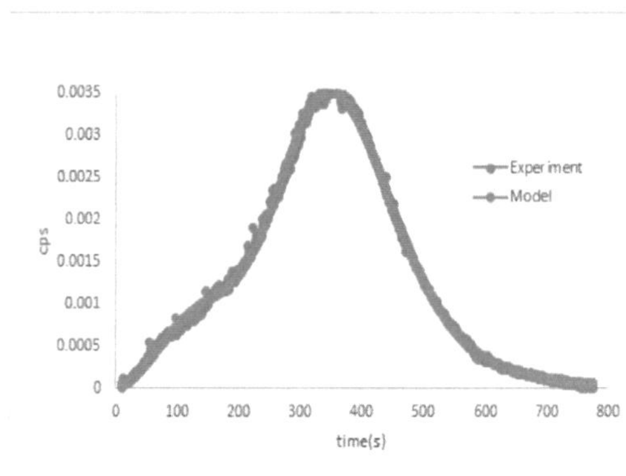

(b) Perfect Mixer with Parallel Model

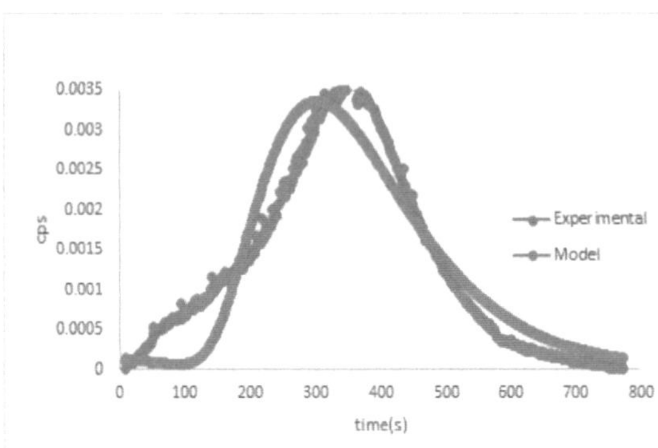

(c) Axial Dispersed Plug Flow Model

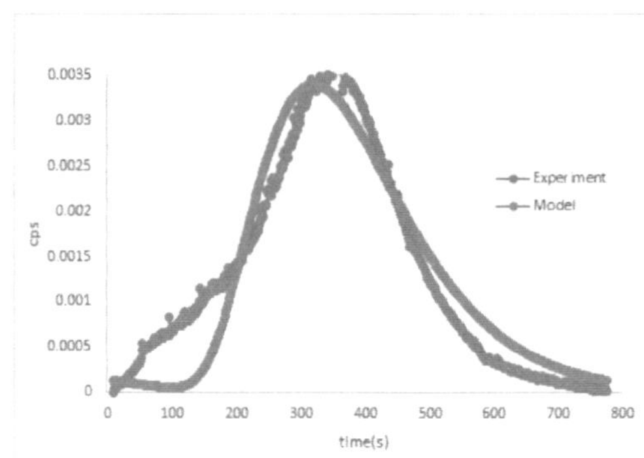

(d) Axial dispersed with Exchanged Model

Figure 4. (a) – (d) RTD Models of the experimental curve.

Figure 4 shows that all RTD Models fit the experimental results well. These models are the perfect mixers in series model, the perfect mixers in parallel model, the axial dispersed plug flow model and the axial dispersed plug flow with exchange model respectively. Nevertheless, according to Table 4, the Perfect Mixer with Parallel Model fits the experimental curve best as the value of Error Sum of Square (SSE) is the minimum.

Table 4. Error Sum of Square (SSE).

Models	Parameters						SSE	Ranks (Minimum value)
	τ_1	τ_2	J_1	J_2	Pe	N		
Perfect Mixer model	360	-	8.2	-	-	-	0.920×10^{-7}	2
Perfect Mixer with Parallel Model	366	295	17	2.9	-	-	0.239×10^{-8}	1
Axial Dispersed Plug Flow Model	365	-	-	-	16	-	0.155×10^{-6}	4
Axial Dispersed Model with Exchange model	352	-	1.0	-	15	1.0	0.131×10^{-6}	3

The results also indicated that there were two series of perfect mixers which were arranged in parallel whereby primary series consists of 17 mixers and secondary series comprises with 3 mixers accordingly. NO short-circuiting or channeling was found in the model, indicating that the sand was well packed inside the column. No long tail was also observed in the RTD curve, which indicates no dead zone discovered inside the column. Moreover, the RTD model was symmetrical that signifies perfect mixer is achieved and radiotracer was mixed homogenously inside the column.

4. Conclusions

Several highlights have been discovered in this work. Ga-68 independently is a sand tracer and does not blend well with water (fluid). On the other hand, Ga-68-DOTA-NHS can represent the fluid flow but the preparation is quite tedious. Moreover, formation water should not be mistaken as saline water because it affects the percentage of oil recovery. Water flooding is proved to be able to extract residual oil from the vertical column excellently up to 67%.

Based on the RTD Model discussion, the sand column was prepared nicely because there was no indication of common process anomalies found such as dead zone, channeling and short-circuiting. The Perfect Mixers in Parallel Model fits the experimental model well due to minimum value of SSE in comparison with other models.

5. References

[1] Zitha P, Felder R, Zornes D, Brown K, and Mohanty K 2011 Increasing hydrocarbon recovery factors (Soc. Pet. Eng.) 1–9.
[2] Khan I H, Farooq M, Ahmad G D M, Gul S and Qureshi R M 2003 Tracer technology to investigate inter-well communications during enhanced oil recovery (Vienna: Research Contract No. PAK-12949/RBF).
[3] Selamat S, Samsuddin S A and Halim N A 2011 Evaluation and optimization of enhanced oil recovery by WAG injection at Tapis and Guntong fields, Malaysia (SPE 145123).
[4] Mohammadian E, Hamidi H, Azdarpour A and Junin R 2015 *Proc. of the Int. Civil and Infrastr. Eng. Conf.* 417-424.
[5] Tayyib D, Al-Qasim A, Kokal S and Huseby O 2019 *SPE Kuwait Oil and Gas Show and Conference* (Soc. of Pet. Eng.).

[6] Melo M A and Almeida A R 2017 *SPE Latin America and Caribbean Mature Fields Symposium* (SPE-184956-MS) https://doi.org/10.2118/184956-MS.
[7] Othman N, Mohamed Hassan N P, Yahya R, Ahmad Khusaini M A, Shari M R, Hassan H and Mahmood A A 2017 *Malaysian J. of Analy. Sci.* **21**(2) 445–451.
[8] Othman N, Saaid I M, Ahmed A, Yusof N H, Yahya R, Yunos M A S M, Engku Chik E M F, Shari M R, Hassan H and Mahmood A A 2020 *ASEAN J. of Chem. Eng.* **20**(1) 49–56.
[9] Kasban H, Zahran O, Arafa H, El-Kordy M, Elaraby S M S and Abd El-Samie F E 2010 Laboratory experiments and modeling for industrial radiotracer applications *Appl. Radiat. and Isot.* **68**(6) 1049–1056.
[10] IAEA 2008 Radiotracer residence time distribution method for industrial and environmental applications *Material for Education and On-the-job Training for Practitioners of Radiotracer Technology, Training Course Series No. 31*.

Acknowledgment
The authors would like to thank National Cancer Institute (NCI), Putrajaya for providing the Ge-68/Ga-68 generator which is used to produce positron-emitting radionuclide Ga-68. The authors also would like to thank Prof Jovan Thereska from IAEA, Vienna Prof Tor Bjornstard from IFE, Norway and Dr Quang from CANTI, Vietnam in providing continuous assistance and knowledge contribution throughout this study. Not to forget all PAT colleagues who participated in this experiment. Finally, the authors would like to express their appreciation for the support of the sponsor IAEA Research Contract No: 22898, IAEA Coordinated Research Project F22069 and FRGS with Project No FRGS/1/2018/TK07/MOSTI/02/3

IOP Conf. Series: Materials Science and Engineering 1106 (2021) 012032 doi:10.1088/1757-899X/1106/1/012032

Density distribution investigation through the geometrical structure for the ^{27}S halo nucleus

J Islam[1,2,3], S Radiman[1,2] and K S. Khoo[1,2*]

[1]Nuclear Technology Research Centre, Faculty of Science and Technology, Universiti Kebangsaan Malaysia (UKM), 43600 Bangi, Selangor, Malaysia.
[2]Department of Applied Physics, Faculty of Science and Technology, Universiti Kebangsaan Malaysia (UKM), 43600 Bangi, Selangor, Malaysia.
[3]Department of Physics, Mawlana Bhashani Science and Technology University, Santosh, Tangail-1902, Bangladesh.

*khoo@ukm.edu.my

Abstract. ^{27}S was predicted as a proton halo nucleus in the $2s_{1/2}$ state due to the abnormally large protons root mean square radius. The presence of weakly bound (S_{2n} = 0.90(20) MeV) two valence protons in the ^{27}S nucleus strongly supported this prediction. The core density, matter density, and halo density for this nucleus (^{27}S) have been investigated in this study. The geometrical halo structure has been used for different parameter calculations to investigate these density distributions since sufficient experimental data are not available for this ^{27}S nucleus. The parameters' values were calculated easily following this geometrical structure. Gaussian-Gaussian (GG) distribution has been applied to investigate these densities. In our present study, the ^{27}S nucleus was investigated as a 2-*p* halo nucleus due to its low valence two-protons separation energy (0.90 MeV). The 2-*p* halo ^{27}S nucleus was described having (^{25}Si) + p + p configuration. These measured parameters' values were then used to investigate the Gaussian-Gaussian (GG) density distributions. MATLAB computational software was also used for these distribution investigations. From these density distributions, a maximum density of 1.042%, 1.001%, and 0.4789% at the centre of the nucleus was obtained for the core density (ρ_c), matter density (ρ_m), and halo density (ρ_h), respectively. The presence of longer tails on the halo and matter distributions also support the prediction as a two-proton halo nucleus.

1. Introduction

The ^{27}S nucleus is located at the proton-drip line in the nuclear landscape, and the nucleus is a proton halo candidate. In 1996, this candidature was confirmed and the ^{27}S nucleus is a proton halo nucleus based on the mean-square radius of proton for the $2s_{1/2}$ state [1]. The low separation energy of the two-valence protons on the ^{27}S nucleus emphasizes the two-proton (2-*p*) halo nucleus demand [2,3]. In addition, its root mean square (rms) halo radius (R_h) is larger than the rms core radius (R_c) and rms matter radius (R_m). Considering the two-proton halo configuration of the nucleus ^{27}S, we have focused our work on the nuclear density distributions (ρ_c, ρ_m and ρ_h) of this nucleus. The Gaussian-Gaussian (GG) density distributions depend on the rms core, matter and halo radii [4]. However, there is a lack of sufficient experimental data to study the nuclear properties, especially the density distributions for the ^{27}S nucleus [5]. Therefore, to investigate these density distributions, we have calculated the density parameters such as distance between core and centre of mass of two valence protons, half distance of the two-valence protons, distance between two-valence protons, etc. using a geometrical halo structure.

We have calculated the nuclear density distributions (ρ_c, ρ_m and ρ_h) for the nucleus ^{27}S theoretically using MATLAB computational software in our present work.

2. Theoretical Frameworks

The simplest shell-model structure consists of $2s_{1/2}$ valence orbital, which is filled up by two protons for $Z = 16$, and the $1d_{3/2}$ valence orbital which is filled up by protons for $Z = 14$ and neutron number $N = 11$ [2]. For this shell-model, the ^{25}Si nucleus was considered as the core of the halo nucleus ^{27}S. To study the density distributions of the halo nucleus (^{27}S), we have applied a geometrical structure for the halo nucleus since this geometrical structure was used for the study of the ^{6}He two-neutron structure [6]. The core nucleus was considered inside a cluster core model (CCM). The geometrical structure for this core and the valence two-proton are shown in Figure 1.

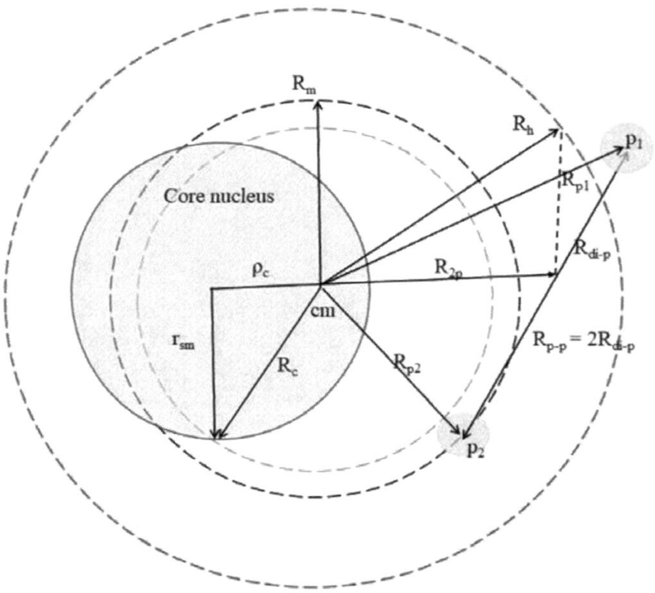

Figure 1. The geometrical structure for the ^{27}S halo nucleus.

Through this geometrical structure, all parameters for the core (^{25}Si) and ^{27}S nucleus can be determined based on the following:

- The distance ρ_c between the nuclear centre of mass and the core centre is [7]:

$$\rho_c = \left(\frac{R_{sm}^2}{A_c} - \frac{R_m^2}{A}\right)^{\frac{1}{2}} \quad (1)$$

where, R_{sm} is rms matter radius of the core (^{25}Si) nucleus, R_m^2 is the rms matter radius of ^{27}S, R_{sm}^2 is the rms radius of the ^{25}Si nucleus and A_c is the mass number of the ^{25}S.

- The core radius of the centre of mass R_c is:

$$R_c = (R_{sm}^2 + \rho_c^2)^{\frac{1}{2}} \quad (2)$$

- The vector R_{2p} joining the nuclear centre of mass and the midpoint of the line connecting the two halo protons is determined from the balancing condition:

$$A_c\rho_c = A_h R_{2p}, \text{where } A_c = 25,\ A_h = 2 \quad (3)$$

- The distance R_{c-2p} from the core centre to the two halo protons is:
  $$R_{c-2p} = \rho_c + R_{2p} \tag{4}$$
- The distance R_{p-p} between the two halo neutrons is given by:
  $$R_{p-p} = 2R_{di-p}, \text{where } R_h^2 = R_{2p}^2 + R_{di-p}^2 \tag{5}$$
- The rms halo radius of the ^{27}S nucleus is [7]:
  $$R_h = \left(\frac{1}{2}\left((A_c + 2)R_m^2 - A_c R_{sm}^2\right)\right)^{1/2} \tag{6}$$

  where, from Equation (3) to Equation (5) are taken from the Ref. [6].

Moreover, the Gaussian-Gaussian (GG) density distributions equations for ρ_c, ρ_m and ρ_h can be written as [4][6]:

$$\rho_{core(halo)}(r) = \left(\frac{3}{2\pi R_{c(h)}^2}\right)^{3/2} \exp\left(-\frac{3r^2}{2\pi R_{c(h)}^2}\right) \tag{7}$$

and,

$$\rho_m(r) = \frac{1}{A}[N_{core}\rho_{core}(r) + N_{halo}\rho_{halo}(r)] \tag{8}$$

where N_{core} and N_{halo} are the number of nucleons in the core and halo respectively and A is the mass number of the halo nucleus, r is the radial distance from the centre of the nucleus' core to infinity. We have carried out all calculations including the core, matter and halo distribution calculations etc. through the MATLAB codes.

3. Result and discussion

Using Equations (1) – (6) through the geometrical halo structure, we obtained the calculated values tabulated in Table 1. From these measurements, the rms core, halo, and matter radii of the ^{27}S nucleus were determined 3.1632 fm, 4.0994 fm, and 3.2347 fm respectively. Using these rms radii in Equations (7) and (8), the core density, halo density, and matter density were calculated. These density distributions are shown in Figure 2.

Table 1. The calculated density distributions parameter values.

Nucleus	Parameter Name	Parameters	Parameter Values
^{25}Si	Matter radius	R_{sm}	3.1553
	Distance between core and centre of mass	ρ_c	0.2232
^{27}S	Core radius	R_c	3.1632
	Matter radius	R_m	3.2347
	Distance between centre of mass and the mid-point of two valence protons	R_{2p}	2.7900
	Halo radius	R_h	4.0994
	Half distance of the two-valence protons	R_{di-p}	3.0035
	Distance between two-valence protons	R_{p-p}	6.0071

The maximum density values of 1.042%, 1.001%, and 0.4789% are noted at the centre of the nucleus respectively for the core density, matter density, and halo density which are shown in Figure 2.

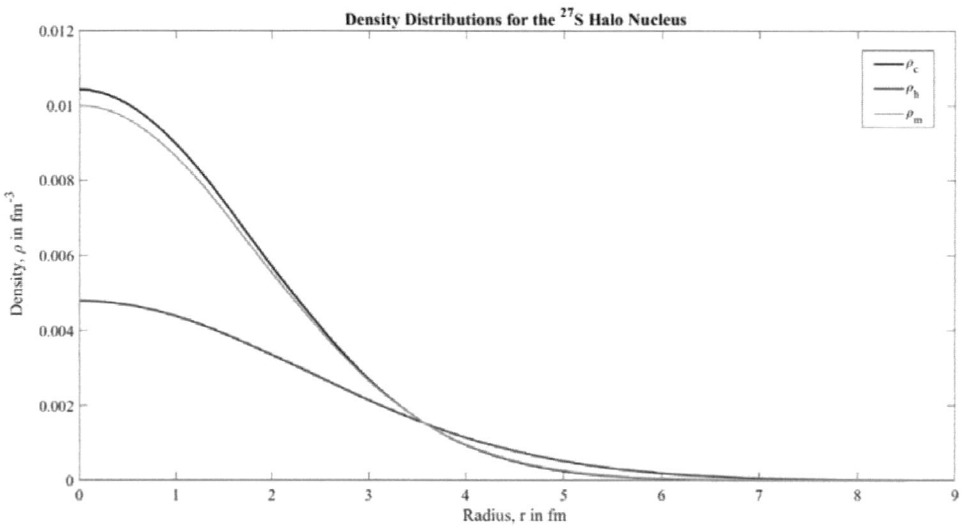

Figure 2. GG density distributions for the ^{27}S halo nucleus.

4. Conclusion

The ^{27}S nucleus core density (ρ_c) is at maximum for all ρ_c, ρ_h and ρ_m and all neutrons are distributed in the nucleus' core. Furthermore, the matter density (ρ_m) is less than ρ_c and the tail of the proton halo density (ρ_h) is larger than ρ_c and ρ_m. This indicates that the protons are distributed at larger radius from the nucleus centre and exist outside the core of the nucleus. Ultimately, ρ_h has the smallest density at the core of the ^{27}S nucleus and the tail of this density is at the largest up to 7 fm in radius which indicates that the protons are lying further away from the nucleus' core. The nucleons' distribution at further distances and the halo density distribution's large tail show that the ^{27}S nucleus is a proton-halo nucleus.

5. References

[1] Ren Z, Chen B, Ma Z and Xu G 1996 *Phys. Rev. C* **53**(2) R572.
[2] Brown B A and Hansen P G 1996 *Phys. Lett. B* **381** 391.
[3] Audi G and Wapstra A H 1995 *Nucl. Phys. A* **595** 409.
[4] Aygün M 2016 *GU J Sci* **29**(2) 263.
[5] Sawhney G and Gupta R K 2016 *Acta Physica Polonica B* **47**(3) 959.
[6] Chung L X, Kiselev O A, Khoa D T and Egelhof P 2015 *Phys. Rev. C* **92** 034608.
[7] Tanihata I, Savajols H, and Kanungo R 2013 *Prog. Part. Nucl. Phys.* **68** 215.

Acknowledgment

Authors would like to thanks Universiti Kebangsaan Malaysia (UKM) for the financial support through grant scheme GGP-2017-003.

IOP Conf. Series: Materials Science and Engineering 1106 (2021) 012033 doi:10.1088/1757-899X/1106/1/012033

Rainfall runoff modeling for the basin in Bukit Kledang, Perak

Azmi Ibrahim*, Norasalwa Zakaria, Nazran Harun, Mohd Muzamil Mohd Hashim

Malaysian Nuclear Agency (Nuclear Malaysia),Ministry of Science, Technology,and Innovation (MOSTI),Bangi, 43000 Kajang, Selangor, Malaysia.

*azmiibrahim@nm.gov.my

Abstract. A fundamental problem in studying a hydrological system is that most of the reaction takes place underground. Precipitation Runoff demonstrating of the Johan River basin was carried out using HEC-HMS technique. The model can improve decision-making about the hydrological problem, whether that be in water resources planning, flood protection and mitigation of contamination. The result shows that the rainfall-runoff model for the Bukit Kledang basin is slow depending on the relative travel time of basin and also associated with soil type, land used, multiple storms and seasonality.

1. Introduction

Rainfall runoff from areas where solid waste is sorted, stored or treated is invariably contaminated to some extent. Waste spilled or lost from waste management areas may build up in soil or on other surface and be transported by runoff after rainfall. According to Semadeni-Devies [1] the storm water pollutant load generated by a specific event depends on storm intensity, the time lapse and the rate of sediment accumulation between storms.

Runoff is one of the significant hydrologic boundaries utilized in the watershed the executives. Appraisal of the surface runoff which fundamentally relies upon the meteorology, geography, topography, soil and land use design is required for legitimate arranging of the hydraulic structures just as relief of natural hazard in the region. There are many methods available for the determination of runoff based on the above factors [2]. Several models of runoff simulations have been proposed and applied to urban runoff prediction, such as the Sacramento model presented by Burnash *et al.* [3], Tank model presented by Sugawara [4], HBV model simulated by Bergstro, [5], MIKE 11/NAM model obtained by Nielsen *et al.* [6] and Havnø *et al.* [7] and the Soil Conservation Service curve number (SCS-CN) method presented by Yousuf *et al.* [8]. According to Yousuf *et al.* [8], the SCS-CN is one of the most enduring methods among these models for estimating the volume of direct surface runoff in ungauged rural catchments.

IOP Conf. Series: Materials Science and Engineering 1106 (2021) 012033 doi:10.1088/1757-899X/1106/1/012033

A mathematical model was used by Jasrotia *et al.* [9] to estimate rainfall runoff in combination with remote sensing data and GIS using SCS CN and runoff potential map method. The runoff from SCS curve number model modified by conventional database and GIS was estimated by Ashish Pandey *et al.* [10]. Amutha *et al.* [11] showed that the GIS-integrated SCS-CN method of runoff estimation can be effectively used in watershed management. Pandey *et al.* [12] concluded that the land use input of the SCS-CN model is an important parameter input. Nayak *et al.* [13] obtained that the measured and intimated runoff depth using GIS and CN has a good correlation.

2. The Kledang watershed

Sg. Kledang is located at the Menglembu area, concentrated with industrial outlet. Sungai Kledang flows for 6.4 km into Menglembu area before entering Sungai Kinta. There are two contributors for this site (Figure 1) which are the residential estate and the industrial area. The Sg. Kledang originating from housing estate of Taman Bukit Kledang Indah will meet-up with the industrial area and then flows into the Kinta River .

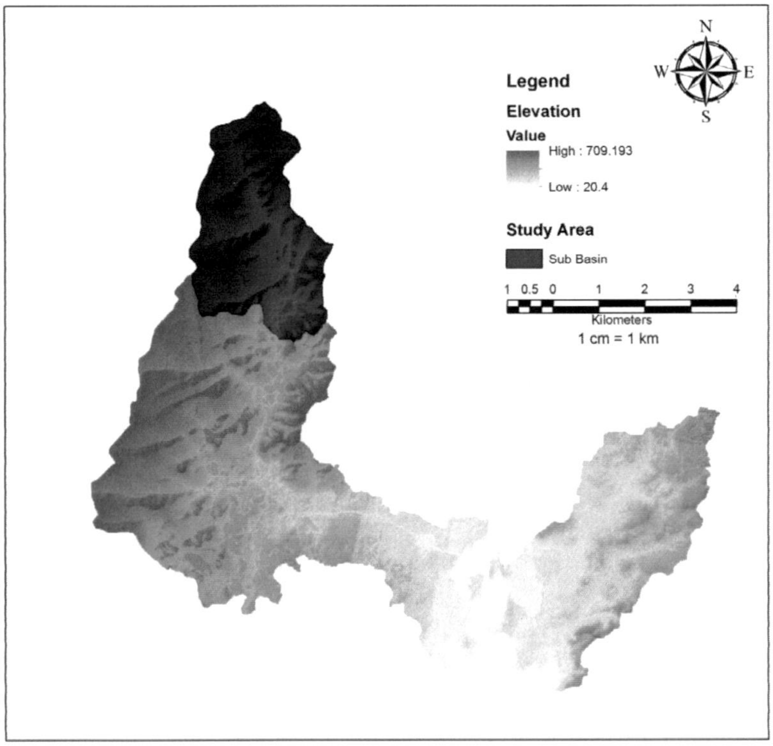

Figure 1. Location of study area.

3. Methodology

The Hydrologic Engineering Center - Hydrologic Modeling System (HEC-HMS) in Figure 2, is design to simulate the precipitation-runoff processes of dendritic drainage basin. It is design to be applicable in wide range of geographic areas for solving the widest possible range of problem. This includes large river basin supply and flood hydrology and small urban or natural watershed runoff. The program is a generalized model system capable of representing many different watersheds. A model of the watershed is constructed by separating the water cycle into manageable pieces and constructing boundaries around the watershed of interest.

The main steps of HEC-HMS methodology adopted in the present study as depicted in Figure 2 is summaries below:

- Create river basin from Digital Elevation Model (DEM) using GIS tool in HEC-HMS. Created Metrological Model from rainfall data of rain gauges located in the basin using HEC-HMS.

- Input time-series rainfall data in HEC-HMS model
- Create Control Model to provide simulation data
- Create Simulation Run model by integrating basin model, metrological model, and control model
- Running of the simulation Model
- Analysis of Result

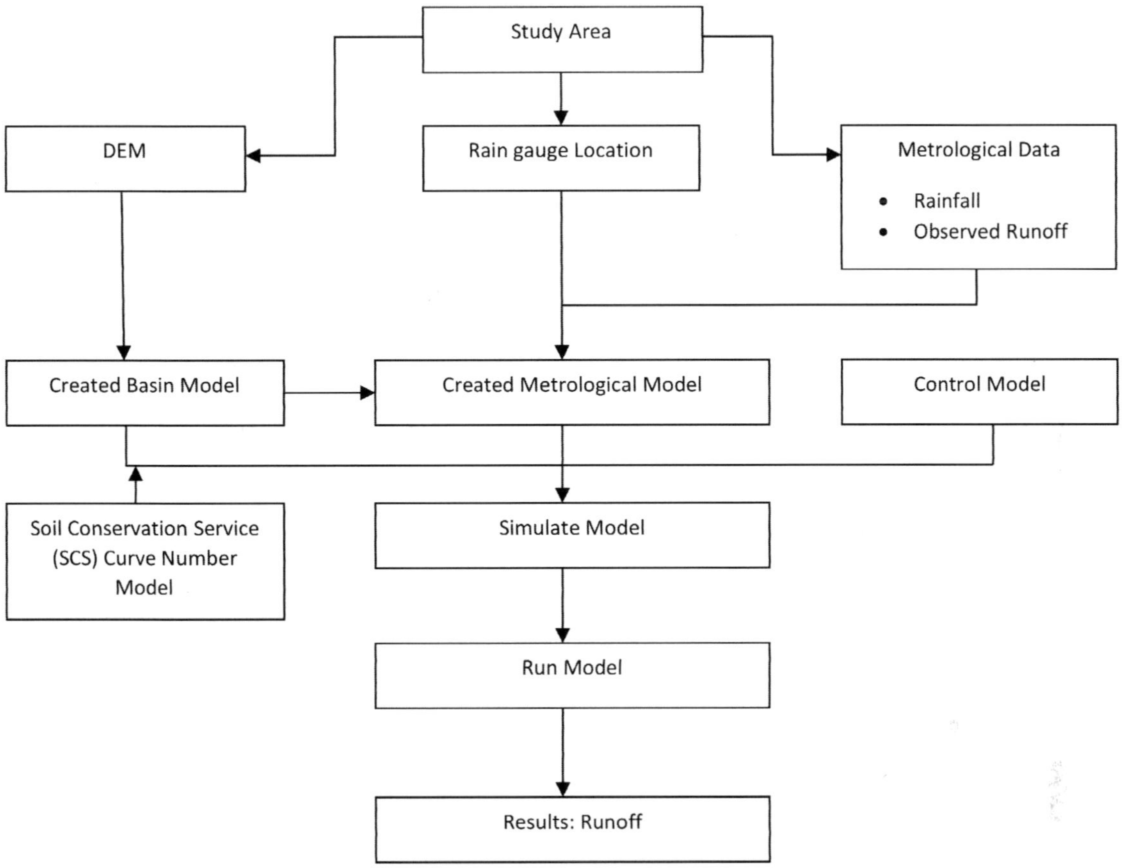

Figure 2. Flow Chart of HEC-HMS model.

3.1. Basin Model

In this study, the basin model was created using GIS tool in HEC-HMS. The DEM was imported into the HEC-HMS and process using GIS tool, the result of basin model as shown in Figure 3. The basin area include 13.73 km^2 and perimeter 19.7 km. Sg. Johan located in the middle of the basin with 2 km length.

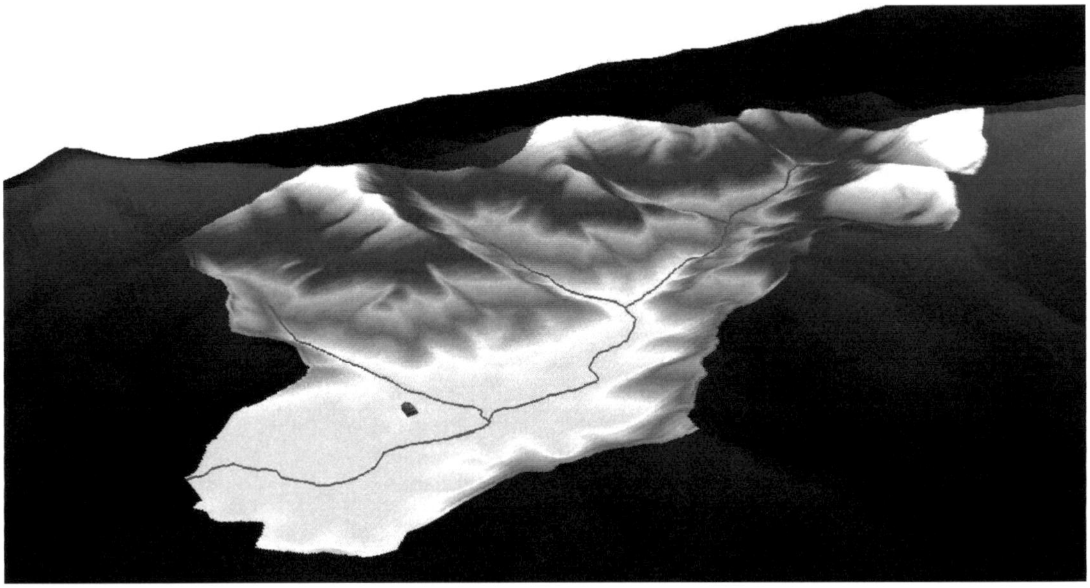

Figure 3. Sub-basin model of Bukit Kledang.

3.2. SCS Curve Number Method (SCS-CN)

The SCS Curve Number is a method of measurement for the surface runoff. This integer or parameter is introduced and developed by the United States Department of Agriculture (USDOA) The object of the curve number is to define the average conditions for the purpose of design. The curve number was initially designed for agricultural watersheds with a land slope of 5% and an initial rainfall loss of 20 % due to infiltration. The initial abstractions consist of lack of interception, surface storage and infiltration prior to runoff.

Table 1. Different soil group classifications.

Soil characteristics	Soil group
Deep sand, deep loess, and aggregated silt	Group A
Shallow loess and sandy loam	Group B
Clay loam, shallow sandy loam, soils low in organic content, and soils usually high in clay	Group C
Soils that swell significantly when wet, heavy plastic clap and certain saline soils	Group D

As per soil characteristics, soils in the study region fall within the hydrological soil groups A, B, C and D as shown in Table 1. In general, soil group A (sandy soil) has the highest infiltration rate and a lower runoff capacity. Conversely, soil group D (clay soil) has the lowest infiltration rate and provides the highest runoff potential. Other groups of soil fall between them.

3.3. Metrological Model

This model is one of the primary Component of the investigation which make metrological limit conditions for the basin. This metrological model can be utilized with various basin models. Result figured by the metrological model were coordinated with the sub-basin in the basin model utilizing the name of the sub-basin

3.4. Control Specification

The function of control specification is to monitor the model regarding to starting and ending simulation at explicit time. The date and timing are imported in the model.

4. Results and discussion

4.1. Evaluation of Model Performance

Lumped models treat the catchment area as a single homogenous unit. Spatial variability of catchment parameters is disregarded in lumped models presented by Moradkhani and Sorooshian [14]; Singh [15]. Average values over the catchment are used such as mean soil storage and uniform precipitation amounts according to Beven [16] and Risema [17]. The catchment characteristics are set as equal for the entire area and often cause over or under parameterization [17]. In these models, a single runoff output value is measured at the outlet of the sub basin area. Lumped models accurately simulate average runoff condition with short computing time. All data, including input, output, and parameters, are constant over time and space in the lumped model. By assuming homogeneity over the sub basin area, lumped models lose the spatial resolution of the results. Empirical and conceptual models typically run spatially as lumped models.

4.2 Evaluation of rainfall-runoff relation

The relationship between rainfall and surface runoff is complex, since it depends on a variety of factors related to the watershed and meteorology. The results obtained in Table 1 below described the peak discharge at study area were 43.4 m^3/s for 13.73km^2 of drainage area. The peak discharge occurs after 18 hours of rainy days and the rainfall data are collected over a period of two days. On the first day of data collection, there was rain all day and no rain on the second day of data collection.

Table 2: Result of the model simulation

Hydrologic Element	Drainage Area (km^2)	Peak Discharge (m^3/s)	Time of Peak	Volume (mm)
Subbasin 1	13.73	43.4	09Aug2020, 00:00	174.74
Sink-1	13.73	43.4	09Aug2020, 00:00	174.74

From the point of view of the management of radioactive waste, this result must be seen in the scope of the mitigation of contamination. In the worst case scenario, if the waste disposal cap infiltrated by meteoric rainfall, the surface contamination will be transported by the surface runoff in the disposal area. According to the simulation in Figure 4, the runoff generation in the study area gives a slow rate depending on the relative travel time of basin and also associated with soil type, land used, multiple storms and seasonality.

The analysis of the findings presents that the runoff yield was in proportion to the precipitation. In additional, runoff has not been a consistent trend. It is obvious that the days of higher rainfall will produce higher runoff and vice versa. Besides, the result obtained showed that the annual runoff yield can be predicted by annual rainfall data with better accuracy using SCS-CN method. The retention in the sub-basin model is attributed by its moderately disturbed forest characteristics and soil characteristic in the study area that represent as soils that swell significantly when wet, heavy plastic clap and certain saline soils. The primary requirement to calculate surface runoff by SCS-CN method involved infiltration rate in each land use element. Base on the land use element in the study area, it's represent a moderate disturbed forest that lowest runoff depth. This is due to the fact that the soil erosion happens in the moderate disturbed forest contributes to land degradation. Lal and Cummings [18] presented shifting cultivation practised in these field may lead to significant reduction in infiltration rates and saturated hydraulic conductivity.

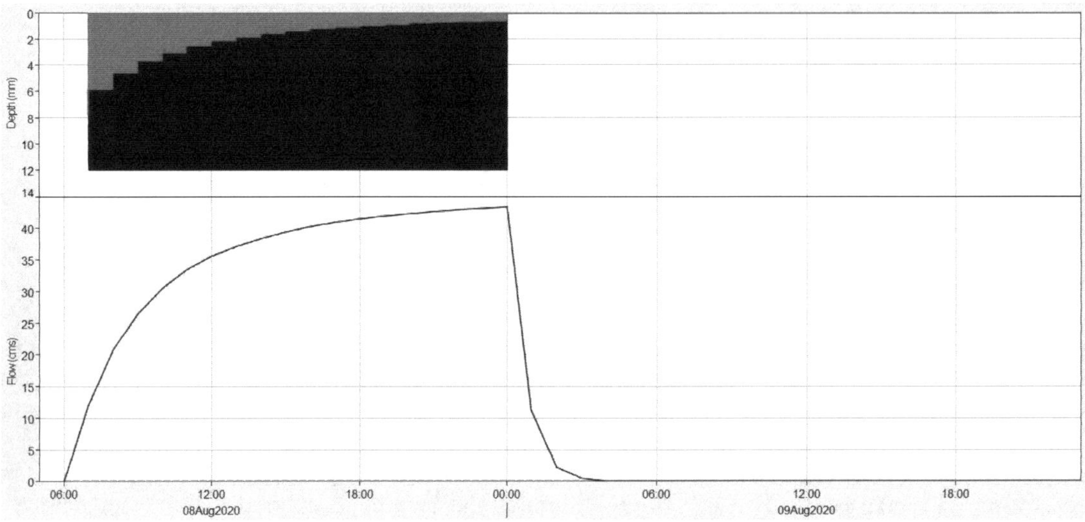

Figure 4. Simulation of rainfall-runoff model at study area.

During the wet season, after storm events and eve in the dry season, a degraded forest discontinues adequate infiltration and groundwater recharge via vertical percolation, despite gain from evapotranspiration. Chappel *et al.* [19] also states that decreased soil infiltration capacity leads to decrease stormflow pathways, particularly on hill slopes, implying negative feedback on the generation of runoff. Ground cover beneath the trees, particularly leaf litter, was more effective than the amount of canopy cover in reducing runoff. During the less intense storms, canopy cover was more effective, but was ineffective when the intensity of rainfall was high. Brabec *et al.* [20], Hatt *et al.* [21], Olivera and DeFee [22] and O'Driscoll *et al.* [23] presented increase in urbanization and impervious surface in watershed strongly impacts the stream hydrology and this linearly increases of the runoff.

5. Conclusion

HEC-HMS 4.5 hydrological model is used to simulate the rainfall runoff in the Bukit Kledang watershed. The area is resistant to flooding and displays high variability in monsoon precipitation. The area of study is characterised by a limited supply of runoff data. The rainfall-runoff simulation is based on hourly rainfall data. It is likely that rainfall values are one of the biggest uncertainties and vary spatially and temporally. Digital Elevation Model (DEM) 1m was used by Geospatial Hydrologic Modeling Extension (HEC-GeoHMS) and ArcHydro extension in ArcGIS 10.6 to delineate the sub-watershed and generate the stream network. HEC-GeoHMS was also used as a meteorological model for use in HEC-HMS to delineate the physical properties and create the input file in the form of sub-basin boundaries. Careful study has been done to assign the CN to the land use. Simulated results should be validated based on field measurement parameters and field observations due to the inadequacy and lack of observed data. Based on the present analysis, rainfall-runoff from basin area is likely to be an insignificant sources of radioactive contamination in the surface area to the vicinity environment. Conceptually, the rainfall-runoff model for the Bukit Kledang basin is slow because the soil type in the basin is sandy clay, which is classified as Group D in the soil group. While land use for the basin area is a type of forest that provides an interpretation of the slow rate of surface water runoff. In conclusion, depending on the relative travel time of the basin, the rainfall-runoff model for the Bukit Kledang basin is slow and also associated with the type of soil land used, multiple storm and seasonality.

6.References

[1] Semadeni-Davies A 2006 Winter performance of an urban stormwater pond in southern Sweden *Hydrol. Process* **20** 165-182.

[2] Subramanya K *Engineeering Hydrology* (New Delhi: TATA Mcgraw Hill Education Private Limited).

[3] Burnash R; Singh V 1995 The NWS River forecast system-catchment modeling. in computer models of watershed hydrology ed. Singh V P (USA: Water Resources Publications, LLC, Littleton Co.) 311–366.

[4] Sugawara M 1995 Tank model. in computer models of watershed hydrology ed. Singh V P (USA: Water Resources Publications, LLC, Littleton Co.) 165–214.

[5] Bergstro m S 1995 The HBV model in computer models of watershed hydrology ed. Singh V P (USA: Water Resources Publications, LLC, Littleton Co.) 443–476.

[6] Nielsen S A, Hansen E 1973 Numerical simulation of the rainfall runoff process on a daily basis *Nordic Hydrol.* **4** 171–190.

[7] Havnø K, Madsen M N, Dørge J 1995 MIKE 11—A generalized river modelling package. in computer models of watershed hydrology ed. Singh V P (USA: Water Resources Publications, LLC, Littleton Co.) 733–782.

[8] Yousuf A, Bhardwaj A, Tiwari A K, Bhatt V K 2015 Modelling runoff and sediment yield from a small forest watershed in Shivalik foot-hills using WEPP model *Int. J. Agric. Sci. Res.* **5** 67-78.

[9] Jasrotia A S, Dhiman S D, and Aggarwal S P 2002 Rainfall-runoff and soil erosion modeling using remote sensing and GIS technique – A case study on tons watershed *J. of Indian Soc. of Remote Sens.* **30**(3) 167–180.

[10] Pandey A and Dabral P P 2004 Estimation of runoff for hilly catchment using satellite data *J. of Indian Soc. of Remote Sens.*

[11] Amutha R, Porchelvan P 2009 Estimation of surface runoff in Malattar sub-watershed using SCSCN method *J. of Indian Soc. of Remote Sens.* **37** 291–304.

[12] Pandey A and Sahu A K 2002 Generation of curve number using remote sensing and geographic information system, htttp://www.GIS development.net.

[13] Nayak T R and Jaiswal R K 2003 Rainfall-runoff modelling using satellite data and GIS for Bebasriver in Madhya Pradesh, *IE (I) J.* **84** 47-50.

[14] Moradkhani H and Sorooshian S 2008 General review of rainfall-runoff modeling: model calibration, data assimilation, and uncertainty analysis eds. Sorooshian S, Hsu K L, Coppola E, Tomassetti B, Verdecchia M, Visconti G *Hydrol. Modelling and the Water Cycle: Coupling the Atmos. and Hydrol. Models* **63** (Berlin: Springer- Verlag) 12-35.

[15] Singh V P 1995 Computer models of watershed hydrology highlands ranch (Water Resources Pub.).

[16] Beven K J 2012 Rainfall-runoff modelling: the primer (2nd ed.)(Society of Remote Sensing, Wiley-Blackwell) **32**(2) 236–240.

[17] Rinsema J G 2014 Comparison of rainfall runoff models for the Florentine Catchment (Univ. of Tasmania).

[18] Lal R and Cummings D J 1979 Clearing a tropical forest: effects on soil and micro-climate. *Field Crops Res.* **2** 91–107.

[19] Chappell N A, Sherlock M, Bidin K, Macdonald R, Najman Y, Davies G 2007. Runoff processes in Southeast Asia: role of soil, regolith, and rock type ed. Swada H, Araki M, Chappell N A, LaFrankie J V, Shimizu *A Forest Environ. in the Mekong River basin* (Tokyo: Springer-Verlag) 3–23.

[20] Brabec E, Schulte S, Richards P L 2002 Impervious surfaces and water quality: a review of current literature and its implications for watershed planning *J. Plann. Lit.* **16** 499–514.

[21] Hatt B, Fletcher T, Walsh C, Taylor S 2004 The influence of urban density and drainage infrastructure on the concentrations and loads of pollutants in small streams. *Environ. Manage.* **34** 112–124.
[22] Olivera F, DeFee B D 2007 Urbanization and its effect on runoff in the Whiteoak Bayou watershed, Texas *J. Am. Water Resour. Assoc. (JAWRA)* **43** (1) 170–182.
[23] O'Driscoll M, Clinton S, Jefferson A, Manda A, McMillan S 2010 Urbanization effects on watershed hydrology and in-stream processes in the southern United States *Water* **2** 605–648.

Acknowledgment
This material is based upon work supported by Malaysia Nuclear Agency under HCD-STI funding. Any opinions, findings, and conclusions or recommendations expressed in this material are those of the author(s) and do not necessarily reflect the views of the Agency.

NDT methods in inspecting road and highway structures

T S T Amran[1,a)], M S M Amin[1], M R Ahmad1, N M Azreen[1], S Sani[1], M A K Adnan[1], N A Razak and S Sayuti[1]

Non-Destructive Testing - Material Structure Integrity group (NDT-MSI), Industrial Technology Division, Malaysian Nuclear Agency, 43000 Kajang, Selangor, Malaysia

[a]sarah@nm.gov.my

Abstract. This paper presents an overview of the non-destructive testing (NDT) methods that can be used to determine the quality and safety of road and highway structures in Malaysia. These NDT methods offer a relatively quick and inexpensive mean for regular inspection, which shall help alert the responsible agencies of critical structural damages even before heavy deterioration is visually apparent.

1. Introduction

The Federal Road network contains bridges with a bridge deck area, tunnels and a large number of other highway structures like retaining walls, noise protection walls and structures for traffic signs. The maintenance programs prepared for this purpose do not only require a high budget, but also influence the traffic infrastructure and, thus, the economy and society as a whole. The safety of the structures has to be ensured under consideration of environmental aspects [1]. At the same time the structure owner has to make sure that the maintenance activities are carried out in the most efficient way. The programs for lasting and systematic maintenance of these structures tie up a large percentage of the budget. This increase is influenced by insufficient durability of most of the structures which were built during the years 1960 to 1980. Further important aspects are the growing volumes of traffic and higher gross weights of trucks which lead to a reduction of the expected lifetime of bridges. Considering the fact that financial resources are restricted, the maintenance costs have to be spent in a way to obtain the greatest possible benefit [2].

Non-destructive testing (NDT) is expected to gain an increasingly important role in the management of structures. Efforts should be made to integrate the NDT data into the systematic maintenance system. This will improve the quality of information by eliminating the subjectivity associated with current visual inspection techniques. NDT methods can be used in future under the condition that results of investigations can be quantified and translated into rating resulting from road and highway inspection [3]. The term NDT used in this connection represents the techniques that are based on the application of physical principles employed for the purpose of determining the characteristics of materials or components or systems and for detecting and assessing the inhomogeneities and harmful defects without impairing the usefulness of such materials or components or systems.

NDT is not an integral part of most inspection procedures. Within the last decade, NDT gets more important for quality control of the workmanship after finishing the construction process and as a non-disruptive tool for the evaluation of current condition [4]. Only a few destructive samples are needed for verification and validation of the result. The condition assessment of the inner structure can localize quite fast hidden construction defects and determine the extent of damage caused by external impact or deterioration. Several NDT techniques have been developed to enable more efficient assessments of road and highway structure. This paper presents an overview on the NDT techniques that have been apply in one of the case studies done for highway inspection that showed that NDT techniques can give faster data acquisition, more contained cost and the capability to be performed in-situ over longer distances.

2. Case study

The development of voids beneath highways can lead to major pavement failures, a serious hazard. Voids typically develop because of subsidence and erosion of the base and subgrade materials. Void-related highway problems have often developed as a result of surface water intrusion into the road base. This may allow fines particles to be carried away, resulting in local base or foundation erosion, and the formation of weak areas, which eventually become voids. Voids continue to increase in size as the load from vehicles using the highway continues. The development of voids beneath highways is a serious hazard, making their early detection an important aspect of infrastructure maintenance and remediation. Early identification of concealed subsurface voids under a highway structure is critical to prevent major failures from occurring.

One of the most effective tools in use is ground penetrating radar (GPR) method. GPR has advanced to a level where the subsurface condition of a highway can be inspected and diagnosed with confidence. GPR is a non-destructive geophysical device used for subsurface exploration and operates by transmitting an electromagnetic pulse from an antenna into the ground and then capturing the partial reflections from subsurface layers [5]. A hyper spectrum (350 MHz) GPR system by GSSI was used in this case study. The highway inspection using GPR data were collected along nine scan lines, parallel to the road with 1-meter spacing. The highway data collection was conducted for 3.7 km length. The interpretation was done manually by identifying the areas showing relatively high amplitude response. This case study was carried out because there is a request from client to do the highway inspection. Therefore, due to the privacy and confidentially agreement of client, the exact location of the case study cannot be disclosed. Other techniques used as a complimentary technique to confirm GPR data was geoelectrical resistivity imaging (GRI) [6]. GRI is used in estimating the resistivity distributions of the subsurface based on several measurements of discrete voltage and current. This paper summarizes the case study, the technology overview, the inspection methodology and results of identified anomalies for the highway.

3. Methodology

The inspection for highway was conducted by quickly survey the area for potential sinkholes that might exist under the road surface. This survey was conducted using a GPR system with a 350 MHz antenna mounted on a survey cart. Subsequently, a more detailed survey of those areas was performed to confirm the existence and to determine the distribution and dimension of the potential sinkholes identified in the preliminary survey using GRI techniques.

3.1. Ground penetrating radar (GPR)

GPR is the general term applied to techniques that employ electromagnetic waves (radio wave range) to profile structures and features in the subsurface [7]. Although typically used for ground (soils) applications, GPR can be used to identify features beneath other surfaces. GPR method is based on transmission, reflection and detection of radio waves. A short pulse of high frequency (250-900 MHz) radar wave is produced and transmitted into the ground or other medium (i.e., concrete). The pulse spreads into the subsurface materials and is affected by the properties of the surrounding material. Waves are reflected at the interface between materials of different dielectric constants (Figure 1). A receiver records the reflected energy at the surface. Processed radar data are plotted as surveyed horizontal distance (meters) versus two-way travel time in nanoseconds (2D mode). When GPR data are collected in grids, 3D data processing and interpretation can be applied. GPR can locate both metallic and non-metallic targets. Penetration depth and detectability of targets will depend on antenna frequency, target orientation and the conductivity of the host material.

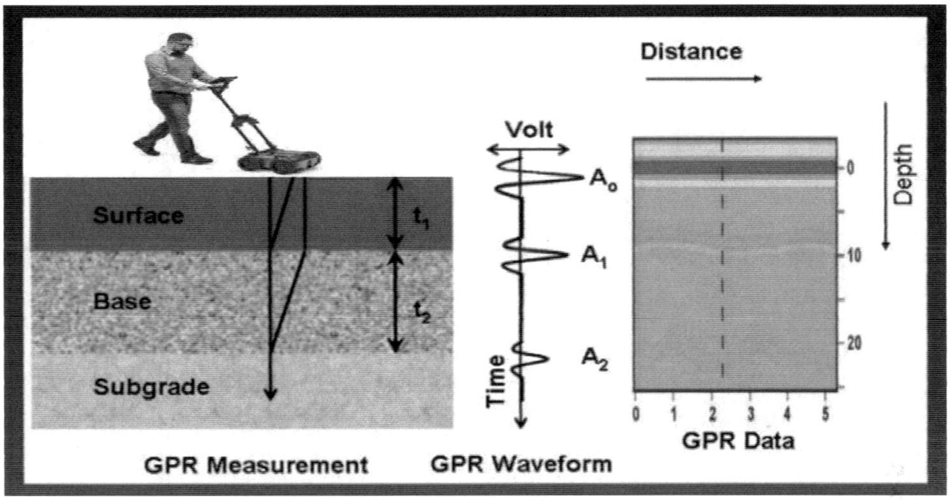

Figure 1: GPR principle and reflection profiling

A GSSI GPR device equipped with a 350 MHz hyper stacking antenna was used (Figure 2). Nine scan lines were selected along the highway with the antenna separation between the lines was 1 m (Figure 3). Data processing was done by the RADAN processing and analysis software package. The interpretation was done manually by identifying the areas showing relatively high amplitude response. Please note that the anomalies can be caused by numerous reasons. Including but not limited to: different electrical properties in the host material (i.e., different water content, different minerals), void space, embedded human-made objects (tank, concrete, metal or plastic pipes etc.).

Figure 2. GSSI GPR system 350 MHz Hyper stacking antennas.

Figure 3. Data collection using GPR.

3.2. Geoelectrical resistivity imaging (GRI)
Geoelectrical resistivity imaging (GRI) technique is a non-invasive technique that can map out subsurface structure based on electrical properties of subsurface material for example resistivity and conductivity which can be used to verify GPR data. Unlike the GPR Technique, GRI can investigate much deeper structure as compared to GPR but with lower resolution. The effective depth of GRI depends on the choice of electrode configuration and spacing. In this project, data acquisition was carried out as 3 sets of electrode spacings which are 40m, 100m and 200m with respect to the effective depth of 5m, 10m and 20m, respectively.

A set of ABEM Terrameter SAS 4000 (Figure 4) and a switcher unit were used to control the induction of current and potential readings from electrodes connected by multicore-cable along the survey line. The depth of the data obtained is depending on the length of the survey line, electrode spacing and the protocol file used in the survey. In this study, Pole-dipole protocol was selected.

Figure 4. ABEM resistivity set used in this survey.

Subsurface resistivity distributions are measured (Figure 5) by applying electrical current into the ground by using two current electrodes. The potential differences caused by the flow of current between any two points in linear line with the current electrodes are then measured by a pair of potential electrodes. From the measured voltage (V) and current (I) values, the resistance at the specified point in the subsurface can be determined. In a homogeneous ground, the penetration depth is directly proportional to electrodes spacing, and changing the electrode's separation gives information on sub surface's stratification [8]. For 2-D resistivity imaging, it is important to have a large set of data recorded along a survey line to effectively map the complex resistivity distribution of subsurface structure. The most practical way to acquire such a large amount of data is by using automated multi-electrode data acquisition system.

Figure 5. Data collection using GRI.

4. Results and discussion

4.1 Ground penetrating radar (GPR)

The 350 MHz hyper stacking GPR system provide sufficient data quality, although the signal penetration depth was limited to approximately 3.0m probably due to soil conditions. Figure 6 illustrated typical payment layers of highway for normal condition as detected by GPR system. It is consisting of asphalt, road base, subbase and subgrade.

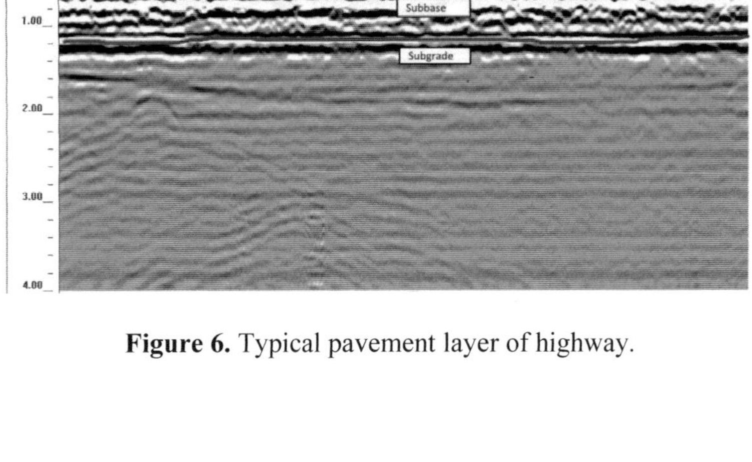

Figure 6. Typical pavement layer of highway.

Ground penetrating radar survey identified 55 anomaly areas which can be categorized into 4 types which are disturbed profile, contaminated profile, displacement and concrete structure profile. Detail of the anomalies profile are summarized in Table 1. GPR survey able to penetrate up to the 3-meter depth and the depth of asphalt, base and sub-grade can clearly identify. Grid scanning improved the quality of GPR data and enable to estimate the size of the defect. The example of GPR anomalies profile is shown in Figure 7. Six potential void-like structures are identified and from the 6-potential void-like structure, GRI test has been carried out at two location and the result shows good agreement with GPR result.

Table 1: Details of GPR anomalies profile

Profile	Definition
Normal Profile	the condition whereby pavement layers such as asphalt layer, sub-base and subgrade can easily be identified from the radargram
Contaminated Profile	the condition whereby different layers beneath the road surface could not easily be identified due to the contamination of the layer by water and clayey materials.
Disturbed Profile	the condition whereby the layer has lost its continuity due to a physical disturbance or modification such as previous maintenance work or the existence of void like structure
Displaced Profile	the condition whereby the continuity of the layers is displaced from the original position
Reinforced Concrete Structure Profile	the condition whereby the radargram is showing the existence of constant tiny hyperbolic signals as a result of the high reflection of radar wave by the equally-spaced metal reinforcement bars.

4.2 Geoelectrical resistivity imaging (GRI)

Resistivity surveys were carried out at two location named profile A and profile B along the highway. The locations were selected based on GPR data which show the most severe location. Resistivity survey lines for profile A were shown in figure 8. The saturated zone was delineated with dotted lines while the weathered and fractured granite bedrock was delineated with a dashed line. Resistivity survey line with a distance of 40 m detailed out the water content that may be filled within the saprolitic soil. Resistivity survey line with a distance of 100 m showed the top of weathered and fractured granite bedrock. The resistivity survey line with a distance of 200 m finally revealed that the weathered and fractured granite bedrock was in the form of the boulder.

Resistivity survey lines for profile B were shown in Figure 9. The saturated zone was delineated with dotted lines while the weathered and fractured granite bedrock was delineated with the dashed line. Resistivity survey line with a distance of 40 m indicated that the weathered fractured granite started to form at the near surface. Resistivity survey line with a distance of 100 m highlighted saprolitic kaolin between the weathered and fractured granite bedrock. The resistivity survey line with a distance of 200 m finally revealed that the subsurface formation herein was mainly composed of weathered and fractured granite bedrock with some saprolitic soil within.

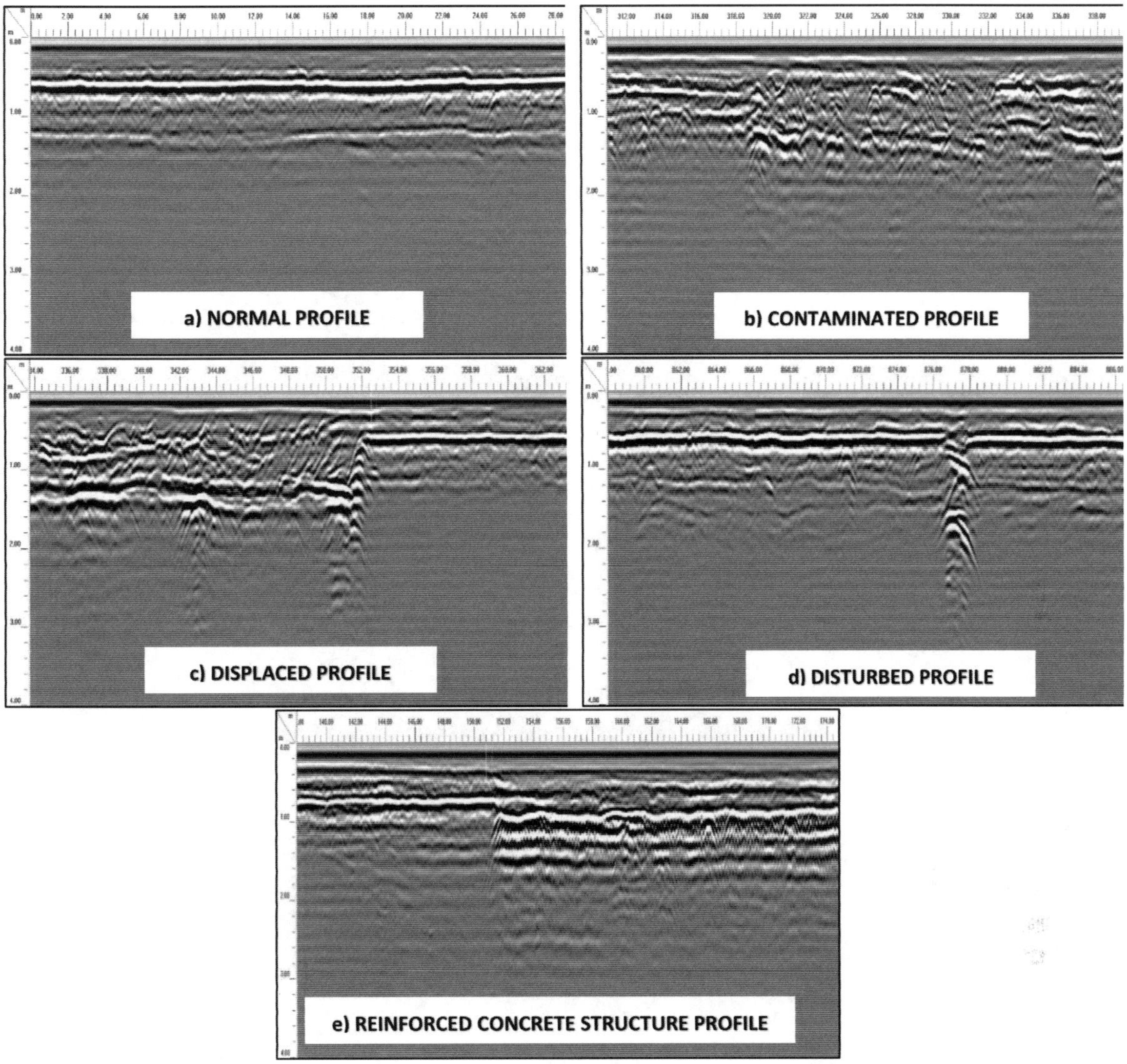

Figure 7. Example of GPR anomalies profiles.

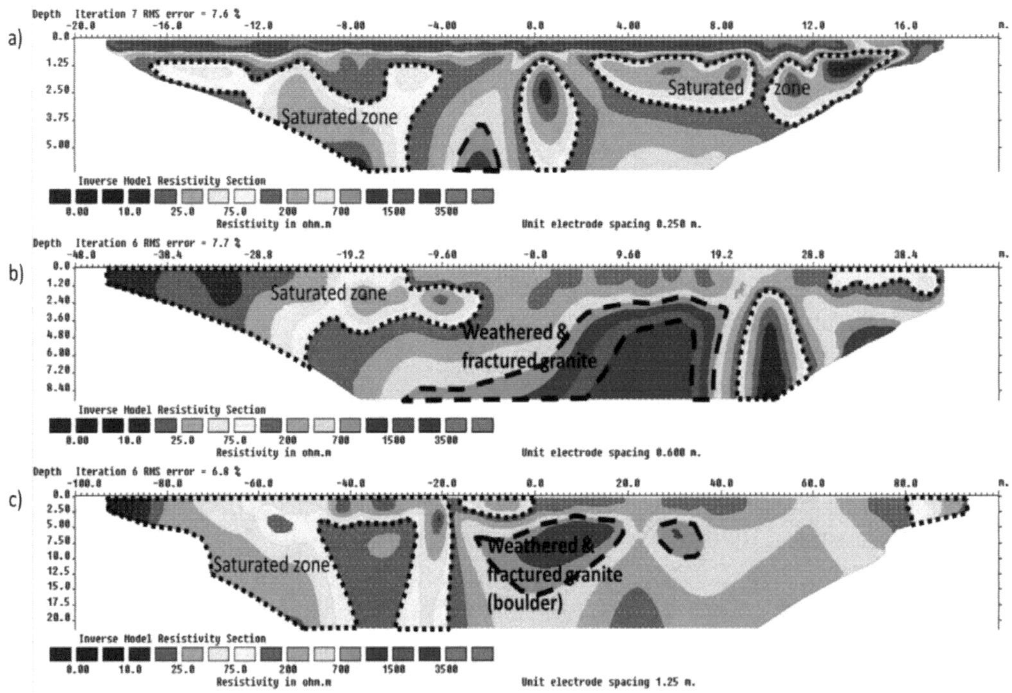

Figure 8. Electrical resistivity tomography at Profile A a) resistivity survey line of 40 m, b) resistivity survey line of 100 m and, c) resistivity survey line of 200 m.

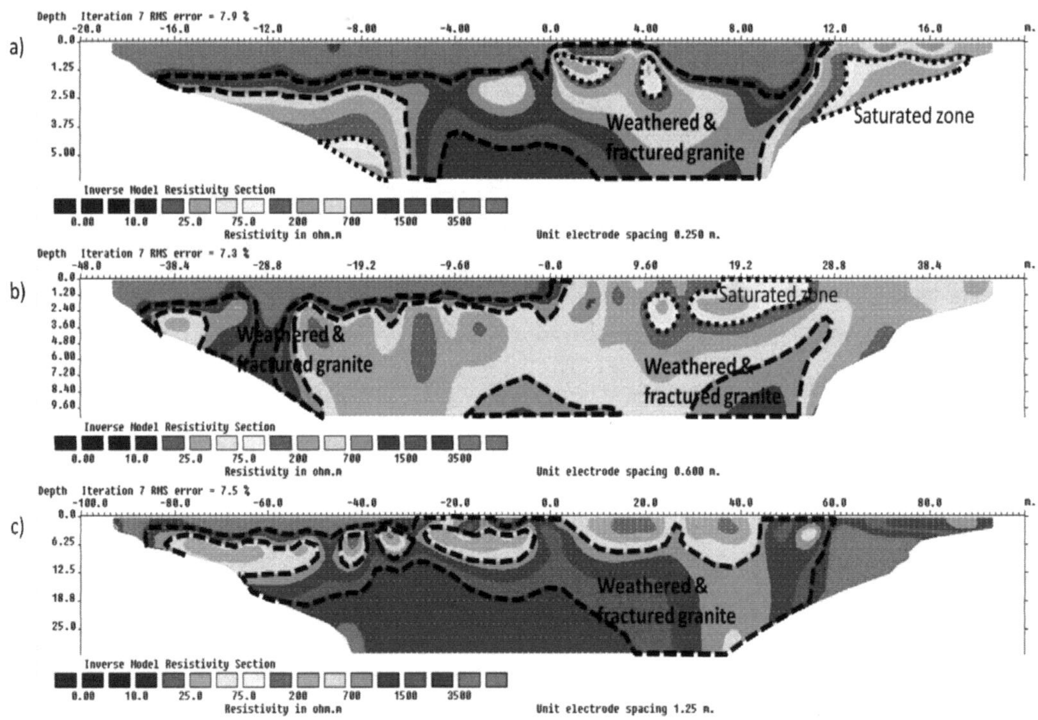

Figure 9. Electrical resistivity tomography at Profile B a) resistivity survey line of 40 m b) resistivity survey line of 100 m and c) resistivity survey line of 200 m

4.3 Comparison GPR-GRI data

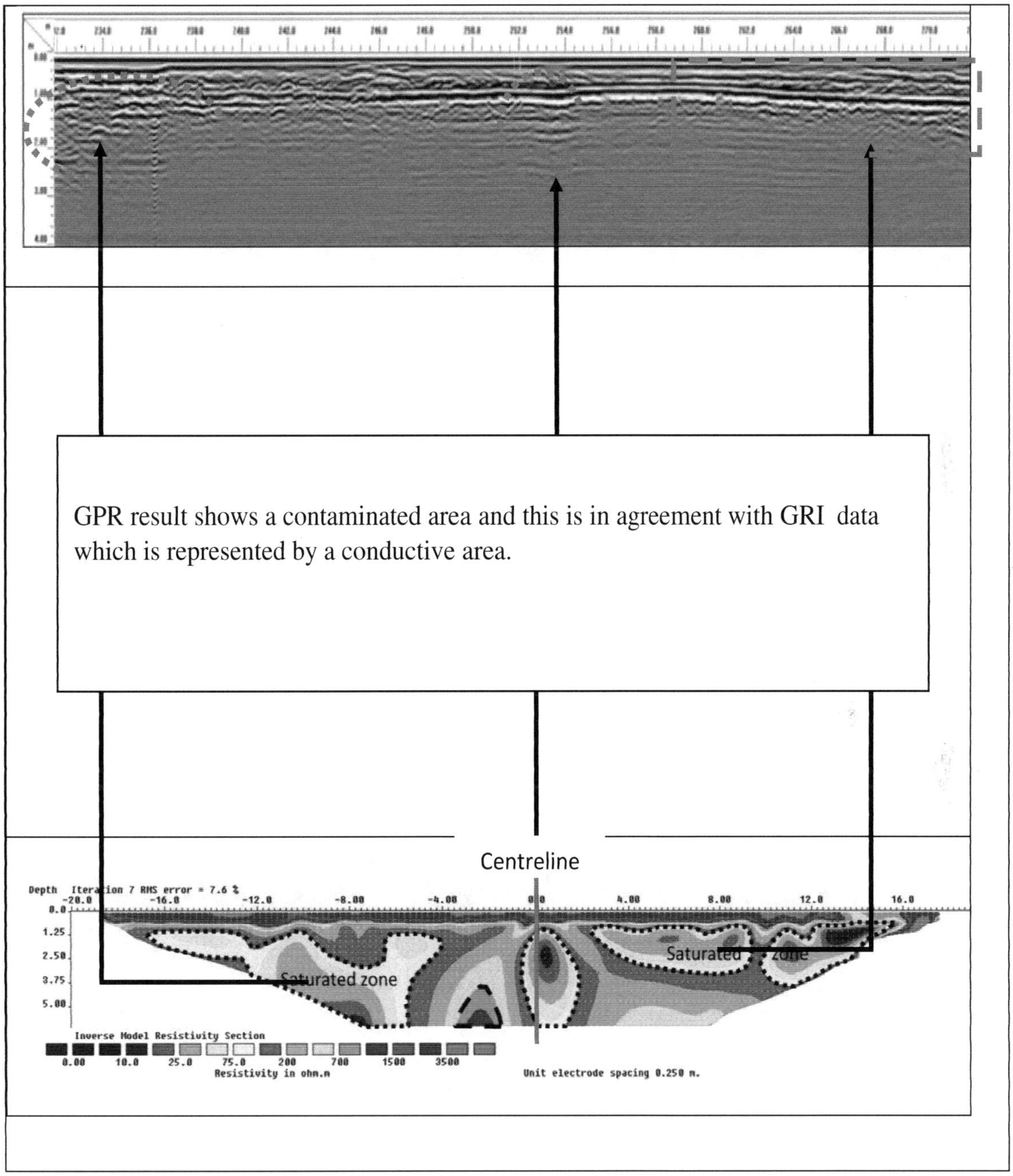

Figure 10. Comparison between GPR radargram and GRI 2-D image at anomalies (red dotted line) which shows in good agreement

IOP Conf. Series: Materials Science and Engineering 1106 (2021) 012034 doi:10.1088/1757-899X/1106/1/012034

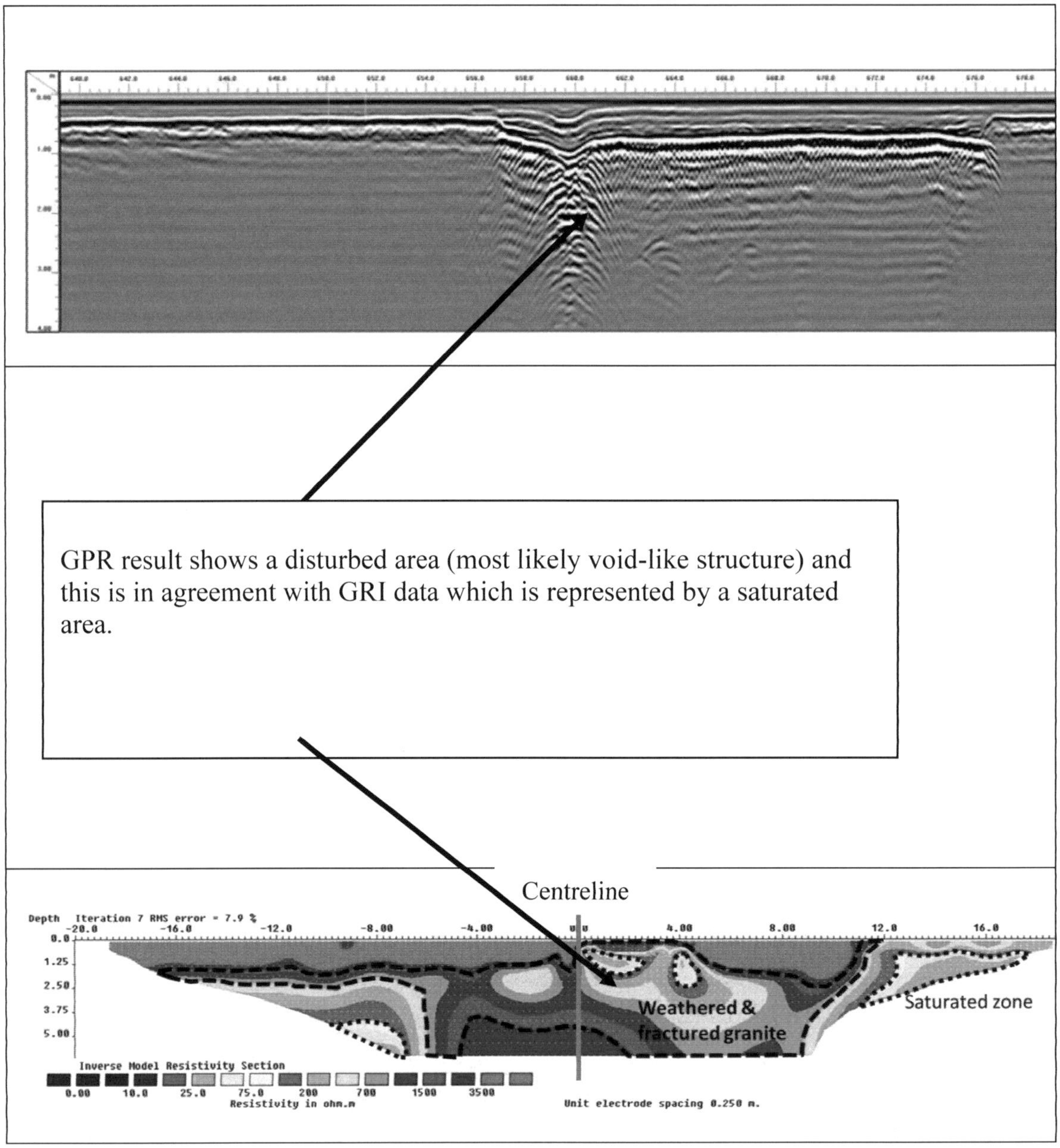

Figure 11. Comparison between GPR radargram and GRI 2-D image at anomaly which shows in good agreement.

5. Conclusion

The work investigates the effectiveness of Ground Penetrating Radar (GPR) and Geoelectrical Resistivity Imaging (GRI) methods for the inspection of road and highway conditions in Malaysia. Based on our measurements, these two methods complement each other, specifically in detecting anomalies beneath the road surface. It is upon these observations that the authors propose the use of these aforementioned NDT methods for the regular inspection of road and highway in Malaysia.

6. References

[1] Rolander D D, Phares B M, Graybeal B A, Moore M E, Washer G A 2001 Highway bridge inspection: state-of-the-practice survey *Transport. Res. Rec.* **1749 (1)** 73-81

[2] Lee S, Kalos N and Shin DH 2014 Non-destructive testing methods in the U.S. for bridge inspection and maintenance *KSCE J. Civil Eng.* **18** 1322-1331

[3] Benedetto A, Tosti F, Ciampoli L B, and D'Amico F 2017 An overview of ground penetrating radar signal processing techniques for road inspections *Signal Process.* **132** 201-209

[4] Willcox M and Downes G 2003 A brief description of NDT techniques *Insight NDT Tech. Paper* 1–22.

[5] Artagan S S, Ciampoli L B, D'Amico F, Calvi A and Tosti F 2020 Non-destructive assessment and health monitoring of railway infrastructures *Surveys in Geophys.* **41** 447-483.

[6] Carriere S D, Chalikakis K, Senechal G, Danquigny C and Emblanch C 2013 Combining electrical resistivity tomography and ground penetrating radar to study geological structuring of karst unsaturated zone *J. App. Geophys.* **94** 31-41.

[7] Baker G S, Jordan T E and Talley J 2007 An introduction to ground penetrating radar (GPR) *The Geological Society of America Special Paper* 432a.

[8] Dahlin T 2001 The Development of DC Resistivity Imaging Techniques *Comp. and Geosci.* **27** 1019-1029.

Acknowledgments

The authors would like to extent their appreciations and sincere thanks to Ministry of Science, Technology and Innovation (MOSTI) and Agensi Nuklear Malaysia as well as the Non-Destructive Testing-Material and Structure Integrity Group (NDT-MSI) for the support and co-operation.

IOP Conf. Series: Materials Science and Engineering 1106 (2021) 012035 doi:10.1088/1757-899X/1106/1/012035

Development of gantry radial arm robotic system for corrosion inspection on wall of reactor tank TRIGA

R Ramli[1,a)], S Sani[2], A S Asri[1]

[1]Institute of Sustainable Energy, Universiti Tenaga Nasional, Kajang, Selangor, Malaysia
[2]Industrial Technology Division, Agensi Nuklear Malaysia, Bangi, Selangor, Malaysia.

[a]Rogemah@uniten.edu.my

Abstract. This paper focuses on the design upgrade of existing scanner of gantry radial arm robotic system for corrosion inspection on wall reactor of a TRIGA tank. The scanner design will be used for under water inspection environment verification. This inspection can be performed by using ultrasonic C-scan that will be designed in this project. This gantry radial arm system is a modular design that allows upgrading, easy dimensional matching and, specifications related to kinematic and technique. The gantry radial arm system is expected to be user-friendly. SolidWorks software was used to design the prototype of the gantry radial arm. Most of the components are made from Aluminium which is much sturdy and low cost. Meanwhile for the rod holder system, spring mechanism was used to prevent rod from bending when it moves down into the water. The test results show that the system designed is safe to be used, as it has been proved by simulation and calculation.

1. Introduction

Non-destructive testing (NDT) is the process of inspecting, testing, or evaluating materials, components or assemblies for discontinuities, or differences in characteristics without destroying the serviceability of the part or system. Today, modern non-destructive tests are used in manufacturing, fabrication and in-service inspections to ensure product integrity and reliability, to control manufacturing processes, lower production costs and to maintain a uniform quality level [1][2][3].

Ultrasonic testing (UT) is a non-destructive inspection method that uses high frequency sound waves (ultrasound) that is above the range of human hearing, to measure geometric and physical properties in materials. Sound is produced by a vibrating body and travels in the form of a wave. One type of sound wave, called longitudinal or compression travels about 330 metres per second in air and about 6,400 metres per second in Aluminium or in steel at approximately 5960 metres per second [4]. A typical UT inspection system consists of several functional units, such as the pulser, receiver, transducer, and display devices. A pulser or receiver is an electronic device that can produce high voltage electrical pulses. Driven by the pulser, the transducer generates high frequency ultrasonic energy [5].

Ultrasonic information can be obtained and presented in various sizes. Three most popular formats are called A-scan, B-scan and C-scan presentations in the NDT world. Every method of presentation offers a different way of evaluating and assessing the region of the product inspected [6]. The C-scan analysis offers an image of features that reflect and scatter the sound in the test piece's surfaces. High-resolution scans can produce extremely detailed images. All images are generated with the transducer scanned on the head side of a method of immersion scanning using a pulse-echo technique [6]. For virtually all touch test use, ultrasound couplers are used to allow sound energy transfer between the transducer and the sample part [7]. Couplant usually are moderately viscous, non-toxic, gels or pastes. In this project, we're going to focus more on C-scan and improvement of the existing design robotic system.

2. Design verification

Design verification is a key step in any product's development. Quality tests ensure that the product planned is the same as the product as expected, often called qualifying checking. During the development cycle, from design creation to post-production, testing will take place at many stages. The tool focuses mainly on prototype testing, but a number of guidelines can be used for all tests [8]:

i. Material, model or sub-assembly development tests are useful in determining the viability of design ideas and in gathering insights that further guide the development.
ii. Prototype experiments for products closely identical to the final product are conducted.
iii. Proof testing is another type of prototyping design testing. The proof tests are designed to test the product for failure instead of checking it for performance.
iv. Acceptance testing is a method of non-destructive testing with manufacturing units. The tests can be conducted on initial production units or on the random or specified samples, or on any single unit produced, depending on the criticality of the defects, test costs and the number of units produced.

2.1. Automatic climbing robot

The climbing robot was used in the modern period for wall corrosion inspection. Figure 1 shows various types of climbing robot used in corrosion inspection.

Figure 1. Automatic climbing robot [9],

The climbing robot has some difficulties, and a particular wheel configuration is required in order to effectively achieve horizontal and vertical motions. In addition, it is also a problem to collect position information, so the Kalman filter technique uses fusion of sensors and position estimation. The adhesion mechanism, which defines the unit sticking ability on the vertical wall or ceiling, is an indispensable feature of the wall-lifting robot [10].

2.2 Existing system for reactor tank TRIGA

The existing system where the current mechanism of ultrasonic probe holder that shown in Figure 2 by using slotted pin to assemble the rod. There was some problem occurred on the ultrasonic rod during the inspection. The ultrasonic probe holder was shaking when it goes inside the reactor tank because of the depth of reactor and density of water. However, there is way to improvise the ultrasonic probe holder by changing the mechanism for slotted pin to the thread to make it work better.

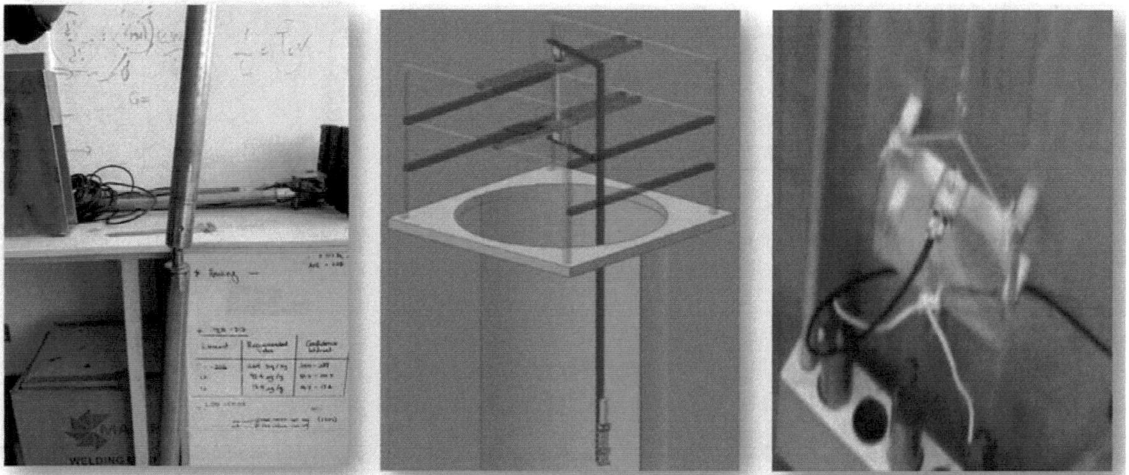
a) b) c)

Figure 2. (a) Ultrasonic probe holder, (b) Gantry radial arm system, (c) Ultrasonic inspection on reactor wall

2.3. Improvement system

The usage of the gantry radial arm system is to ensure the sensor not moving when the ultrasonic inspection on wall reactor is performed. Aluminium is the only material that can enter inside the reactor tank because it has long term stability, strength and resilience for high temperature reactor application such as the cladding of nuclear fuel pellets and the reactor tank surface also made from Aluminium material. The advantage of using Aluminum material is that it has high impact resistant and extremely strong and also low corrosion rates compared to rust, especially if anodized.

The function of the whole system is also the same as automatic climbing robot. The existing design system, which only can be operate 180 degrees because of limited design itself. Another type of ultrasonic inspection on wall reactor tank is the one that used by others country where the system with fully hydraulic and also used robot crawler as their inspection devices. Table 1 shows the advantages and disadvantages of the designing gantry radial arm robotic system and the existing system of the ultrasonic inspection on wall reactor.

Table 1. Existing system versus new gantry radial arm robotic system.

Existing Ultrasonic Inspection System	New Gantry Radial Arm Robotic System
The movement of the system was limited	The system designed was free moving as it use x-y table
The assembly rod using a slotted pin	User friendly because the rod having thread as it can easily to detach
High maintenance	Low maintenance

3. Project design method

3.1. Requirement
To identify all the information and requirement such as software, planning must be done in decent manner. The planning phase have two vital points which is data collection and the requirements of software.

3.1.1. Data collection
Data collection is the first stage in planning. At this stage about the project's resources and requirements, literature studies and schedule to get more information in this study. All the materials are collected based on the previous design from Malaysian Nuclear Agency. Within the data collection period, some research about ultrasonic probe holder was done and the product was designed.

3.1.2. Software Requirement
Data collection is the first stage in planning. This stage is about the project's resources and requirements, literature studies and schedule in order to get more information for this study. All the materials are collected based on the previous design from Malaysian Nuclear Agency. Within the data collection period, some research about ultrasonic probe holder was done and the product was designed.

3.2. Design

3.2.1. Sketching draft of product's design
Before starting the implementing phase, a draft of sketching must be produced for checking the design progress and approving one of the designs. The design is loosely based on the existing design at Malaysian Nuclear Agency. During designing phase, all factors are taken into consideration to make sure the design user friendly and it will function as the objectives of this project. The sketching drafts that manage to brainstorm before proceeding to the next step is shown in Figure 3.

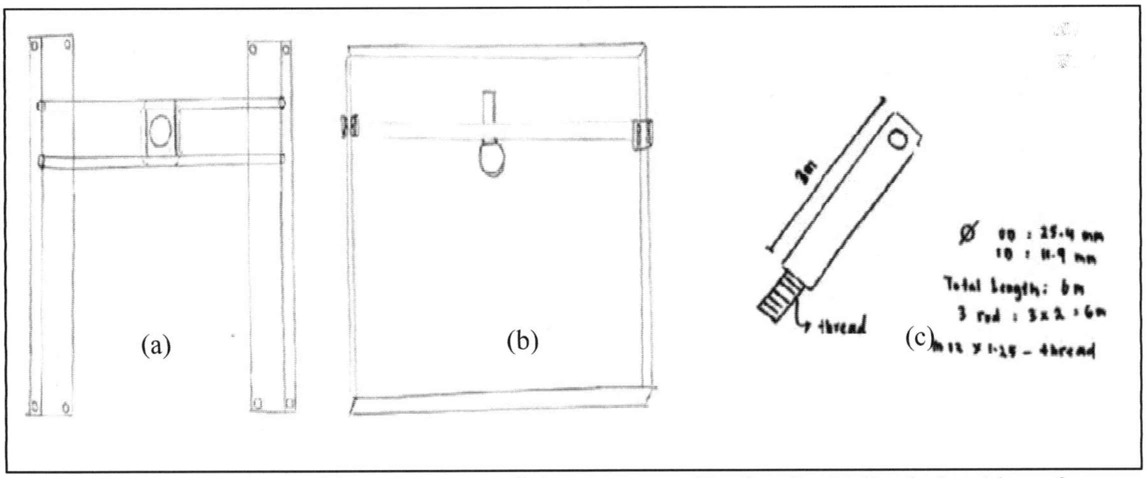

Figure 3 (a) The draft sketching of gantry radial arm system (Design 1), **(b)** Draft sketching of gantry radial arm system (Design 2), **(c)** Ultrasonic probe holder for corrosion inspection.

IOP Conf. Series: Materials Science and Engineering 1106 (2021) 012035 doi:10.1088/1757-899X/1106/1/012035

3.2.2. Finalize design

In all the drafts provided, every design created was discussed, compared and improvised for the design to function more efficiently and easy to operate. The final design that has been improvised is shown in Figure 4.

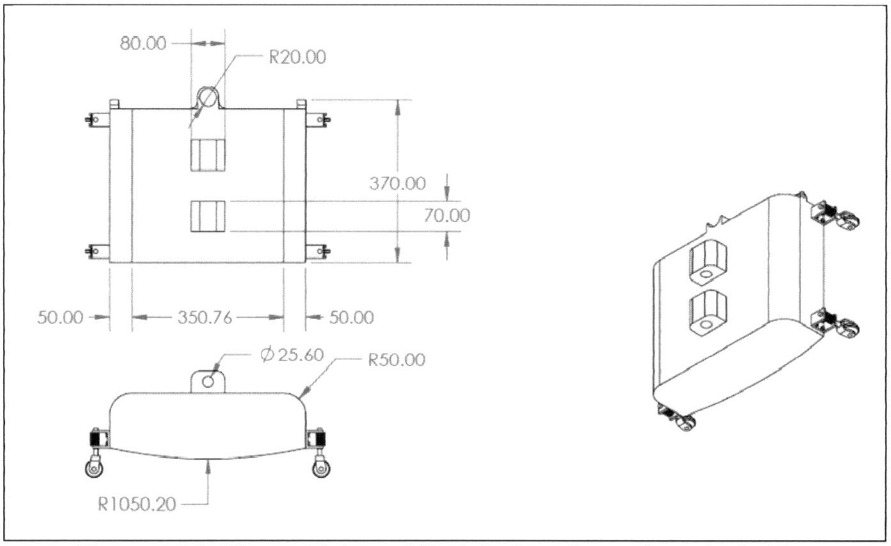

Figure 4 (a). The multiview of final design ultrasonic probe holder.

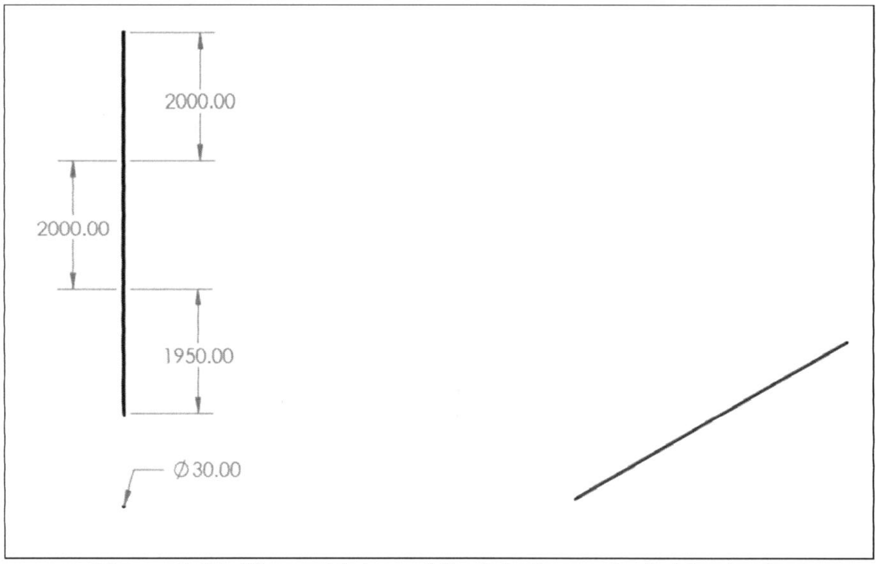

Figure 4 (b). The multiview of final design probe holder (rod).

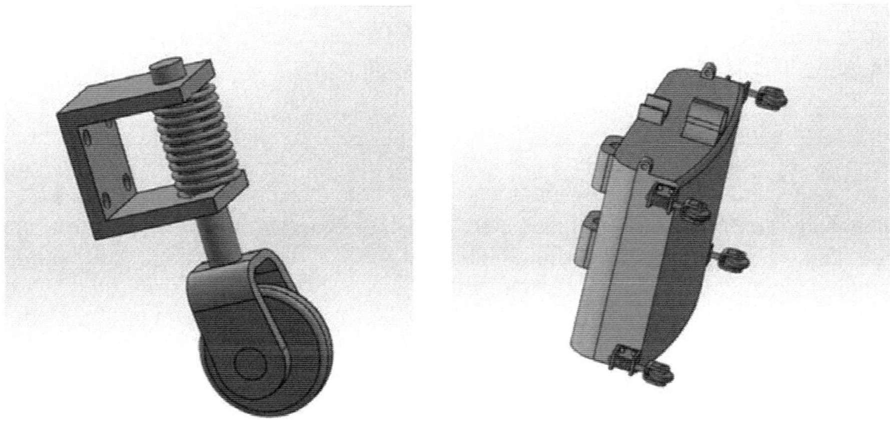

Figure 4 (c). The Full View of Final Design of Ultrasonic Probe Holder; (Left) spring roller, (right) ultrasonic probe holder.

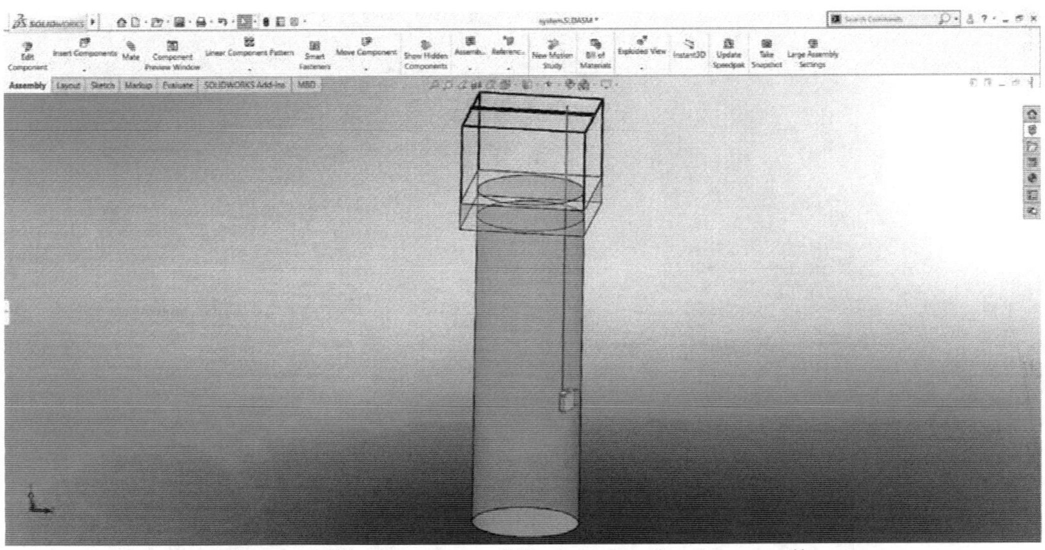

Figure 4 (d). The view of final design for the overall system.

3.3. Implementing

In the implementing phase, right after the design of the product is finalized, the design was simulated to ensure that there will be no failure happens during the fabrication process. Testing is needed in order to know whether it is functional or not, and how strong the probe holder can operate in a reactor tank with the density of water and the height of the reactor which is 6 meters. The design of the gantry radial arm robotic system for corrosion inspection on wall reactor tank must be running during testing. The point of undergoing this phase is to know whether it is efficient or not, and how accurate the data display once the scanner in mounted on the wall of the reactor tank. This was simulated by using SolidWorks, buckling test and calculation of buoyancy force.

IOP Conf. Series: Materials Science and Engineering 1106 (2021) 012035 doi:10.1088/1757-899X/1106/1/012035

4. Results and discussion

The design was test and it has been simulated using SolidWorks to check whether the design is safe to be used or not, and make some comparison with the previous system and the high-end technology.

4.1. Stress and strain analysis

Stress analysis is an important part of engineering research, because stress is typically caused by the failure of most engineering components. The stress analysis is based on the mechanical characteristics of a material and its ability to survive predicted loads. It generally includes tension, stress, shear, performance, UTS, young modulus, Poisson ratio, and others. Based on the result, the deformation scale of stress and strain analysis are 6.045 (Table 2). These values are used to ensure the rigors of life can be met with whatever the design has been created and built. Trust and safety are the two main focus points which enhance the need for stress analysis. Figure 5 shows the stress and strain analysis done by using SolidWorks.

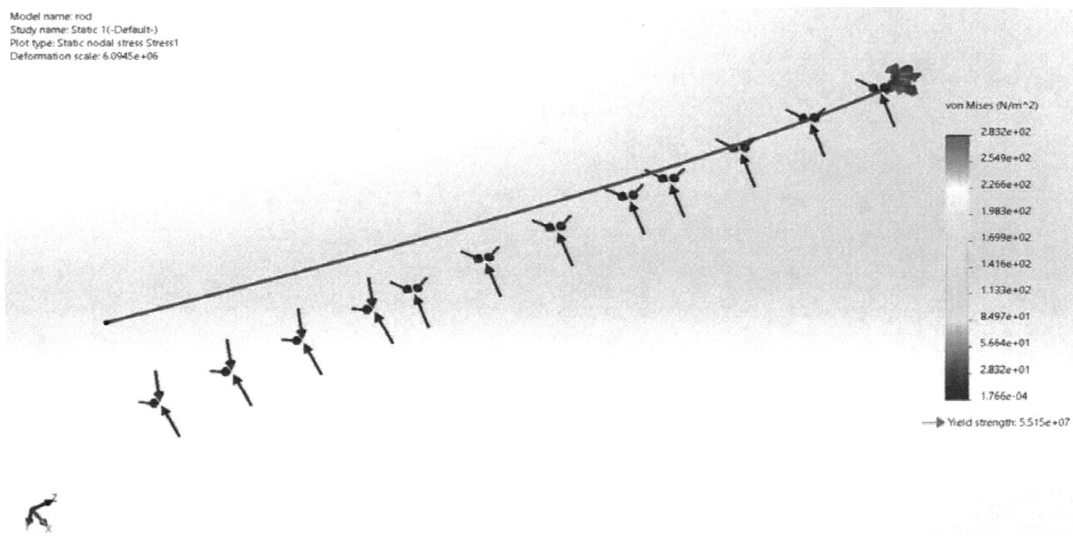

(a)

(b)

Figure 5. Stress and strain analysis.

IOP Conf. Series: Materials Science and Engineering 1106 (2021) 012035 doi:10.1088/1757-899X/1106/1/012035

4.2. Buckling test and factor of safety (FOS)

The buckling load factor (BLF) is a measure of the safety factor against buckling or the buckling load ratio. Today, stiffness usually refers to the finite element stiffness matrix. Analytic buckling studies are used to identify additional classes of instability besides Euler buckling. Based on the result shown in Figure 6, the deformation scale was 9.23112 (Table 2). This result shows that the rod will not buckle because the applied loads are less than the estimated critical load.

Figure 6. Buckling test.

The material safety factor (FOS) is defined as the material output stress divided by the effective stress of von Mises for Aluminium material used in this project. The FOS is not too high which is 1.9 (Table 2), ranging from a low value of about 1 to a high value of about 100. The result in Figure 7 shows that this design was safe to be used and without any failure on fabrication. Besides that, this probably suggests that a simple design will save material and money.

Figure 7. Factor of safety (SOF).

Table 2: Results summary.

Analysis	Deformation Scale
Stress	6.045
Strain	6.045
	9.23112
Buckling	**Load Factor**
	699.14
Factor of Safety (FOS)	
1.9	

4.3. Buoyancy force calculation

The buoyancy force on an object is directly proportional to the volume of the submerged body [32]. In other words, the more a solid object is immersed, the greater the booming power. This means that even objects falling into liquid have an uplifting booming power. Based on the water inside the reactor, the density of distilled water at 30-degree Celsius is 995.67 kg/m^3. The length of the rod is 6m and the volume of the ultrasonic probe holder:

$$\boldsymbol{F_b = V_s \times D \times g} \tag{1}$$

where,

F_b = Buoyancy force that is acting on the object
V_s = Submerged volume of the object
D = Density of the fluid the object is submerged in
g = force of gravity.

With

$$V_s = \frac{\pi}{4} \times D^2 \times h \tag{2}$$

$$V = \frac{\pi}{4} \times 0.0254^2 \times 6 = \mathbf{3.04 \times 10^{-3}}\ \mathrm{m^3} \tag{3}$$

Therefore,

$$F_b = 3.04 \times 10^{-3} \times 995.67 \times 6 = \mathbf{29.6933}\ \boldsymbol{N} \tag{4}$$

The mass m of probe holder calculated based on the density ρ of the material used which is Aluminium 6061 alloy, $\rho = 2700 kg/m^3$:

$$V = \frac{\pi}{4} \times 2^2 \times 6 = \mathbf{18.8496}\ \mathrm{m^3} \tag{5}$$

$$\rho = \frac{m}{V} \tag{6}$$

$$m = \rho V = 2700 \times 3.04 \times 10^{-3} = \mathbf{8.208}\ \mathbf{kg} \tag{7}$$

Assuming ordinary gravity and water with ordinary density, so that the force of buoyancy on the reactor:

$$F = 18.8496 \times 995.67 \times 9.81 = \mathbf{184113.8959}\ \boldsymbol{N}. \tag{8}$$

Therefore,

$$G = 8.208 \times 9.81 = 80.8205\ N \tag{9}$$

The value of force of gravity on the reactor is much more than the buoyancy force. So that, the rod will sink into the distilled water.

5. Conclusion

The result that was obtained showed that this objective has been achieved. Due to the usage of Aluminium alloy 6061, the prototype is reliable and has high melting temperature – with the factor of safety 1.9 and 9.23112 deformation for buckling. Also, the system of gantry radial arm having a free movement that is easily used to inspect on reactor under water inspection. Apart from that, the thread mechanism on the rod can be user-friendly without the difficult learning and, easy to be detached after the inspection process is done.

6. References

[1] Amza G S Ultrasunete aplicaŃii active 2006 ed. Agir (Bucuresti).
[2] Amza G S 1989 Sisteme ultraacustice ed. Tehnică (Bucureşti).
[3] Hillger W 2002 Inspection of CFRP components by ultrasonic imaging with air coupling *8th European Conf. on Non-destruc. Testing* (Barcelona).
[4] Vipparthy S T Inspection of defects in rails using ultrasonic probe
[5] *Basic Principles of Ultrasonic Testing* https://www.ndeed.org/EducationResources/CommunityCollege/Ultrasonics/Introduction/description.htm (NDT Resource Center) [Accessed: 8 November 2019].
[6] https://www.nde-ed.org/EducationResources/CommunityCollege/Ultrasonics/EquipmentTrans/DataPres.htm. (NDT Resource Centre) [Accessed: 8 December 2019].
[7] *Olympus Industrial Resources* https://www.olympus-ims.com/en/applications/ultrasonic-couplant/. [Accessed: 8 December 2019].
[8] Burgess J A 1984 Design assurance for engineers and managers (New York: Marcel Dekker), 150-165.
[9] . Xu K Z X Z A H S Z 2015 Design and optimization of magnetic wheel for wall climbing robot *Robotic Welding, Intelligence and Automation, Advances in Intelligent Systems and Computing*, (Switzerland: Pringer Int. Pub.) 363.
[10] Chen Y C A X 2015Design of a scalable wall climbing robot for inter-plane *Robotic Welding, Intelligence and Automation, Advances in Intelligent Systems and Computing*, (Switzerland: Pringer Int. Pub.) 363.

AUTHOR INDEX

Institute of Physics
Dirac House, Temple Back
Bristol BS1 6BE UK

ISSN: 1757-8981
ISBN 978-1-7138-3399-4

CURRAN ASSOCIATES INC.
proceedings.com